IDENTIFICATION AND CONTROL OF MECHANICAL SYSTEMS

Vibration is a significant issue in the design of many structures including aircraft, spacecraft, bridges, and high-rise buildings. This book discusses the *control of vibrating systems*, integrating structural dynamics, vibration analysis, modern control, and system identification. Integrating these subjects is an important feature in that engineers will need only one book, rather than several texts or courses, to solve vibration/control problems.

The book begins with a review of the fundamentals in mathematics, dynamics, and control that are needed for understanding subsequent materials. Chapters then cover recent developments in aerospace control and identification theory, including virtual passive control, observer and state–space system identification, and data-based controller synthesis. Many practical issues and applications are addressed, with examples showing how various methods are applied to real systems. Some methods show the close integration of system identification and control theory from the state–space perspective, rather than from the traditional input–output model perspective of adaptive control.

This text will be useful for advanced undergraduate and beginning graduate students in aerospace, mechanical, and civil engineering, as well as for practicing engineers.

Dr. Jer-Nan Juang is the Principal Scientist at the Structural Dynamics Branch, NASA Langley Research Center. He is also an Adjunct Professor at George Washington University, Duke University, and Virginia Polytechnic Institute and State University.

Dr. Minh Q. Phan is Associate Professor of Engineering at Thayer School of Engineering, Dartmouth College.

Identification and Control of Mechanical Systems

JER-NAN JUANG
NASA Langley Research Center

MINH Q. PHAN
Dartmouth College

CAMBRIDGE
UNIVERSITY PRESS

CAMBRIDGE UNIVERSITY PRESS
Cambridge, New York, Melbourne, Madrid, Cape Town, Singapore, São Paulo

Cambridge University Press
The Edinburgh Building, Cambridge CB2 2RU, UK

Published in the United States of America by Cambridge University Press, New York

www.cambridge.org
Information on this title: www.cambridge.org/9780521783552

First published 2001
This digitally printed first paperback version 2006

A catalogue record for this publication is available from the British Library

Library of Congress Cataloguing in Publication data
Juang, Jer-Nan.
    Identification and control of mechanical systems / Jer-Nan Juang, Minh Q. Phan.
        p. cm.
    Includes bibliographical references.
    ISBN 0-521-78355-0
        1. Vibration. 2. Structural control (Engineering) 3. Damping (Mechanics) I. Phan,
    Minh Q. II. Title.
    TA355 .J83 2001
    620.3 – dc21   00-063087

ISBN-13 978-0-521-78355-2 hardback
ISBN-10 0-521-78355-0 hardback

ISBN-13 978-0-521-03190-5 paperback
ISBN-10 0-521-03190-7 paperback

To Lily, Philo, Derek, and my parents
JNJ

To Suzu, Daniel, my mother, and to the memory of my father
MQP

Contents

Preface

This book is based on a series of lecture notes developed by the authors. The first author has used part of the notes for two graduate-level classes in System Identification and Control of Large Aerospace Systems at the Joint Institute for Advancement of Flight Sciences, George Washington University at NASA Langley Research Center for the past 10 years. The second author has used part of the notes for senior and first year graduate-level courses in Dynamics and Control of Mechanical Systems and System Identification at Princeton University and Dartmouth College since 1995. There are many reasons that motivated the writing of this book; some of them are outlined below.

First, the lecture notes received overwhelming response from the students taking these courses, with many urging us to turn these materials into a textbook. When developing the notes, we tried to place emphasis on the fundamentals and clarity of presentation. Second, the subject matter is important in practice, but it is challenging both for students to learn and for us to teach because it is an integration of several disciplines: structural dynamics, vibration analysis, modern control, and system identification. The primary goal is for students to learn what these tools are without having to take a separate course for each subject and how they are brought together to solve a vibration control problem. Third, although there are many excellent textbooks dedicated to each of the individual disciplines, there are none that bring all of the disciplines together in this specific area of system identification and control of vibrating systems. Recently, there have been several textbooks that integrate dynamics and control, and others that deal with vibrations and control. In all cases, little attention, if any, is paid to system identification, which is still viewed as an advanced and separate topic. If one takes a course in system identification, it is likely that the course's emphasis will be on adaptive estimation and control for single-input single-output models for which there are many textbooks available. In this book, the emphasis is on the multi-input multi-output state–space models that are common in aerospace vibration control practice. We believe that the topic of state–space system identification has reached the critical point of maturity at which it has become useful as a practical tool and should no longer be treated as an advanced topic. Fourth, the book contains a number of recent and useful developments in aerospace control and identification theory. These include virtual passive control, observer and state–space identification, and data-based controller synthesis. Some of these new developments show the close integration of system identification and control theory as done from the state–space perspective, not from the traditional input–output model perspective of adaptive control.

The book begins with a review of basic mathematics (Chaps. 1 and 2) that is needed for understanding the subsequent materials. Often, the source of difficulty for some students is the lack of knowledge of certain specific areas of mathematics. We wish to eliminate this kind of difficulty up front by making the mathematical review part of the course, rather than including it as an appendix as in most other textbooks. Next, we describe basic modeling techniques in Chap. 3, including Newton's laws, D'Alembert's principle, the principle of virtual work, Hamilton's principle, Lagrange's equations, Gibbs–Appell equations, and Kane's equations. Of course, in a single chapter it is not possible to go into depth on each of the above topics, but we do not think that this is necessary for our present purpose. By having many modeling techniques in one place, we hope to bring out the essence of each of the techniques and, more importantly, show how one modeling technique is related to another. This kind of global perspective is important, and a beginner should not be biased toward one or another. Dedicated texts addressing each of the modeling techniques are available and should be used for further studies. The finite-element method presented in Chap. 4 is a very popular technique for the numerical solution of complex problems in engineering. In continuous systems, the formulations describing the system response are governed by partial differential equations as described in Chap. 3. The exact solutions of the partial differential equations satisfying all boundary conditions are possible for only relatively simple systems such as a uniform beam. Numerical techniques must be introduced to discretize the partial differential equations to turn them into linear ordinary differential equations to predict the system response approximately, i.e., the finite-element method. Next, in Chap. 5 we present the core materials of any standard linear vibrations text. We do not create separate chapters for single- and multiple-degree-of-freedom systems, and we do not place any undue emphasis on single-variable systems. Instead, we move quickly from the single-degree-of-freedom to the multiple-degree-of-freedom case and show how both cases can be treated within a common framework. In Chap. 6 the subject of control is introduced, starting with designs that have clear physical interpretation by taking advantage of the second-order differential equations of system dynamics. Modern control is often criticized for its abstract tendency. This chapter shows what can be done if one wishes to stay as close as possible to the physics and what the advantages and disadvantages are with this approach. The next three chapters contain some of the most basic results of modern control theory. Chapter 7 introduces the notion of the state–space model that provides a common platform for the treatment of a wide variety of systems. Chapter 8 addresses the topic of state-feedback control. Some of the key results associated with classical optimal control are presented here. Chapter 9 focuses on dynamic feedback control including the observer-based state-feedback approach. For the most part we do not make a great deal of distinction between continuous-time and discrete-time representations. There are cases in which the mathematics are simpler in continuous time than in discrete time, but the most basic results in continuous-time control have their equivalent discrete-time counterparts. One should see this similarity right away, not after taking a course in linear system theory, which is usually taught in continuous time, and then see it again in a course in digital control theory some time later. By showing how the two approaches are actually similar, we hope that the readers will be able to see when they are different and why. Chapter 10 gives basic concepts and properties of state–space system identification. Computation of system Markov parameters (pulse-response time history) is described. Most time-domain methods in modal testing are based on the pulse-response time history to identify modal parameters such as system frequencies, damping ratio, and mode shapes at the sensor points. Here identification techniques are derived that provide the state–space models needed for the design of a modern controller. In Chap. 11, the problem of predictive control is addressed. Originating in the chemical process community, predictive control has found its way into aerospace control problems. Some of the latest results in the integration of system identification and predictive control to produce the so-called data-based controller synthesis are presented here.

This book can be used in a junior-, senior-, or first-year-graduate-level course. Very minimal background is required other than elementary differential equations and linear algebra. When used at the graduate level, certain specific directions can be dealt with in more detail at the discretion of the instructor. The materials here provide a firm formulation so that the instructor's own research can easily be incorporated into such a course. Thus, the students will get exposed to the instructor's research with proper understanding and insight to the solution procedure for a general vibration/control problem. In our opinion, the ability to see complicated things in the simplest way across disciplinary boundaries is a prerequisite in creating new and fundamental knowledge. We believe that, when properly presented, many complicated techniques will look simple, and, being far from perfect, we hope that our attempt at this is a useful one.

We probably would not have been able to write this book or develop a number of techniques presented here without considerable support from our organizations including NASA Langley Research Center, George Washington University, Princeton University, and Dartmouth College. We sincerely acknowledge the influence of our colleagues John L. Junkins, Richard W. Longman, Lucas G. Horta, Leonard Meirovitch, Daniel J. Inman, Earl H. Dowell, Raymond G. Kvaternik, and our undergraduate and graduate students over the years. Their friendship and encouragement provided the motivation for writing this text.

We want to thank our family members, Lily Juang, Philo Juang, Derek K. Juang, Suzu Phan, and Daniel Phan for their support and understanding during those numerous dislocations in both time and space. Last, but not least, we would like to thank our parents for their love and support.

1

Ordinary Differential Equations

1.1 Introduction

The formulation of many problems in physical and social sciences involves differential equations that express the relationship among the derivatives of one or more unknown functions with respect to the independent variables (Refs. [1–7]). In an ordinary differential equation (ODE), all derivatives are with respect to a single independent variable. The order of a differential equation is the order of the *highest* derivative that appears in the equation. For example, Newton's law describing the angular position of an oscillating pendulum consisting of a massless string of length ℓ and a point mass m under the influence of gravity (see Fig. 1.1) takes the form of a second-order differential equation:

$$\frac{d^2\theta}{dt^2} + \frac{g}{\ell}\sin\theta = 0,$$

where $\theta = \theta(t)$ and t denotes the time variable. Because of the term $\sin\theta$, the above is a nonlinear differential equation whose solution has been extensively studied. Unfortunately, to date, there is no comprehensive theory to solve general nonlinear differential equations analytically. This stands in contrast to the well-developed theory of linear differential equations. In the above example, if the angle of oscillation is small, then $\sin\theta \approx \theta$, and the solution can be approximated from the following linear differential equation:

$$\frac{d^2\theta}{dt^2} + \frac{g}{\ell}\theta = 0.$$

In most cases, a linear differential equation can be obtained by linearization of a nonlinear one. We are fortunate that many physical systems can be adequately described by linear differential equations.

Perhaps one of the most famous ODE is that of a linear spring–mass–damper system (see Fig. 1.2) whose dynamics is governed by

$$m\frac{d^2w}{dt^2} + \zeta\frac{dw}{dt} + kw = f,$$

where w denotes the position of the mass from its equilibrium position, $f(t)$ is the forcing function, and m, ζ, and k denote the mass, damping, and stiffness coefficients, respectively. In practice, many vibration problems are treated as linear even if they

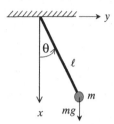

Figure 1.1. An oscillating pendulum.

involve a large number of degrees of freedom. This is because they typically deal with small-amplitude motion about an equilibrium position. The control of such linear systems can be handled conveniently by linear control, which represents a major portion of control theory in general.

In this chapter, we will show how to solve homogeneous ODEs with constant coefficients. The characteristic equation is derived and its solutions are discussed including all possible cases such as distinct roots, repeated roots, etc. In the section of the non-homogeneous ODEs with constant coefficients, the solution of the nonhomogeneous equations and its properties are discussed. The last section briefly introduces coupled differential equations and the definition of a matrix differential equation.

1.2 Homogeneous ODE with Constant Coefficients

An nth-order linear homogeneous ODE has the form

$$a_0 \frac{d^n y}{dt^n} + a_1 \frac{d^{n-1} y}{dt^{n-1}} + \cdots + a_{n-1} \frac{dy}{dt} + a_n y = 0, \tag{1.1}$$

where $a_0, a_1, \ldots, a_n$ are real constant coefficients. Note that the right-hand side is zero in this case. In a physical system, a homogeneous ODE may correspond to the situation in which the dynamic system does not have any input or force applied to it and its response is due to some nonzero initial conditions.

EXAMPLE 1.1

The following equation is a homogeneous ODE:

$$\frac{d^2 y}{dt^2} + 3 \frac{dy}{dt} + 2y = 0.$$

This may represent a spring–mass–dashpot system with mass $m = 1$, damping coefficient $\zeta = 3$, and spring constant $k = 2$.

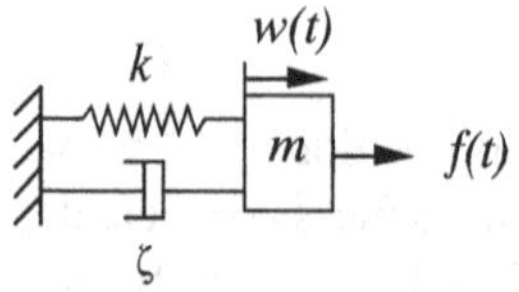

Figure 1.2. A spring–mass–damper system.

1.2.1 General Solution

Functions of $y(t)$ that satisfy Eq. (1.1) are called the homogeneous solutions of the ODE. It is anticipated that the solutions are of the form $y = e^{\gamma t}$ with appropriate values of γ. To find these values, we substitute $y = e^{\gamma t}$ into the differential equations and simplify the resultant expression to obtain

$$(a_0\gamma^n + a_1\gamma^{n-1} + \cdots + a_{n-1}\gamma + a_n)e^{\gamma t} = 0. \tag{1.2}$$

The equation

$$a_0\gamma^n + a_1\gamma^{n-1} + \cdots + a_{n-1}\gamma + a_n = 0. \tag{1.3}$$

is called the characteristic equation of the ODE. A polynomial of degree n has n roots, say, $\gamma_1, \gamma_2, \ldots, \gamma_n$. Therefore, the characteristic equation can be written in the form

$$a_0(\gamma - \gamma_1)(\gamma - \gamma_2)\cdots(\gamma - \gamma_n) = 0. \tag{1.4}$$

Solving the roots of characteristic equation (1.3) yields the values of $\gamma_1, \gamma_2, \ldots, \gamma_n$. Each value of γ represents one solution to the ODE. The general solution is a linear combination of these solutions:

$$y = c_1 e^{\gamma_1 t} + c_2 e^{\gamma_2 t} + \cdots + c_n e^{\gamma_1 n}. \tag{1.5}$$

The simplest situation is that in which all the characteristic roots $\gamma_1, \gamma_2, \ldots, \gamma_n$ are real and distinct. Minor complexities will occur if there are some complex roots or repeated roots. The following will summarize various possible cases.

CASE 1: REAL AND NONREPEATED ROOTS

If all the roots of characteristic equation (1.3) are distinct and real, the general solution is

$$y = c_1 e^{\gamma_1 t} + c_2 e^{\gamma_2 t} + \cdots + c_n e^{\gamma_1 n}.$$

The coefficients $c_1, c_2, \ldots, c_n$ are determined by initial conditions, which will be addressed later in this chapter.

EXAMPLE 1.2

The characteristic equation for the differential equation

$$\frac{d^2y}{dt^2} + 3\frac{dy}{dt} + 2y = 0$$

is

$$\gamma^2 + 3\gamma + 2 = 0,$$

which has two roots, $\gamma_1 = -1$ and $\gamma_2 = -2$. The general (homogeneous) solution is

$$y = c_1 e^{-1t} + c_2 e^{-2t}.$$

CASE 2: COMPLEX ROOTS

If the characteristic equation has complex roots, they must occur in complex-conjugate pairs, $\sigma \pm i\omega$, as the coefficients $a_0, a_1, \ldots, a_n$ are real numbers. Provided that the roots are not repeated, the corresponding solutions that make up the general solution will have the form $e^{(\sigma+i\omega)t}$, $e^{(\sigma-i\omega)t}$. We can carry the mathematics one step further by considering the linear combination

$$
\begin{aligned}
Ae^{(\sigma+i\omega)t} + Be^{(\sigma-i\omega)t} &= Ae^{\sigma t}e^{i\omega t} + Be^{\sigma t}e^{-i\omega t} \\
&= Ae^{\sigma t}(\cos\omega t + i\sin\omega t) + Be^{\sigma t}(\cos\omega t - i\sin\omega t) \\
&= (A+B)e^{\sigma t}\cos\omega t + i(A-B)e^{\sigma t}\sin\omega t \\
&= c_1 e^{\sigma t}\cos\omega t + c_2 e^{\sigma t}\sin\omega t.
\end{aligned}
\tag{1.6}
$$

The last equality is possible because of the expectation that the solution of a physical system is real, so that A and B will be such that the combinations $(A+B)$ and $i(A-B)$ are indeed real numbers. For convenience, $(A+B)$ and $i(A-B)$ are denoted by the real coefficients c_1 and c_2, respectively. For this reason, we normally use the real-valued solutions

$$
e^{\sigma t}\cos\omega t, \quad e^{\sigma t}\sin\omega t
$$

as solutions that make up the general solution of the ODE.

EXAMPLE 1.3

Find the general solution of

$$
\frac{d^2 y}{dt^2} + \frac{dy}{dt} + y = 0.
$$

The characteristic equation is $\gamma^2 + \gamma + 1 = 0$, which has two roots:

$$
\gamma_1 = -\frac{1}{2} + i\frac{\sqrt{3}}{2}, \quad \gamma_2 = -\frac{1}{2} - i\frac{\sqrt{3}}{2}.
$$

The general solution is

$$
y(t) = c_1 e^{-\frac{1}{2}t}\cos\frac{\sqrt{3}}{2}t + c_2 e^{-\frac{1}{2}t}\sin\frac{\sqrt{3}}{2}t.
$$

CASE 3: REPEATED REAL ROOTS

If the roots of the characteristic equation are not distinct, i.e., some of the roots are repeated, then some additional solutions to the ODE must be found to make up the general solution. Fortunately, it can be shown that they have simple forms. If the real root γ is repeated s times, then the corresponding solutions are not only

$$
e^{\gamma t},
\tag{1.7}
$$

as known before, but also

$$
te^{\gamma t}, \quad t^2 e^{\gamma t}, \ldots, \quad t^{s-1}e^{\gamma t}.
\tag{1.8}
$$

It is easy to verify that the above solutions do indeed satisfy the homogeneous ODE.

EXAMPLE 1.4

Find the general solution of

$$\frac{d^2y}{dt^2} + 2\frac{dy}{dt} + y = 0.$$

The characteristic equation is $\gamma^2 + 2\gamma + 1 = (\gamma + 1)^2 = 0$, which has two repeated roots, $\gamma_1 = -1$ and $\gamma_2 = -1$. The general solution is

$$y(t) = c_1 e^{-t} + c_2 t e^{-t}.$$

CASE 4: REPEATED COMPLEX ROOTS

If the complex root $\sigma + i\omega$ is repeated s times, then we have $2s$ solutions because the complex roots always appear as conjugate pairs (for characteristic equations with real coefficients). These roots can be shown to be

$$e^{\sigma t} \cos \omega t, \quad t e^{\sigma t} \cos \omega t, \quad t^2 e^{\sigma t} \cos \omega t, \ldots, \quad t^{s-1} e^{\sigma t} \cos \omega t,$$
$$e^{\sigma t} \sin \omega t, \quad t e^{\sigma t} \sin \omega t, \quad t^2 e^{\sigma t} \sin \omega t, \ldots, \quad t^{s-1} e^{\sigma t} \sin \omega t. \tag{1.9}$$

Again, the above solutions can be easily shown to satisfy the homogeneous ODE.

EXAMPLE 1.5

Find the general solution of

$$\frac{d^4y}{dt^4} + 2\frac{d^2y}{dt^2} + y = 0.$$

The characteristic equation is $\gamma^4 + 2\gamma^2 + 1 = (\gamma + i)^2(\gamma - i)^2 = 0$, which has two repeated roots:

$$\gamma_1 = i, \quad \gamma_2 = i, \quad \gamma_3 = -i, \quad \gamma_4 = -i.$$

The general solution is

$$y(t) = c_1 \cos t + c_2 t \cos t + c_3 \sin t + c_4 t \sin t.$$

1.2.2 Multiple Roots of $\gamma^s - \alpha = 0$

In solving for the characteristic roots, it is sometimes necessary to solve for the multiple roots of a number α, which can be real or complex. In general, α may be written as

$$\alpha = R e^{i\theta}, \tag{1.10}$$

where R is the amplitude and θ is the phase angle. For a real number α, θ is zero or integer multipliers of 2π. Therefore, we need to solve for the s roots of the equation

$$\gamma^s = R e^{i\theta}. \tag{1.11}$$

One root is obviously,

$$\gamma_1 = R^{1/s} e^{i\theta/s}. \tag{1.12}$$

To find the remaining $s - 1$ roots, first realize that the angle of a complex number is determined up to only a multiple of 2π. This means that the number α can be represented by $\alpha = R e^{i(\theta + 2n\pi)}$, where n can be zero or any positive or negative number. Thus, to be more general, the equation to be solved becomes

$$\gamma^s = R e^{i(\theta + 2n\pi)}, \tag{1.13}$$

whose roots are

$$\gamma = R^{1/s} e^{i(\theta + 2n\pi)/s}. \tag{1.14}$$

Setting $n = 0$ gives us the root γ_1, as before. Setting $n = 1, 2, \ldots, s - 1$ will produce the remaining $s - 1$ roots for a total of s roots. It can easily be verified that setting n equal to any other integer value, either positive or negative, will reproduce one of these s roots.

EXAMPLE 1.6

Find the four roots of $\gamma^4 + 1 = 0$. Application of Eq. (1.13) yields

$$\gamma^4 = -1 = e^{i(\pi + 2n\pi)}.$$

Hence,

$$\gamma = e^{i(\pi/4 + n\pi/2)}.$$

We obtain the four roots by setting $n = 0, 1, 2, 3$, which give $(1 + i)/\sqrt{2}, (1 - i)/\sqrt{2}, (-1 + i)/\sqrt{2}$ and $(-1 - i)/\sqrt{2}$. The four roots are located on the complex plane shown in Fig. 1.3.

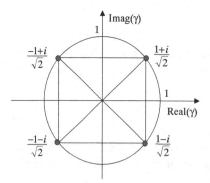

Figure 1.3. Four complex roots.

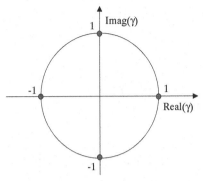

Figure 1.4. Two real and two pure imaginary roots.

EXAMPLE 1.7

Find the four roots of $\gamma^4 - 1 = 0$. Since $\gamma^4 = 1 = e^{i(2n\pi)}$ we have

$$\gamma = e^{i(n\pi/2)}.$$

Setting $n = 0, 1, 2, 3$, yields the four roots:

$$1, \; i, \; -1, \; -i.$$

The four roots are located on the complex plane shown in Fig. 1.4. Note that in both Examples 1.6 and 1.7 the roots are equally spaced in the complex plane. This is a general property of the complex roots of a number.

1.2.3   Determination of Coefficients

The general form of the solution to a homogeneous ODE is known from the roots of the characteristic equation. However, the general solution contains the unknown coefficients $c_1, c_2, \ldots, c_n$, which need to be determined. To be able to solve for the unknown coefficients, new information is needed. The new information comes in the form of the initial conditions that must be specified. The initial conditions are the specified values of

$$y(0), \; \left.\frac{dy}{dt}\right|_{t=0}, \; \left.\frac{d^2 y}{dt^2}\right|_{t=0}, \ldots, \; \left.\frac{d^{n-1} y}{dt^{n-1}}\right|_{t=0}. \tag{1.15}$$

Satisfying n initial conditions gives us n linear algebraic equations with n unknowns $c_1, c_2, \ldots, c_n$. Solving this set of algebraic equations will give us the values of $c_1, c_2, \ldots, c_n$. In a physical system, the initial conditions can be the values of its initial position and initial velocity. It is clear that the response of a system can be uniquely determined only after its initial conditions are specified.

EXAMPLE 1.8

Find the solution of the following ODE:

$$\frac{d^2 y}{dt^2} + 5\frac{dy}{dt} + 6y = 0$$

subject to the initial conditions $y(0) = 0$ and $y'(0) = 1$, where $y'(0) = \frac{dy}{dt}\big|_{t=0} = 1$. The general solution is

$$y(t) = c_1 e^{-2t} + c_2 e^{-3t}.$$

To satisfy the initial conditions, the coefficients must satisfy

$$c_1 + c_2 = 0,$$
$$2c_1 + 3c_2 = -1.$$

Solving these equations gives $c_1 = 1$ and $c_2 = -1$. Thus, the solution that satisfies the initial conditions is

$$y(t) = e^{-2t} - e^{-3t}.$$

1.3 Nonhomogeneous ODE with Constant Coefficients

When the right-hand-side term is not zero, the ODE is said to be nonhomogeneous. Thus, a general nonhomogeneous ODE with constant coefficients has the form

$$a_0 \frac{d^n y}{dt^n} + a_1 \frac{d^{n-1} y}{dt^{n-1}} + \cdots + a_{n-1} \frac{dy}{dt} + a_n y = f(t). \tag{1.16}$$

In a physical system, this may correspond to the case in which the dynamic system is subjected to some input or forcing function $f(t)$.

1.3.1 General Solution

The general solution of a nonhomogeneous ODE is the sum of the solution of the homogeneous part of the ODE and a particular solution of the nonhomogeneous ODE. It is simple to show mathematically why this is the case. Let the solution of the homogeneous part of the ODE be denoted by $y_h(t)$ and the particular solution by $y_p(t)$,

$$a_0 \frac{d^n y_h}{dt^n} + a_1 \frac{d^{n-1} y_h}{dt^{n-1}} + \cdots + a_{n-1} \frac{dy_h}{dt} + a_n y_h = 0,$$

$$a_0 \frac{d^n y_p}{dt^n} + a_1 \frac{d^{n-1} y_p}{dt^{n-1}} + \cdots + a_{n-1} \frac{dy_p}{dt} + a_n y_p = f(t). \tag{1.17}$$

Adding the two equations and recognizing the property

$$\frac{d^i (y_h + y_p)}{dt^i} = \frac{d^i y_h}{dt^i} + \frac{d^i y_p}{dt^i} \tag{1.18}$$

for any i, we have

$$a_0 \frac{d^n (y_h + y_p)}{dt^n} + a_1 \frac{d^{n-1}(y_h + y_p)}{dt^{n-1}} + \cdots + a_{n-1} \frac{d(y_h + y_p)}{dt}$$

$$+ a_n(y_h + y_p) = f(t), \tag{1.19}$$

Table 1.1: Forcing Functions and Particular Solutions

Forcing Function $f(t)$	Particular Solution $y_p(t)$ to Try
Constant	a
t	$at + b$
t^2	$at^2 + bt + c$
$\sin \omega t$	$a \sin \omega t + b \cos \omega t$
$e^{\sigma t}$	$ae^{\sigma t}$
$t^2 e^{\sigma t} \cos \omega t$	$at^2 e^{\sigma t} \cos \omega t + bt^2 e^{\sigma t} \sin \omega t$

which implies that

$$y(t) = y_h(t) + y_p(t) \tag{1.20}$$

is the general solution to the nonhomogeneous ODE. From previous sections, we know how to find the general form of the homogeneous solution. It is very important to realize that the unknown coefficients $c_1, c_2, \ldots, c_n$ in the homogeneous solution must not be determined at this stage. These coefficients can be determined only after a particular solution has been found and the general solution is constructed. This is because any initial conditions of the system are for $y_h(t) + y_p(t)$, not $y_h(t)$ alone.

1.3.2   Particular Solution

For a simple forcing function $f(t)$, it is sometimes possible to guess the form of the particular solution with a certain number of undetermined coefficients that will make this candidate solution satisfy the nonhomogeneous ODE. This method is known as the method of undetermined coefficients. Table 1.1 gives some simple cases that are commonly encountered in practice (for control applications).

EXAMPLE 1.9
Consider the nonhomogeneous ODE

$$\frac{d^2y}{dt^2} + 2\frac{dy}{dt} + y = e^t.$$

We try $y_p(t) = ae^t$ as a candidate particular solution. Substituting ae^t back to the differential equation will yield $a = 1/4$. Thus $y_p(t) = e^t/4$ is a particular solution.

In the following, we provide a justification for the above process. Because we know how to handle a homogeneous ODE, we try to find a way to turn our nonhomogeneous ODE into a homogeneous one by looking for a differential operator L such that

$$L\left\{\frac{d^2y}{dt^2} + 2\frac{dy}{dt} + y\right\} = L\{e^t\} = 0.$$

Because it is required that e^t be a solution to the homogeneous problem $L\{e^t\} = 0$, $\gamma = 1$ must be a characteristic root. A corresponding characteristic equation is then $\gamma - 1 = 0$, which implies that the desired differential operator is $L = d(.)/dt - 1$. Having found L, we now apply it to the original nonhomogeneous ODE to turn it into a homogeneous one:

$$\left(\frac{d}{dt} - 1\right)\left(\frac{d^2y}{dt^2} + 2\frac{dy}{dt} + y\right) = 0.$$

The solution to this homogeneous ODE, which we know how to solve, will contain a particular solution to the original nonhomogeneous problem. The characteristic equation for this homogeneous ODE is

$$(\gamma - 1)(\gamma^2 + 2\gamma + 1) = (\gamma - 1)(\gamma + 1)^2 = 0.$$

Thus, the general solution has the form

$$y(t) = ae^t + c_1e^{-t} + c_2te^{-t}.$$

The constant a can be determined by substituting the above general solution back into the original nonhomogeneous ODE. Note that the part $c_1e^{-t} + c_2te^{-t}$ is simply the solution to the homogeneous part of the ODE and will be eliminated automatically. Completing this procedure will yield $a = 1/4$. Hence,

$$y_p(t) = \frac{1}{4}e^t$$

is a particular solution and $c_1e^{-t} + c_2te^{-t}$ is the solution to the homogeneous part of the ODE:

$$y_h(t) = c_1e^{-t} + c_2te^{-t}.$$

The general solution is the sum of the homogeneous solution and a particular solution, as claimed. The constants c_1, c_2 can be determined if the general solution $y(t) = y_h(t) + y_p(t)$ is made to satisfy the initial conditions

$$y(0) \quad \text{and} \quad \frac{dy}{dt}\Big|_{t=0},$$

which must be specified.

1.4 Coupled Ordinary Differential Equations

The earlier sections in this chapter address the ODE of only a single variable. This is the case when single-input–single-output (SISO) systems are considered. The dynamics of a multiple-input–multiple-output (MIMO) system can be described by a set of coupled ODEs. The set of equations

$$\frac{dx_1}{dt} = a_{11}x_1 + a_{12}x_2 + f_1(t),$$

$$\frac{dx_2}{dt} = a_{21}x_1 + a_{22}x_2 + f_2(t) \tag{1.21}$$

is one such example. Here the inputs (or forcing functions) are $f_1(t)$ and $f_2(t)$. To handle this type of problem, it is mathematically convenient to express it in matrix form:

$$\frac{d}{dt}\begin{bmatrix} x_1 \\ x_2 \end{bmatrix} = \begin{bmatrix} a_{11} & a_{12} \\ a_{21} & a_{22} \end{bmatrix}\begin{bmatrix} x_1 \\ x_2 \end{bmatrix} + \begin{bmatrix} f_1(t) \\ f_2(t) \end{bmatrix}. \tag{1.22}$$

This brings us to the study of linear algebra in the next chapter, which deals with matrix operations.

EXAMPLE 1.10

The set of two coupled differential equations

$$\frac{dx_1}{dt} = -2x_1 + x_2 + 2e^{-t},$$

$$\frac{dx_2}{dt} = x_1 - 2x_2 + 3t$$

can be represented in matrix form as

$$\frac{d}{dt}\begin{bmatrix} x_1 \\ x_2 \end{bmatrix} = \begin{bmatrix} -2 & 1 \\ 1 & -2 \end{bmatrix}\begin{bmatrix} x_1 \\ x_2 \end{bmatrix} + \begin{bmatrix} 2e^{-t} \\ 3t \end{bmatrix}.$$

When we say coupled, we really mean that $x_1(t)$ in the first differential equation cannot be solved without knowing $x_2(t)$ nor can $x_2(t)$ in the second differential equation be solved without knowing $x_1(t)$. In fact, both $x_1(t)$ and $x_2(t)$ must be solved simultaneously.

1.5 Concluding Remarks

In this chapter, we have focused on the mathematics of scalar linear differential equations (with constant coefficients) for single-degree-of-freedom systems. The solutions of linear differential equations are discussed, including homogeneous solutions and particular solutions. In the next chapter, this elementary knowledge will help us handle equations that describe multiple-degree-of-freedom systems by using matrix theory. These first two chapters will provide the basic mathematical foundation for studying in subsequent chapters the dynamics and control of such systems.

1.6 Problems

1.1 Assume that $y_{h1}(t)$ and $y_{h2}(t)$ satisfy the homogeneous part of the following linear equation,

$$a_0\frac{d^n y}{dt^n} + a_1\frac{d^{n-1} y}{dt^{n-1}} + \cdots + a_{n-1}\frac{dy}{dt} + a_n y = h(t),$$

and that $y_{p1}(t)$ and $y_{p2}(t)$ are two linearly independent particular solutions. Prove that both $c_1 y_{h1}(t) + c_2 y_{h2}(t)$ (for any constant values of c_1 and c_2) and $y_{p1}(t) - y_{p2}(t)$ satisfy the homogeneous part of the above equation.

1.2   Solve the following differential equations:

(a) $\dfrac{d^2y}{dt^2} - \dfrac{dy}{dt} - 2y = 0,$

(b) $\dfrac{d^3y}{dt^3} - \dfrac{d^2y}{dt^2} - \dfrac{dy}{dt} + y = 0,$

(c) $\dfrac{d^2y}{dt^2} - 2\dfrac{dy}{dt} + 2y = 0,$

(d) $\dfrac{d^3y}{dt^3} - y = 0.$

1.3   Solve the following differential equations that describe the structural problems noted:

(a) $\dfrac{d^2y}{dt^2} + \omega^2 y = 0$ (free vibration of a string under tension),

(b) $\dfrac{d^4y}{dt^4} - \omega^4 y = 0$ (free vibration of a beam),

(c) $\dfrac{d^4y}{dt^4} + 16\omega^4 y = 0$ (beam on an elastic foundation),

(d) $\dfrac{d^4y}{dt^4} - 2\omega^2\dfrac{dy^2}{dt^2} + \omega^4 y = 0$ (bending of an elastic plate),

where ω is a nonzero constant.

1.4   Find the complete solution of each of the following differential equations:

(a) $\dfrac{d^2y}{dt^2} + \omega^2 y = a e^{i\omega_f t}, \quad (i = \sqrt{-1}),$

(b) $\dfrac{d^4y}{dt^4} - \omega^4 y = a e^{i\omega_f t},$

(c) $\dfrac{d^4y}{dt^4} - 2\omega^2\dfrac{dy^2}{dt^2} + \omega^4 y = a e^{i\omega_f t},$

where a, ω, and ω_f are nonzero constants.

1.5   Solve the following differential equation:

$$\dfrac{d^4 y_1}{dt^4} + 2\dfrac{d^2 y_1}{dt^2} = (1 - \omega^2)\sin\omega t,$$

where ω is a constant. Prove that the solution also satisfies the simultaneous equations

$$\dfrac{d^2 y_1}{dt^2} - y_2 + y_1 = \sin\omega t,$$

$$\dfrac{d^2 y_2}{dt^2} + y_2 - y_1 = 0.$$

1.6 Show that the following equation,

$$\frac{d^4y}{dt^4} + 2\frac{d^2y}{dt^2} = \frac{d^2u}{dt^2} + u,$$

can be represented in matrix form as

$$\frac{d}{dt}\begin{bmatrix} x_1 \\ x_2 \\ x_3 \\ x_4 \end{bmatrix} = \begin{bmatrix} 0 & 0 & 0 & 0 \\ 1 & 0 & 0 & 0 \\ 0 & 1 & 0 & -2 \\ 0 & 0 & 1 & 0 \end{bmatrix}\begin{bmatrix} x_1 \\ x_2 \\ x_3 \\ x_4 \end{bmatrix} + \begin{bmatrix} 1 \\ 0 \\ 0 \\ 0 \end{bmatrix}u$$

$$y = \begin{bmatrix} 0 & 1 & 0 & -1 \end{bmatrix}\begin{bmatrix} x_1 \\ x_2 \\ x_3 \\ x_4 \end{bmatrix},$$

where x_1, x_2, x_3, and x_4 are commonly called state variables.

BIBLIOGRAPHY

[1] Arnold, V. I., *Ordinary Differential Equations*, Translated by Richard A. Silverman, Cambridge University Press, U.K., 1991.

[2] Coddington, E. A. and Landin J., *An Introduction to Ordinary Differential Equations*, Dover, New York, 1989.

[3] Hildebrand, F. B., *Advanced Calculus for Applications*, Prentice-Hall, Englewood Cliffs, NJ, 1962.

[4] Hurewicz, W., *Lectures on Ordinary Differential Equations*, Dover, New York, 1990.

[5] Ross, S. L., *Introduction to Ordinary Differential Equations*, 3rd ed., John Wiley & Sons, New York, 1980.

[6] Sanchez, D. A., *Ordinary Differential Equations & Stability Theory: An Introduction*, Dover, New York, 1979.

[7] Wilson, H. K., *Ordinary Differential Equations*, Addison-Wesley, New York, 1971.

2

Elementary Matrix Algebra

2.1 Introduction

In the previous chapter, we paid attention to the solution of a scalar ODE. A scalar ODE describes the dynamics of a single-input single output (SISO) dynamic system. In general, a system can have several inputs and several outputs. Such a multi-input multi-output (MIMO) dynamic system can be described by a set of coupled ODEs involving several input and output variables. We must solve these equations simultaneously to obtain the dynamic response of the system. We can find such a solution by using matrix theory, which conveniently rearranges the set of coupled equations in a compact form. Instead of having several coupled scalar ODEs, we now have one single *matrix* ODE. Matrix operations can be performed on the matrix differential equations, and the final solution can be expressed in a simple form. It is important to realize that applying such matrix operations is equivalent to operating on the scalar ODEs individually, although with the scalar approach it is very easy to miss the general picture. Thus, matrix theory is one fine example in which one gains by simply rewriting an old problem in a new form that can be analyzed more effectively. Another reason why the matrix formulations are so useful is that a set of coupled differential equations of any order can be rewritten as a single matrix differential equation of *first* order. The same statement applies for a single scalar ODE of any order as well. What this means in our problem is that we need to focus only on the mathematics of a single first-order matrix ODE, knowing that any higher-order problem can be reduced to this one.

In this chapter, we first review the basic concepts of matrix algebra (Ref. [1–2]). This will form the basis of handling first-order matrix differential equations. This review is adequate for our present purpose which is dynamics and control of (linear) flexible structures. It is not a complete review of linear algebra or matrix differential equations per se.

2.2 Vectors and Matrices

A matrix is an array of elements arranged in rows and columns. The elements that make up a matrix can be real, complex, or functions of single or multiple variables. An $m \times n$ matrix has m times n elements arranged as m rows and n columns:

$$A = [a_{ij}] = \begin{bmatrix} a_{11} & a_{12} & \cdots & a_{1n} \\ a_{21} & a_{22} & \cdots & a_{2n} \\ \vdots & \vdots & \ddots & \vdots \\ a_{m1} & a_{m2} & \cdots & a_{mn} \end{bmatrix}. \tag{2.1}$$

A column vector can be viewed as a matrix with only one column. Similarly, a row vector can be viewed as a matrix with only one row. If the number of rows of a matrix is equal to the number of columns, the matrix is called a square matrix. Otherwise, it is a rectangular matrix. A diagonal matrix is a special square matrix in which all elements not on the main diagonal are zero:

$$A = \begin{bmatrix} a_{11} & 0 & \cdots & 0 \\ 0 & a_{22} & \cdots & 0 \\ \vdots & \vdots & \ddots & \vdots \\ 0 & 0 & \cdots & a_{nn} \end{bmatrix} = \text{diag}[a_{11}, a_{22}, \ldots, a_{nn}]. \tag{2.2}$$

A special diagonal matrix in which all the diagonal elements are the same and equal to 1 is called an identity matrix and is normally denoted by I:

$$I = \begin{bmatrix} 1 & 0 & \cdots & 0 \\ 0 & 1 & \cdots & 0 \\ \vdots & \vdots & \ddots & \vdots \\ 0 & 0 & \cdots & 1 \end{bmatrix}. \tag{2.3}$$

A matrix in which all of its elements are zero is called a zero (or null) matrix.

2.3 Basic Matrix Operations

The basic matrix operations reviewed in this section are matrix addition, multiplication, transpose, complex-conjugate transpose, matrix trace, and inner and outer products.

MATRIX ADDITION

Matrix addition can be performed only on matrices of the same dimensions. The sum of two matrices is a matrix whose elements is the sum of the corresponding elements of the two constituent matrices. Mathematically, we write $C = A + B$, where

$$[c_{ij}] = [a_{ij}] + [b_{ij}] \tag{2.4}$$

for all i and j.

MATRIX MULTIPLICATION

Matrix multiplication is defined for only matrices of compatible dimensions. A matrix A of dimensions $m \times n$ multiplied by a matrix B of dimensions $n \times q$ will produce a matrix C of dimensions $m \times q$. Mathematically, we write $C = AB$, where

$$[c_{ij}] = \left[\sum_{k=1}^{n} a_{ik} b_{kj} \right]. \tag{2.5}$$

A matrix multiplied by a scalar has all of its elements multiplied by the scalar, that is,

$$\alpha[a_{ij}] = [\alpha a_{ij}]. \tag{2.6}$$

Note that if a scalar is thought of as a 1×1 matrix, the compatibility rule for matrix multiplication does not apply here.

Matrix multiplication satisfies the associative law

$$(AB)C = A(BC) \tag{2.7}$$

and distributive law

$$A(B + C) = AB + AC, \tag{2.8}$$

but generally does not commute:

$$AB \neq BA. \tag{2.9}$$

MATRIX TRANSPOSE

The transpose of A is commonly denoted by A^T, which can be obtained simply by interchanging rows and columns of A, i.e.,

$$[a_{ij}]^T = [a_{ji}].$$

Some basic properties of the transpose operation are

$$(A + B)^T = A^T + B^T,$$
$$(cA)^T = cA^T,$$
$$(AB)^T = B^T A^T,$$
$$(A^T)^T = A,$$

where A and B are matrices but c is a scalar.

MATRIX COMPLEX-CONJUGATE TRANSPOSE

The analogy of the transpose of a real-valued matrix is its complex-conjugate transpose. This operation is performed by taking the regular transpose first and then taking the complex conjugate of the resultant matrix. The complex-conjugate transpose of a matrix is sometimes denoted by A^*.

MATRIX TRACE

The trace of a square matrix is simply the sum of its diagonal elements:

$$\text{trace}(A) = a_{11} + a_{22} + \cdots + a_{nn} = \sum_{k=1}^{n} a_{kk}.$$

INNER PRODUCT

The inner product of two vectors x and y of the same dimensions (say $n \times 1$) is a single value obtained from

$$x^T y = y^T x = \sum_{k=1}^{n} x_k y_k.$$

OUTER PRODUCT

The outer product of two vectors, say $n \times 1$ and $m \times 1$, is an $n \times m$ matrix defined as

$$xy^T = \begin{bmatrix} x_1 y_1 & x_1 y_2 & \cdots & x_1 y_m \\ x_2 y_1 & x_2 y_2 & \cdots & x_2 y_m \\ \vdots & \vdots & \ddots & \vdots \\ x_n y_1 & x_n y_2 & \cdots & x_n y_m \end{bmatrix}.$$

2.4 Special Matrices

There are a number of matrices that are particular useful because they carry special properties. Let A be an $n \times n$ real matrix and x be an $n \times 1$ real vector. The matrix A is said to be

symmetric	if $A^T = A$,
skew symmetric (or antisymmetric)	if $A^T = -A$,
positive definite	if $x^T A x > 0$ for all nonzero x,
positive semidefinite	if $x^T A x \geq 0$ for all nonzero x,
orthogonal	if $A^T A = D = $ diagonal matrix,
orthonormal (unitary)	if $A^T A = I = $ identity matrix.

If A is an $n \times n$ complex matrix and x is an $n \times 1$ complex vector, the matrix A is said to be

Hermitian (or self-adjoint)	if $A^* = A$,
skew (or anti-) Hermitian	if $A^* = -A$,
positive definite	if $x^* A x > 0$ for all nonzero x,
positive semidefinite	if $x^* A x \geq 0$ for all nonzero x,
orthogonal	if $A^* A = D = $ diagonal matrix,
orthonormal (or unitary)	if $A^* A = I = $ identity matrix.

EXAMPLE 2.1

A few examples are given below to illustrate the definitions shown above.

(1) $\begin{bmatrix} 1 & 4 \\ 4 & 2 \end{bmatrix}$ is symmetric.

(2) $\begin{bmatrix} 1 & 1+3i \\ 1-3i & 2 \end{bmatrix}$ is Hermitian.

(3) $\begin{bmatrix} 0 & 1 \\ -1 & 0 \end{bmatrix}$ is skew symmetric.

(4) $\begin{bmatrix} 2 & 1 \\ 1 & 4 \end{bmatrix}$ is positive definite because

$$\begin{bmatrix} x_1 & x_2 \end{bmatrix} \begin{bmatrix} 2 & 1 \\ 1 & 4 \end{bmatrix} \begin{bmatrix} x_1 \\ x_2 \end{bmatrix} = 2x_1^2 + 2x_1 x_2 + 4x_2^2$$
$$= x_1^2 + 3x_2^2 + (x_1 + x_2)^2 > 0$$

for all $x_1, x_2 \neq 0$.

(5) $\begin{bmatrix} 1 & 0 \\ 0 & 0 \end{bmatrix}$ is positive semidefinite because

$$\begin{bmatrix} x_1 & x_2 \end{bmatrix} \begin{bmatrix} 1 & 0 \\ 0 & 0 \end{bmatrix} \begin{bmatrix} x_1 \\ x_2 \end{bmatrix} = x_1^2 \geq 0$$

for any x_2. For example, the quadratic form is zero for $x_1 = 0, x_2 = 1$.

(6) $\begin{bmatrix} 1 & 1 \\ -1 & 1 \end{bmatrix}$ is orthogonal because

$$\begin{bmatrix} 1 & 1 \\ -1 & 1 \end{bmatrix}\begin{bmatrix} 1 & -1 \\ 1 & 1 \end{bmatrix} = \begin{bmatrix} 2 & 0 \\ 0 & 2 \end{bmatrix}$$

is a diagonal matrix.

(7) $\frac{1}{\sqrt{2}}\begin{bmatrix} 1 & 1 \\ -1 & 1 \end{bmatrix}$ is orthonormal because

$$\frac{1}{\sqrt{2}}\begin{bmatrix} 1 & 1 \\ -1 & 1 \end{bmatrix} \times \frac{1}{\sqrt{2}}\begin{bmatrix} 1 & -1 \\ 1 & 1 \end{bmatrix} = \begin{bmatrix} 1 & 0 \\ 0 & 1 \end{bmatrix}$$

is an identity matrix.

2.5   Determinants and Matrix Inverse

The determinant of an $n \times n$ square matrix is the sum of the signed products of all possible combinations of n elements, in which each element is taken from a different row and column. The determinant of A, denoted by $\det(A)$, is

$$\det(A) = \sum^{n!}(-1)^p a_{1p_1} a_{1p_2} \cdots a_{1p_n}, \tag{2.10}$$

where $p_1, p_2, \ldots, p_n$ is a permutation of $1, 2, \ldots, n$ and the sum is taken over all possible permutations. A permutation is a rearrangement of $1, 2, \ldots, n$ into some other order, such as $n, 1, \ldots, 2$, that is obtained by successive transpositions. A transposition is the interchange of places of two numbers in the list $1, 2, \ldots, n$. The exponent p of -1 in Eq. (2.10) is the number of transpositions it takes to go from the natural order to $p_1, p_2, \ldots, p_n$. There are $n!$ possible permutations of n numbers, so each determinant is the sum of $n!$ products.

EXAMPLE 2.2

The determinant of the matrix

$$A = \begin{bmatrix} 1 & 2 \\ 3 & 4 \end{bmatrix}$$

is computed by

$$\det(A) = 1 \times 4 - 2 \times 3 = -2.$$

A matrix containing a row or column of zeros has a zero determinant. If two rows or columns are identical or multiples of one another, then the determinant is also zero.

Useful properties of the determinant include

$$\det(AB) = \det(A)\det(B),$$
$$\det(A^T) = \det(A),$$
$$\det(cA) = c^n \det(A).$$

An $n \times n$ matrix B is referred to as the inverse of an $n \times n$ matrix A if

$$AB = BA = I$$

and is denoted by $B = A^{-1}$. If A has an inverse, it is said to be nonsingular; otherwise it is singular. A matrix is nonsingular if and only if its determinant is not zero.

EXAMPLE 2.3

The inverse of the matrix

$$A = \begin{bmatrix} 1 & 2 \\ 3 & 4 \end{bmatrix}$$

is

$$A^{-1} = \begin{bmatrix} -2 & 1 \\ 1.5 & -0.5 \end{bmatrix}.$$

It is easy to verify that

$$\begin{bmatrix} 1 & 2 \\ 3 & 4 \end{bmatrix}\begin{bmatrix} -2 & 1 \\ 1.5 & -0.5 \end{bmatrix} = \begin{bmatrix} -2 & 1 \\ 1.5 & -0.5 \end{bmatrix}\begin{bmatrix} 1 & 2 \\ 3 & 4 \end{bmatrix} = \begin{bmatrix} 1 & 0 \\ 0 & 1 \end{bmatrix}.$$

MATRIX INVERSION LEMMA

Suppose that A is an $n \times n$ nonsingular matrix and that x and y are both $n \times 1$ column vectors. If $y^T A^{-1} x \neq -1$, then

$$(A + xy^T)^{-1} = A^{-1} - \frac{A^{-1}xy^T A^{-1}}{1 + y^T A^{-1}x}. \tag{2.11}$$

This is know as the Sherman–Morrison formula. Let X and Y be $n \times m$ matrices and let B be an $m \times m$ matrix. A useful generalization of Eq. (2.11) is

$$(A + XBY^T)^{-1} = A^{-1} - A^{-1}X(B^{-1} + Y^T A^{-1}X)^{-1}Y^T A^{-1}, \tag{2.12}$$

where A, B and $(B^{-1} + Y^T A^{-1} X)$ are assumed to be nonsingular. This is known as the Sherman–Morrison–Woodbury formula or simply the matrix inversion lemma. This lemma is very useful in deriving formulation for system identification methods, particularly in the frequency domain.

2.6 Subspaces and Rank

A set of $m \times 1$ vectors $\{x_1, x_2, \ldots, x_n\}$ is linearly dependent if there exist scalars $\alpha_1, \alpha_2, \ldots, \alpha_n$, not all zero, such that $\sum_{i=1}^{n} \alpha_i x_i = 0$. On the other hand, the $m \times 1$ vectors $\{x_1, x_2, \ldots, x_n\}$ are linearly independent if $\sum_{i=1}^{n} \alpha_i x_i = 0$ implies that all scalars $\alpha_1, \alpha_2, \ldots, \alpha_n$ are zero.

The set of all possible linear combinations of $\{x_1, x_2, \ldots, x_n\}$ is a subspace $\mathcal{S}_x$ referred to as the span of $\{x_1, x_2, \ldots, x_n\}$:

$$\mathcal{S}_x = \text{span}\{x_1, x_2, \ldots, x_n\} = \text{the set of } \{\alpha_1 x_1 + \alpha_2 x_2 + \cdots + \alpha_n x_n\}$$

for all possible scalars $\alpha_1, \alpha_2, \ldots, \alpha_n$. If the set of vectors $\{x_1, x_2, \ldots, x_n\}$ is linearly independent, then the space $\mathcal{S}_x$ is said to have dimension n. For any space $\mathcal{S}_x$ of dimension r, there always exist independent vectors $x_1, x_2, \ldots, x_r$ in $\mathcal{S}_x$ such that $\mathcal{S}_x = \text{span}\{x_1, x_2, \ldots, x_r\}$.

There are four important subspaces associated with an $m \times n$ matrix A:

(1) The column space of A is the space spanned by the n columns of A, with each column considered a separate vector. The column rank of A is the maximum number of linearly independent vectors formed by the columns of A.

(2) The row space of A is the space spanned by the m rows of A, with each row considered a separate vector. The row rank of A is the maximum number of linearly independent vectors formed by the rows of A. The row rank and column rank of A are equal. Therefore the rank of A equals the maximum number of independent rows or columns.

(3) The column null space of A is the space spanned by the set of nonzero column vectors $\{y_1, y_2, \ldots, y_q\}$ of dimension $m \times 1$ that satisfy

$$A^T\{y_1, y_2, \ldots, y_q\} = 0.$$

The dimension of the column null space is the maximum number of nonzero independent column vectors y_i that satisfy $A^T y_i = 0$.

(4) The row null space of A is the space spanned by the set of nonzero row vectors $\{x_1, x_2, \ldots, x_p\}$ of dimension $1 \times n$ that satisfy

$$A\{x_1^T, x_2^T, \ldots, x_p^T\} = 0.$$

The dimension of the row null space is the maximum number of independent nonzero row vectors x_i that satisfy $Ax_i^T = 0$.

For a rectangular matrix, the dimensions of row null space and column null space are generally not equal, but they are equal for a square matrix. For a square matrix, $Ax = 0$ has a solution $x \neq 0$ if and only if the columns or rows of A are not independent, i.e., A is a singular matrix and thus $\det(A) = 0$.

2.7   Quadratic Form of a Matrix

The (scalar) product $x^T Ax$ is known as the quadratic form of A and x. A matrix can be expressed as the sum of a symmetric matrix and a skew-symmetric matrix:

$$A = \frac{1}{2}(A + A^T) + \frac{1}{2}(A - A^T).$$

It can be easily shown that

$$A_1 = \frac{1}{2}(A + A^T)$$

is symmetric because $A_1 = A_1^T$ and that

$$A_2 = \frac{1}{2}(A - A^T)$$

is skew symmetric because $A_2 = -A_2^T$. Note that the quadratic form of a skew-symmetric matrix is zero and therefore the quadratic form of a matrix is the same as the quadratic form of its symmetric component.

EXAMPLE 2.4

The skew-symmetric (or antisymmetric) part of the matrix

$$\begin{bmatrix} 2 & 6 \\ 4 & 2 \end{bmatrix}$$

is

$$\frac{1}{2}\begin{bmatrix} 2 & 6 \\ 4 & 2 \end{bmatrix} - \frac{1}{2}\begin{bmatrix} 2 & 4 \\ 6 & 2 \end{bmatrix} = \begin{bmatrix} 0 & 1 \\ -1 & 0 \end{bmatrix},$$

whose quadratic form is zero:

$$[x_1 \quad x_2]\begin{bmatrix} 0 & 1 \\ -1 & 0 \end{bmatrix}\begin{bmatrix} x_1 \\ x_2 \end{bmatrix} = [-x_2 \quad x_1]\begin{bmatrix} x_1 \\ x_2 \end{bmatrix} = 0.$$

2.8 Matrix Functions

The elements of a matrix do not have to be real or complex numbers. They can be functions of a variable, such as time. For example,

$$A(t) = \begin{bmatrix} a_{11}(t) & a_{12}(t) & \cdots & a_{1n}(t) \\ a_{21}(t) & a_{22}(t) & \cdots & a_{2n}(t) \\ \vdots & \vdots & \ddots & \vdots \\ a_{m1}(t) & a_{m2}(t) & \cdots & a_{mn}(t) \end{bmatrix}.$$

Many rules of elementary calculus apply to matrix functions, for example,

$$\frac{d}{dt}[CA(t)] = C\frac{dA}{dt} = C\left[\frac{da_{ij}}{dt}\right],$$

$$\frac{d}{dt}[A(t) + B(t)] = \frac{dA}{dt} + \frac{dB}{dt} = \left[\frac{da_{ij}}{dt}\right] + \left[\frac{db_{ij}}{dt}\right],$$

$$\frac{d}{dt}[A(t)B(t)] = \frac{dA}{dt}B + A\frac{dB}{dt} = \left[\frac{da_{ij}}{dt}\right]B + A\left[\frac{db_{ij}}{dt}\right].$$

2.9 Solving Linear Algebraic Equations

In this section, we are concerned with the set of simultaneous linear algebraic equations of the form

$$a_{11}x_1 + a_{12}x_2 + \cdots + a_{1n}x_n = b_1,$$
$$a_{21}x_1 + a_{22}x_2 + \cdots + a_{2n}x_n = b_2$$
$$\vdots \quad \vdots \quad \vdots$$
$$a_{n1}x_1 + a_{n2}x_2 + \cdots + a_{nn}x_n = b_n, \tag{2.13}$$

where the unknowns are $x_1, x_2, \ldots, x_n$ and the other coefficients are assumed to be known. The above set of equations can be put in matrix form as

$$Ax = b, \tag{2.14}$$

where

$$A = \begin{bmatrix} a_{11} & a_{12} & \cdots & a_{1n} \\ a_{21} & a_{22} & \cdots & a_{2n} \\ \vdots & \vdots & \ddots & \vdots \\ a_{n1} & a_{n2} & \cdots & a_{nn} \end{bmatrix}, \quad x = \begin{bmatrix} x_1 \\ x_2 \\ \vdots \\ x_n \end{bmatrix}, \quad b = \begin{bmatrix} b_1 \\ b_2 \\ \vdots \\ b_n \end{bmatrix}.$$

There are various situations that may apply, which are summarized below:

(1) If A is a full-rank matrix, then the solution for x is unique and is given by

$$x = A^{-1}b.$$

For example, the set of equations

$$x_1 + x_2 = 1,$$
$$x_1 + 2x_2 = 1$$

produces a unique solution, $x_1 = 1, x_2 = 0$.

(2) If A is not a full-rank matrix, then one of the two following cases will apply: If b is such that the set of equations is consistent, then there are an infinite number of solutions for x. For example, the set of equations

$$x_1 + x_2 = 1,$$
$$2x_1 + 2x_2 = 2$$

implies that there are an infinite number of solutions $x_2 = \alpha$, $x_1 = 1 - \alpha$ for any α. On the other hand, if b is such that the set of equations is inconsistent, then there is no solution at all. For example, the set of equations

$$x_1 + x_2 = 1,$$
$$2x_1 + 2x_2 = 4$$

has no solution because they are inconsistent.

2.10   Eigenvalues and Eigenvectors

The product Ax can be thought of as the result of a (linear) transformation that converts x into a new vector $y = Ax$. (In the language of linear transformation, solving the problem $Ax = b$ is finding x such that it is turned into b when transformed by A). Given a general square matrix A, we now ask the following question: Is there a special vector v such that the transformation Av produces a new vector that is just a multiple of v itself? The general answer is yes. In fact, given a general matrix A of dimensions $n \times n$, there can be up to n such special vectors. These vectors are called the eigenvectors of A, and the proportionality constants are called the eigenvalues. Mathematically, to find these eigenvectors and their corresponding eigenvalues, we look for solutions of the following problem:

$$Av = \lambda v \tag{2.15}$$

or

$$(A - \lambda I)v = 0.$$

Obviously, a null vector v will satisfy the equation $Av = \lambda v$, but this is not really what we are looking for. For $v \neq 0$, λ must be such that the matrix $A - \lambda I$ is singular. If $A - \lambda I$ is nonsingular, then its inverse exists and we have $v = (A - \lambda I)^{-1}0 = 0$. The requirement that $A - \lambda I$ be singular is satisfied provided that

$$\det(A - \lambda I) = 0, \tag{2.16}$$

which is the equation to be solved for the eigenvalues.

In general, the eigenvalues of a matrix may be real, complex, or zero. They can even be repeated. If all of the eigenvalues of an $n \times n$ matrix are distinct then there are n independent eigenvectors. However, if the matrix has one or more repeated eigenvalues, then there may be fewer than n linearly independent eigenvectors.

For a symmetric matrix, we can make some definite statements about its eigenvalues and eigenvectors:

(1) All eigenvalues are real.
(2) There is always a full set of linearly independent eigenvectors. This is true whether there are any repeated eigenvalues or not.
(3) Eigenvectors that correspond to different eigenvalues are orthogonal to each other. Recall that the inner product of two orthogonal vectors is zero.
(4) For each eigenvalue that repeats m times, it is possible to find m eigenvectors that are mutually orthogonal. It is also possible to find m eigenvectors that are not mutually orthogonal.
(5) When statements (3) and (4) are combined, it can be shown that, for a symmetric matrix, it is always possible to find a full set of eigenvectors that are not only linearly independent but also mutually orthogonal.

In fact, all the above properties also hold true for not only a symmetric matrix but also for a Hermitian matrix.

EXAMPLE 2.5

The procedure to find the eigenvalues and eigenvectors of an matrix can be illustrated by the following example. Consider the following matrix:

$$A = \begin{bmatrix} 1 & 1 \\ 4 & 1 \end{bmatrix}.$$

First, the eigenvalues are computed from the expression $\det(A - \lambda I) = 0$. Here, we have

$$\det(A - \lambda I) = \begin{vmatrix} 1 - \lambda & 1 \\ 4 & 1 - \lambda \end{vmatrix} = \lambda^2 - 2\lambda - 3 = 0,$$

which yields two eigenvalues $\lambda_1 = 3, \lambda_2 = -1$. Second, to find an eigenvector or eigenvectors corresponding to each eigenvalue, we write $Av = \lambda v$ for each eigenvalue found. This will determine the relationship among the elements of an eigenvector, from which we can choose one or more independent eigenvectors, depending on whether there is more than one free parameter. In this example, for the first eigenvalue $\lambda_1 = 3$, write

$$\begin{bmatrix} 1 & 1 \\ 4 & 1 \end{bmatrix} \begin{bmatrix} v_1^{(1)} \\ v_2^{(1)} \end{bmatrix} = 3 \begin{bmatrix} v_1^{(1)} \\ v_2^{(1)} \end{bmatrix},$$

which reduces to $v_2^{(1)} = 2v_1^{(1)}$. We can therefore choose

$$v^{(1)} = \begin{bmatrix} v_1^{(1)} \\ v_2^{(1)} \end{bmatrix} = \alpha \begin{bmatrix} 1 \\ 2 \end{bmatrix}$$

for any free parameter $\alpha \neq 0$ as one eigenvector, say $\alpha = 1$ for simplicity. Note that a multiple of an eigenvector is still an eigenvector. Associated with another value for α is another eigenvector, but this eigenvector is not an independent one. In fact, it is just a multiple of the first one. Therefore, out of the relationship $v_2^{(1)} = 2v_1^{(1)}$, we can extract only one independent eigenvector. Similarly, an eigenvector corresponding to the second eigenvalue $\lambda_2 = -1$ can be chosen to be

$$v^{(2)} = \begin{bmatrix} 1 \\ -2 \end{bmatrix}.$$

Finally, any eigenvector (or indeed any vector) can be scaled so that it has unit length. The vector is said to be normalized, and the process is called normalization. For example, the vector

$$\begin{bmatrix} 1 \\ 2 \end{bmatrix}$$

can be normalized by its length to become

$$\frac{1}{\sqrt{1^2 + 2^2}} \begin{bmatrix} 1 \\ 2 \end{bmatrix} = \begin{bmatrix} \dfrac{1}{\sqrt{5}} \\ \dfrac{2}{\sqrt{5}} \end{bmatrix}.$$

Note that normalizing a vector changes only its length, not its direction.

2.11 Diagonalization of a Matrix

In solving a set of coupled algebraic or differential equations, it is sometimes useful to transform a matrix into a diagonal matrix so that the variables of interest may become decoupled and then can be solved one at a time. If a matrix has a full set of linearly independent eigenvectors, then this transformation can be done as follows.

Let the eigenvalues and eigenvectors of an $n \times n$ matrix A be denoted by λ_k and $v^{(k)}$, respectively:

$$Av^{(k)} = \lambda_k v^{(k)}, \quad k = 1, 2, \ldots, n. \tag{2.17}$$

Let T be a (square) matrix of the (column) eigenvectors stacked side by side,

$$T = [\, v^{(1)} \quad v^{(2)} \quad \cdots \quad v^{(n)} \,], \tag{2.18}$$

and let Λ be a diagonal matrix of the eigenvalues,

$$\Lambda = \begin{bmatrix} \lambda_1 & & & \\ & \lambda_2 & & \\ & & \ddots & \\ & & & \lambda_n \end{bmatrix}. \tag{2.19}$$

It can be verified that

$$\begin{bmatrix} a_{11} & a_{12} & \cdots & a_{1n} \\ a_{21} & a_{22} & \cdots & a_{2n} \\ \vdots & \vdots & \ddots & \vdots \\ a_{n1} & a_{n2} & \cdots & a_{nn} \end{bmatrix} \begin{bmatrix} v_1^{(1)} & v_1^{(2)} & \cdots & v_1^{(n)} \\ v_2^{(1)} & v_2^{(2)} & \cdots & v_2^{(n)} \\ \vdots & \vdots & \ddots & \vdots \\ v_n^{(1)} & v_n^{(2)} & \cdots & v_n^{(n)} \end{bmatrix}$$

$$= \begin{bmatrix} v_1^{(1)} & v_1^{(2)} & \cdots & v_1^{(n)} \\ v_2^{(1)} & v_2^{(2)} & \cdots & v_2^{(n)} \\ \vdots & \vdots & \ddots & \vdots \\ v_n^{(1)} & v_n^{(2)} & \cdots & v_n^{(n)} \end{bmatrix} \begin{bmatrix} \lambda_1 & & & \\ & \lambda_2 & & \\ & & \ddots & \\ & & & \lambda_n \end{bmatrix},$$

or simply

$$AT = T\Lambda. \tag{2.20}$$

If the matrix has a full set of eigenvectors, then T is full rank and its inverse exists. We then have the following relationships:

$$A = T\Lambda T^{-1} \tag{2.21}$$

or

$$\Lambda = T^{-1}AT. \tag{2.22}$$

Matrix A is said to be similar to matrix Λ and it is diagonalizable. This process is called a similarity transformation. If matrix A has less than n linearly independent eigenvectors, then it is not diagonalizable.

EXAMPLE 2.6

The matrix

$$A = \begin{bmatrix} -2 & 1 \\ 1 & -2 \end{bmatrix}$$

has two eigenvalues and two associated eigenvectors:

$$\lambda_1 = -3, \quad v^{(1)} = \begin{bmatrix} 1 \\ -1 \end{bmatrix} \quad \text{and} \quad \lambda_2 = -1, \quad v^{(2)} = \begin{bmatrix} 1 \\ 1 \end{bmatrix}.$$

The eigenvectors of A can be used to diagonalize A:

$$A = \begin{bmatrix} 1 & 1 \\ -1 & 1 \end{bmatrix} \begin{bmatrix} -3 & 0 \\ 0 & -1 \end{bmatrix} \begin{bmatrix} 1 & 1 \\ -1 & 1 \end{bmatrix}^{-1} = \begin{bmatrix} -2 & 1 \\ 1 & -2 \end{bmatrix},$$

or

$$\Lambda = \begin{bmatrix} 1 & 1 \\ -1 & 1 \end{bmatrix}^{-1} \begin{bmatrix} -2 & 1 \\ 1 & -2 \end{bmatrix} \begin{bmatrix} 1 & 1 \\ -1 & 1 \end{bmatrix} = \begin{bmatrix} -3 & 0 \\ 0 & -1 \end{bmatrix}.$$

EXAMPLE 2.7

This example shows that it is still possible to diagonalize a matrix with complex eigenvalues and eigenvectors. The matrix

$$A = \begin{bmatrix} -0.5 & 1 \\ -1 & -0.5 \end{bmatrix}$$

has two complex-conjugate eigenvalues and two complex-conjugate eigenvectors:

$$\lambda_1 = -0.5 + i, \quad v^{(1)} = \begin{bmatrix} 1 \\ i \end{bmatrix} \quad \text{and} \quad \lambda_2 = -0.5 - i, \quad v^{(2)} = \begin{bmatrix} 1 \\ -i \end{bmatrix}.$$

We can use the matrix of the eigenvectors to diagonalize A:

$$A = \begin{bmatrix} 1 & 1 \\ i & -i \end{bmatrix} \begin{bmatrix} -0.5 + i & 0 \\ 0 & -0.5 - i \end{bmatrix} \begin{bmatrix} 1 & 1 \\ i & -i \end{bmatrix}^{-1} = \begin{bmatrix} -0.5 & 1 \\ -1 & -0.5 \end{bmatrix}.$$

or

$$\Lambda = \begin{bmatrix} 1 & 1 \\ i & -i \end{bmatrix}^{-1} \begin{bmatrix} -0.5 & 1 \\ -1 & -0.5 \end{bmatrix} \begin{bmatrix} 1 & 1 \\ i & -i \end{bmatrix} = \begin{bmatrix} -0.5 + i & 0 \\ 0 & -0.5 - i \end{bmatrix}.$$

If A is a symmetric matrix, then it is very simple to determine the inverse of T. If the eigenvectors are normalized and T is formed by these normalized eigenvectors, then it can be shown that the inverse of T is the same as its complex-conjugate transpose.

2.12 Singular-Value Decomposition

Matrix decomposition techniques that are aimed at reducing a matrix to simpler or canonical form have established themselves as key computational tools in a wide variety of applications in engineering. One of the most important decompositions in matrix computations is the singular-value decomposition (SVD) [3, 4].

2.12.1 Basic Equations

If A is an $m \times n$ real matrix of rank r with $m \geq n$, there exist two orthonormal matrices U $(m \times m)$ and V $(n \times n)$ such that

$$A = U\Sigma V^T \quad \text{or} \quad U^T A V = \Sigma, \tag{2.23}$$

where

$$\Sigma = \begin{bmatrix} S & 0 \\ 0 & 0 \end{bmatrix}, \quad U^T U = I_m, \quad V^T V = I_n,$$

in which 0 is a zero matrix with proper dimensions and I_k is a $k \times k$ matrix for $k = m$ or n and $S = \mathrm{diag}[\sigma_1, \sigma_2, \ldots, \sigma_r]$, with

$$\sigma_1 \geq \sigma_2 \geq \cdots \sigma_r > 0.$$

The positive numbers $\sigma_1 \geq \sigma_2 \geq \cdots \sigma_r > 0$ together with $\sigma_{r+1} = \sigma_{r+2} = \ldots = \sigma_n = 0$ are called the singular values of A. The nonzero singular values are unique, but U and V are not unique. The columns of U are called the left singular vectors, and the columns of V are called the right singular vectors of A.

The SVD given by Eq. (2.23) is closely related to the eigensolution of the symmetric positive-semidefinite matrices $A^T A$ and AA^T. Note that

$$A^T A = V\Sigma^T U^T U\Sigma V^T = V\Sigma^T \Sigma V^T, \tag{2.24}$$

$$AA^T = U\Sigma V^T V\Sigma^T U^T = U\Sigma\Sigma^T U^T. \tag{2.25}$$

The dimension of the diagonal matrices $\Sigma^T \Sigma$ and $\Sigma\Sigma^T$ may not be identical, but they share the same number of nonzero singular values. The nonzero singular values of A are the positive square roots of the nonzero eigenvalues of $A^T A$ or AA^T. The columns of U are the eigenvectors corresponding to the eigenvalues of AA^T, and the columns of V are the eigenvectors corresponding to the eigenvalues of $A^T A$.

The case in which $m \geq n$ is the one most likely to be encountered in practice. However, if $m < n$, the SVD can be applied to A^T instead, and then the transpose of the result can be taken. If A is a complex matrix, Eq. (2.23) is still applicable, with the transpose replaced with the complex-conjugate transpose.

One interpretation of the singular values is as follows. If we take a unit sphere in n-dimensional space and multiply each vector in it by an $m \times n$ matrix A, an ellipsoid will be produced in m-dimensional space. The singular values give the lengths of the principal axes of the ellipsoid. If the matrix A is singular in some way, this will be reflected in the shape of the ellipsoid. In fact, the ratio of the largest singular value of a matrix to the smallest one gives a condition number of the matrix, which determines, for example, the accuracy of numerical matrix inverses.

The SVD is also applicable to a complex matrix. The SVD of an $m \times n$ complex matrix A is

$$A = U\Sigma V^*, \tag{2.26}$$

where

$$\Sigma = \begin{bmatrix} S & 0 \\ 0 & 0 \end{bmatrix}, \quad U^*U = I_m, \quad V^*V = I_n.$$

The matrix S is diagonal, and U and V are unitary. Equation (2.26) for a complex matrix is identical to Eq. (2.23) for a real matrix, with the transpose of V replaced with the complex-conjugate transpose of V.

EXAMPLE 2.8

The SVD of the matrix

$$A = \begin{bmatrix} -0.5 & 1 \\ -1 & -0.5 \end{bmatrix}$$

is

$$\begin{bmatrix} -0.5 & 1 \\ -1 & -0.5 \end{bmatrix} = \begin{bmatrix} -0.4472 & -0.8944 \\ -0.8944 & 0.4472 \end{bmatrix} \begin{bmatrix} 1.118 & 0 \\ 0 & 1.118 \end{bmatrix} \begin{bmatrix} 1 & 0 \\ 0 & -1 \end{bmatrix}.$$

From Example 2.7 in Section 2.11, the eigenvalues of A are $-0.5 + i$ and $-0.5 - i$. The square roots of the eigenvalues of $A^T A$ or $A A^T$ are both $\sqrt{(-0.5 + i)(-0.5 - i)}$, which is 1.118.

2.12.2 Some Areas of Application

In system identification, the SVD is very useful in determining the various ranks that a data matrix may have if subject to errors (measuring and/or modeling). There are many other applications, and only some are noted here.

EFFECTIVE RANK AND CONDITIONED NUMBER

The condition number of a matrix is given by the ratio of the maximum singular value to the minimum singular value, i.e., σ_1/σ_r. If the reciprocal of this ratio approaches a computer's floating-point precision, the matrix is said to be ill conditioned. The rank of a matrix is equal to the number of nonzero singular values. Although the question is not nearly as clearcut within the context of computation on a digital computer, it is now acknowledged that the SVD is the only generally reliable method of determining the rank numerically. Customarily a rank tolerance equal to the square of the machine precision is chosen and the singular values above it are counted to determine the rank numerically.

FOUR FUNDAMENTAL SUBSPACES

With appropriate partitioning of U and V, Eq. (2.23) can be written as

$$A = [U_1 \ U_0] \begin{bmatrix} S & 0 \\ 0 & 0 \end{bmatrix} [V_1 \ V_0]^T = U_1 S V_1^T \tag{2.27}$$

or, equivalently,

$$\begin{bmatrix} U_1^T A \\ U_0^T A \end{bmatrix} = \begin{bmatrix} S V_1^T \\ 0 \end{bmatrix}, \qquad [A V_1 \ A V_0] = [U_1 S \ 0],$$

which imply that

$$U_1^T A V_1 = S,$$
$$A V_0 = 0,$$
$$U_0^T A = 0.$$

The matrix U_1 contains the maximum number (i.e., r) of independent column vectors that span the column space of A. The matrix V_1^T contains the maximum number (r) of independent row vectors that span the row space of A. The matrix U_0 contains the maximum number ($m - r$) of independent column vectors that span the column null space of A. The matrix V_0^T contains the maximum number ($n - r$) of independent row vectors that span the row null space of A.

LEAST SQUARES AND PSEUDOINVERSE

The pseudoinverse of an $m \times n$ matrix A, denoted by $A^\dagger$, is the matrix that satisfies the equation $A A^\dagger A = A$. Using Eq. (2.27), we can compute $A^\dagger$ by

$$A^\dagger = V_1 S^{-1} U_1^T = V_1 \ \mathrm{diag} \left[\sigma_1^{-1}, \sigma_2^{-1}, \ldots, \sigma_r^{-1} \right] U_1^T.$$

The pseudoinverse $A^\dagger$ is a generalization of the matrix inverse A^{-1} and is applicable to rectangular as well as square matrices. When A^{-1} fails to exist, $A^\dagger$ provides a natural substitute. The pseudoinverse solves the least-squares problem of the linear equation $Ax = b$, when it does not have a solution or has more than one solution. For the overdetermined case that has more independent equations than unknowns, the least-squares solution $\hat{x} = A^\dagger b$ minimizes the sum of squared errors of individual equations. Note that $\hat{x}$ does not satisfy the linear equation, i.e., $A\hat{x} \neq b$. If there is more than one solution, i.e., an underdetermined case that has more unknowns than independent equations, the pseudoinverse solution $\hat{x} = A^\dagger b$ minimizes the norm (the length) of x. Note that $\hat{x}$ for the underdetermined case satisfies the linear equation, i.e., $A\hat{x} = b$.

Although SVD is the most reliable method, it is also computationally the most expensive one. If the matrix is known to have full rank and its condition number is relatively small in comparison with the uncertainty in the data, other factorization methods may be used. Such matrix factorization methods include Cholesky factorization, the LU decomposition with partial or complete pivoting, and orthogonal QR factorization with column pivoting. See Ref. [2] for more information.

EXAMPLE 2.9

Consider the 3×2 rectangular matrix

$$A = \begin{bmatrix} 1 & 0 \\ 0 & 1 \\ 1 & 1 \end{bmatrix}.$$

The SVD of A is

$$A = U \Sigma V^T$$

$$= \begin{bmatrix} \frac{1}{\sqrt{6}} & -\frac{1}{\sqrt{2}} & -\frac{1}{\sqrt{3}} \\ \frac{1}{\sqrt{6}} & \frac{1}{\sqrt{2}} & -\frac{1}{\sqrt{3}} \\ \frac{2}{\sqrt{6}} & 0 & \frac{1}{\sqrt{3}} \end{bmatrix} \begin{bmatrix} \sqrt{3} & 0 \\ 0 & 1 \\ 0 & 0 \end{bmatrix} \begin{bmatrix} \frac{1}{\sqrt{2}} & -\frac{1}{\sqrt{2}} \\ \frac{1}{\sqrt{2}} & \frac{1}{\sqrt{2}} \end{bmatrix}^T$$

$$= \begin{bmatrix} \frac{1}{\sqrt{6}} & -\frac{1}{\sqrt{2}} \\ \frac{1}{\sqrt{6}} & \frac{1}{\sqrt{2}} \\ \frac{2}{\sqrt{6}} & 0 \end{bmatrix} \begin{bmatrix} \sqrt{3} & 0 \\ 0 & 1 \end{bmatrix} \begin{bmatrix} \frac{1}{\sqrt{2}} & -\frac{1}{\sqrt{2}} \\ \frac{1}{\sqrt{2}} & \frac{1}{\sqrt{2}} \end{bmatrix}^T$$

$$= U_1 S V^T.$$

The square roots of the nonzero eigenvalues of $A^T A$ or $A A^T$ are the nonzero singular values of A, which are $\sqrt{3}$ and 1. The matrix S is a diagonal matrix with the nonzero singular values as its diagonal elements, and U_1 is a rectangular matrix formed by the first two columns of U associated with two nonzero singular values. The rank of A is 2 because there are two nonzero singular values. The pseudoinverse $A^\dagger$ is

$$A^\dagger = V S^{-1} U_1^T = V \operatorname{diag}[\sigma_1^{-1}, \quad \sigma_2^{-1}] U^T$$

$$= \begin{bmatrix} \frac{1}{\sqrt{2}} & -\frac{1}{\sqrt{2}} \\ \frac{1}{\sqrt{2}} & \frac{1}{\sqrt{2}} \end{bmatrix} \begin{bmatrix} \frac{1}{\sqrt{3}} & 0 \\ 0 & 1 \end{bmatrix} \begin{bmatrix} \frac{1}{\sqrt{6}} & -\frac{1}{\sqrt{2}} \\ \frac{1}{\sqrt{6}} & \frac{1}{\sqrt{2}} \\ \frac{2}{\sqrt{6}} & 0 \end{bmatrix}^T$$

$$= \begin{bmatrix} \frac{2}{3} & -\frac{1}{3} & \frac{1}{3} \\ -\frac{1}{3} & \frac{2}{3} & \frac{1}{3} \end{bmatrix}.$$

2.13   Homogeneous First-Order Matrix Differential Equations

The first-order matrix differential equation with constant coefficients is not only a simpler problem but also it is one of the most useful tools as far as linear time-invariant systems are concerned. We will show later that any set of higher-order linear matrix differential equations can be put in first-order form and solved with first-order theory.

Consider the following problem,

$$\frac{dx}{dt} = Ax, \tag{2.28}$$

subject to the initial condition $x(0) = x_0$, where x is an $n \times 1$ column vector and A is an $n \times n$ square matrix with real constant coefficients. By analogy with the first-order scalar differential equation, we attempt to look for solutions of the form

$$x = ve^{\lambda t}, \tag{2.29}$$

where v is some $n \times 1$ column vector(s) and λ is some number(s) to be determined. Substituting Eq. (2.29) into Eq. (2.28) yields

$$\lambda v e^{\lambda t} = A v e^{\lambda t}, \tag{2.30}$$

or

$$(A - \lambda I)ve^{\lambda t} = 0,$$

where 0 is a zero matrix of the same dimensions as v. Note that the identity matrix I of dimension $n \times n$ is used so that $ve^{\lambda t}$ can be factored out. Because $e^{\lambda t}$ is nonzero for all finite t, the vector v and the number λ must satisfy

$$(A - \lambda I)v = 0. \tag{2.31}$$

We encountered this situation earlier: v and λ are simply an eigenvector and an eigenvalue of the matrix A, respectively. Depending on the nature of the matrix A, there are three possibilities for the eigenvalues of A.

REAL AND DISTINCT EIGENVALUES

This case is the simplest possible. For each eigenvalue there is an associated independent eigenvector. The general solution is a linear combination of these eigensolutions:

$$x = c_1 v^{(1)} e^{\lambda_1 t} + c_2 v^{(2)} e^{\lambda_2 t} + \cdots + c_n v^{(n)} e^{\lambda_n t}. \tag{2.32}$$

COMPLEX-CONJUGATE PAIRS OF EIGENVALUES

As long as the complex eigenvalues are not repeated, the (real) matrix A has n linear independent eigenvectors. The corresponding eigensolutions are complex. For a physical problem, however, we expect the solution to be real. Analogous to the scalar differential equation case, it is possible to obtain a full set of real solutions. Let us examine this case in more detail. Consider the special case in which the system has a pair of complex-conjugate eigenvalues $\lambda^{(1)} = \sigma + i\omega$ and $\lambda^{(2)} = \sigma - i\omega$. The corresponding eigenvectors are also complex conjugates, i.e., $v^{(1)} = v_r + iv_i$ and $v^{(2)} = v_r - iv_i$. The linear combination

$$
\begin{aligned}
x &= \tilde{c}_1(v_r + iv_i)e^{(\sigma+i\omega)t} + \tilde{c}_2(v_r - iv_i)e^{(\sigma-i\omega)t} \\
&= \tilde{c}_1 e^{\sigma t}(v_r + iv_i)(\cos\omega t + i\sin\omega t) + \tilde{c}_2 e^{\sigma t}(v_r - iv_i)(\cos\omega t - i\sin\omega t) \\
&= (\tilde{c}_1 + \tilde{c}_2)e^{\sigma t}(v_r\cos\omega t - v_i\sin\omega t) + i(\tilde{c}_1 - \tilde{c}_2)e^{\sigma t}(v_r\sin\omega t + v_i\cos\omega t) \\
&= c_1 e^{\sigma t}(v_r\cos\omega t - v_i\sin\omega t) + c_2 e^{\sigma t}(v_r\sin\omega t + v_i\cos\omega t)
\end{aligned} \tag{2.33}
$$

has $e^{\sigma t}(v_r\cos\omega t - v_i\sin\omega t)$ and $e^{\sigma t}(v_r\sin\omega t + v_i\cos\omega t)$ as real solutions associated with the complex-conjugate pair of eigenvalues. Note that in the above derivation, v_r and

v_i are vectors and $c_1 = \tilde{c}_1 + \tilde{c}_2$ and $c_2 = i(\tilde{c}_1 - \tilde{c}_2)$ are scalars (real) to be determined by initial conditions.

REPEATED EIGENVALUES

If some eigenvalues are repeated but the system has a full set of eigenvectors, then no difficulties are expected. The solution can be written down, one for each eigenvector–eigenvalue combination. If the system does not have a full set of eigenvectors, however, then the number of linear independent solutions of the form $ve^{\lambda t}$ will be smaller than n. Similar to the scalar differential equation case, the additional solutions that are associated with the repeated eigenvalue take the form $te^{\lambda t}, t^2 e^{\lambda t}, \ldots$, each multiplied by an appropriate eigenvector or generalized eigenvector. The situation is quite complicated. The following summarizes the case in which there is an eigenvalue $\lambda = \rho$ that is repeated twice, yet there is only one eigenvector $v = \xi^{(1)}$ associated with it. One solution is

$$x^{(1)} = \xi^{(1)} e^{\rho t}. \tag{2.34}$$

A second solution will take the form

$$x^{(2)} = \xi^{(1)} t e^{\rho t} + \xi^{(2)} e^{\rho t}. \tag{2.35}$$

Substituting Eq. (2.35) into matrix differential equation (2.28) yields

$$
\begin{aligned}
\frac{dx^{(2)}}{dt} &= \rho \xi^{(1)} t e^{\rho t} + \xi^{(1)} e^{\rho t} + \rho \xi^{(2)} e^{\rho t} \\
&= A(\xi^{(1)} t e^{\rho t} + \xi^{(2)} e^{\rho t}),
\end{aligned} \tag{2.36}
$$

which requires that

$$A\xi^{(1)} = \rho \xi^{(1)}, \tag{2.37}$$

$$A\xi^{(2)} = \rho \xi^{(2)} + \xi^{(1)}. \tag{2.38}$$

The first of the two requirements, Eq. (2.37), is already satisfied because $\rho, \xi^{(1)}$ is an eigenvalue–eigenvector pair. The second of the two requirements, Eq. (2.38), can be rewritten as

$$(A - \rho I)\xi^{(2)} = \xi^{(1)}, \tag{2.39}$$

from which $\xi^{(2)}$ is determined. The vector $\xi^{(2)}$ is called a generalized eigenvector of A corresponding to the eigenvector $\xi^{(1)}$. Because the matrix $A - \rho I$ in Eq. (2.39) is rank deficient, the solution for $\xi^{(2)}$ is not unique. The one that is commonly used is the minimum-norm solution, i.e.,

$$\xi^{(2)} = (A - \rho I)^{\dagger} \xi^{(1)}, \tag{2.40}$$

where † means the pseudoinverse. Among all possible solutions, the vector $\xi^{(2)}$ obtained from Eq. (2.40) has the minimum norm.

If the eigenvalue is repeated three times but there is only one eigenvector associated with it, then the first two solutions will be $x^{(1)} = \xi^{(1)} e^{\rho t}$ and $x^{(2)} = \xi^{(1)} t e^{\rho t} + \xi^{(2)} e^{\rho t}$, as before. The third solution will assume the form

$$x^{(3)} = \xi^{(1)} t^2 e^{\rho t} + \xi^{(2)} t e^{\rho t} + \xi^{(3)} e^{\rho t}, \tag{2.41}$$

where $\xi^{(3)}$ is determined from the condition

$$(A - \rho I)\xi^{(3)} = \xi^{(2)}. \tag{2.42}$$

Again, the vectors $\xi^{(2)}$ and $\xi^{(3)}$ are called the generalized eigenvectors of A corresponding to the eigenvector $\xi^{(1)}$. The general solution is a linear combination of these individual solutions that are associated with the repeated eigenvalues and those associated with other eigenvalues, if any.

Finally, after all the solutions that make up the general solution are found, regardless of whether the eigenvalues are real, complex, or repeated, the unknown coefficients $c_1, c_2, \ldots, c_n$ in the general solution are determined from the initial conditions specified for the system, $x(0) = x_0$.

EXAMPLE 2.10

Consider the matrix differential equation

$$\frac{dx}{dt} = Ax,$$

where

$$A = \begin{bmatrix} 0 & 1 \\ 0 & 0 \end{bmatrix},$$

subject to some initial condition $x(0) = x_0$. The matrix A has only one eigenvalue and one eigenvector:

$$\rho = 0, \qquad \xi^{(1)} = \begin{bmatrix} 1 \\ 0 \end{bmatrix}.$$

The eigenvalue is repeated twice. The generalized eigenvector can be computed with Eq. (2.39):

$$(A - \rho I)\xi^{(2)} = \begin{bmatrix} 0 & 1 \\ 0 & 0 \end{bmatrix} \xi^{(2)} = \begin{bmatrix} 1 \\ 0 \end{bmatrix}.$$

The obvious solution for $\xi^{(2)}$ is

$$\xi^{(2)} = \begin{bmatrix} 0 \\ 1 \end{bmatrix}.$$

From Eq. (2.34), the first solution becomes

$$x^{(1)} = \xi^{(1)}e^{\rho t} = \begin{bmatrix} 1 \\ 0 \end{bmatrix}.$$

From Eq. (2.35), the second solution is

$$x^{(2)} = \xi^{(1)}te^{\rho t} + \xi^{(2)}e^{\rho t} = \begin{bmatrix} 1 \\ 0 \end{bmatrix} t + \begin{bmatrix} 0 \\ 1 \end{bmatrix} = \begin{bmatrix} t \\ 1 \end{bmatrix}.$$

Thus, the general solution to the matrix differential equation is

$$x(t) = c_1 x^{(1)} + c_2 x^{(2)} = c_1 \begin{bmatrix} 1 \\ 0 \end{bmatrix} + c_2 \begin{bmatrix} t \\ 1 \end{bmatrix}.$$

The constants c_1 and c_2 are determined from the initial condition specified for the problem, $x(0) = x_0$.

2.14 Matrix Exponential

At this point, we have a general idea about how to solve a first-order matrix differential equation, and it seems to be tedious. There is another way to approach this problem by making an analogy with the first-order scalar differential equation $\dot{y} = \alpha y$ having the initial condition $y(0) = y_0$, which has the solution of the form $y(t) = y_0 e^{\alpha t}$. Similarly, the solution of the first-order matrix differential equation

$$\frac{dx}{dt} = Ax, \quad x(0) = x_0 \tag{2.43}$$

can be conveniently written as

$$x(t) = e^{At} x_0, \tag{2.44}$$

where e^{At} is a square matrix of the same dimension as A and $x(t)$ is a column vector. Equation (2.44) is derived as follows.

Assume that Eq. (2.43) has a (vector) power-series solution about $t = 0$,

$$x(t) = z_0 + z_1 t + z_2 t^2 + \cdots. \tag{2.45}$$

Note that $x(0) = x_0 = z_0$. Substituting Eq. (2.45) into Eq. (2.43) yields

$$\frac{dx}{dt} = z_1 + 2z_2 t + 3z_3 t^2 + \cdots$$

$$= A(z_0 + z_1 t + z_2 t^2 + \cdots).$$

Equating equal powers of t produces

$$z_1 = Az_0 = Ax_0,$$

$$z_2 = \frac{1}{2} Az_1 = \frac{1}{2} A^2 x_0,$$

$$z_3 = \frac{1}{3} Az_2 = \frac{1}{3} \times \frac{1}{2} A^3 x_0. \tag{2.46}$$

Equation (2.45) can then be written as

$$x(t) = \left(I + At + \frac{1}{2} A^2 t^2 + \frac{1}{3} \times \frac{1}{2} A^3 t^3 + \cdots \right) x_0$$

$$= \left(I + \sum_{k=1}^{\infty} \frac{1}{k!} A^k t^k \right) x_0$$

$$= e^{At} x_0. \tag{2.47}$$

Here the initial condition is defined at time $t = 0$. If the initial condition $x(t_0)$ is given at some other time instead, say at $t = t_0$, then the solution is simply

$$x(t) = e^{A(t-t_0)} x(t_0). \tag{2.48}$$

The matrix $e^{A(t-t_0)}$ is commonly called the state-transition matrix if $x(t)$ is defined as the state vector. Given that t and τ are scalars and A and B are square matrices of the same size, a few properties of the state transition matrix are

(1) composition rule: $e^{At}e^{A\tau} = e^{A(t+\tau)}$,
(2) inverse: $[e^{At}]^{-1} = e^{-At}$,
(3) determinant: $\det[e^{At}] = e^{tr(A)t}$,
(4) product: $e^{At}e^{Bt} = e^{(A+B)t}$ if and only if A and B commute, i.e., $AB = BA$,
(5) similarity: If $A = T\Lambda T^{-1}$, then $e^{At} = Te^{\Lambda t}T^{-1}$.

The following are the matrix exponentials of a few common matrices:

$$\exp\left(\begin{bmatrix} \lambda_1 & & & \\ & \lambda_2 & & \\ & & \ddots & \\ & & & \lambda_n \end{bmatrix} t\right) = \begin{bmatrix} e^{\lambda_1 t} & & & \\ & e^{\lambda_2 t} & & \\ & & \ddots & \\ & & & e^{\lambda_n t} \end{bmatrix},$$

$$\exp\left(\begin{bmatrix} \lambda_1 & 1 & & \\ & \lambda_2 & \ddots & \\ & & \ddots & 1 \\ & & & \lambda_n \end{bmatrix} t\right) = \begin{bmatrix} 1 & t & \cdots & \frac{t^{n-1}}{(n-1)!} \\ & 1 & \ddots & \vdots \\ & & \ddots & t \\ & & & 1 \end{bmatrix},$$

$$\exp\left(\begin{bmatrix} 0 & \omega \\ -\omega & 0 \end{bmatrix} t\right) = \begin{bmatrix} \cos\omega t & \sin\omega t \\ -\sin\omega t & \cos\omega t \end{bmatrix},$$

$$\exp\left(\begin{bmatrix} \sigma & \omega \\ -\omega & \sigma \end{bmatrix} t\right) = e^{\sigma t}\begin{bmatrix} \cos\omega t & \sin\omega t \\ -\sin\omega t & \cos\omega t \end{bmatrix}.$$

Property (5) states that if A is diagonalizable, i.e., $A = T\Lambda T^{-1}$, then $e^{At} = Te^{\Lambda t}T^{-1}$ can be used conveniently to compute the matrix exponential. This property can be verified as follows. Multiplying both sides of Eq. (2.43), i.e, $\dot{x} = Ax$, by T^{-1} and making use of the identity $TT^{-1} = I$ produces

$$T^{-1}\dot{x} = T^{-1}ATT^{-1}x$$
$$= \Lambda T^{-1}x.$$

Defining $z = T^{-1}x$, the above equation becomes

$$\dot{z} = \Lambda z,$$

which has the solution

$$z(t) = e^{\Lambda t}z(0),$$

from which the solution for $x(t)$ is

$$x(t) = Te^{\Lambda t}T^{-1}x(0).$$

We already know that $x(t) = e^{At}x(0)$; therefore $e^{At} = Te^{\Lambda t}T^{-1}$, as claimed.

If complex eigenvalues are present, the matrix of eigenvectors T may contain complex numbers. This is sometimes inconvenient for computation. The following factorization avoids complex numbers altogether. Instead of having a purely diagonal matrix Λ, which may contain complex elements, we will have diagonal blocks corresponding to the complex eigenvalues. Suppose that the $n \times n$ matrix A has m complex eigenvalues ($m/2$ conjugate pairs),

$$\lambda_k = \sigma_k + i\omega_k,$$

$$\lambda_{k+1} = \sigma_k - i\omega_k, \quad k = 1, 3, 5, \ldots, m-1, \tag{2.49}$$

with the corresponding linearly independent eigenvectors

$$v^{(k)} = v_r^{(k)} + i\, v_i^{(k)},$$

$$v^{(k+1)} = v_r^{(k)} - i\, v_i^{(k)}, \quad k = 1, 3, 5, \ldots, m-1, \tag{2.50}$$

and $n - m$ real eigenvalues λ_j, $j = m+1, m+2, \ldots, n$, with the corresponding linear independent eigenvectors,

$$v^{(j)}, \quad j = m+1, m+2, \ldots, n.$$

Then the following real-valued matrix,

$$\Psi = \begin{bmatrix} v_r^{(1)} & v_i^{(1)} & v_r^{(3)} & v_i^{(3)} & \cdots & v_r^{(m-1)} & v_i^{(m-1)} & v^{(m+1)} & \cdots & v^{(n)} \end{bmatrix}, \tag{2.51}$$

is nonsingular and can be used to transform A into the following form:

$$\Gamma = \Psi^{-1} A \Psi = \begin{bmatrix} \Gamma_1 & & & & \\ & \Gamma_3 & & & \\ & & \ddots & & \\ & & & \Gamma_{m-1} & \\ & & & & \Gamma_{m+1} \end{bmatrix}, \tag{2.52}$$

where

$$\Gamma_k = \begin{bmatrix} \sigma_k & \omega_k \\ -\omega_k & \sigma_k \end{bmatrix}, \quad k = 1, 3, 5, \ldots, m-1,$$

$$\Gamma_{m+1} = \begin{bmatrix} \lambda_{m+1} & & & \\ & \lambda_{m+2} & & \\ & & \ddots & \\ & & & \lambda_n \end{bmatrix}.$$

If the eigenvalues are distinct, then the eigenvectors must be linearly independent. This ensures that the inverse of the matrix Ψ exists and is unique.

EXAMPLE 2.11

Consider the case in which A is given as

$$\begin{bmatrix} -0.5 & 1 & 0 \\ -1 & -0.5 & 0 \\ 0 & 1 & -1 \end{bmatrix}.$$

The matrix A has three eigenvalues and three corresponding eigenvectors, as follows:

$$\lambda_1 = -0.5 + i, \quad v^{(1)} = \begin{bmatrix} -i \\ 1 \\ 0.4 - 0.8i \end{bmatrix},$$

$$\lambda_2 = -0.5 - i, \quad v^{(2)} = \begin{bmatrix} i \\ 1 \\ 0.4 + 0.8i \end{bmatrix},$$

$$\lambda_3 = -1, \quad v^{(3)} = \begin{bmatrix} 0 \\ 0 \\ 1 \end{bmatrix}.$$

The real-valued matrix Ψ formed from the eigenvectors,

$$\Psi = \begin{bmatrix} 0 & -1 & 0 \\ 1 & 0 & 0 \\ 0.4 & -0.8 & 1 \end{bmatrix},$$

transforms A into the block diagonal form $\Gamma = \Psi^{-1} A \Psi$,

$$\Gamma = \begin{bmatrix} 0 & -1 & 0 \\ 1 & 0 & 0 \\ 0.4 & -0.8 & 1 \end{bmatrix}^{-1} \begin{bmatrix} -0.5 & 1 & 0 \\ -1 & -0.5 & 0 \\ 0 & 1 & -1 \end{bmatrix} \begin{bmatrix} 0 & -1 & 0 \\ 1 & 0 & 0 \\ 0.4 & -0.8 & 1 \end{bmatrix}$$

$$= \begin{bmatrix} -0.5 & 1 & 0 \\ -1 & -0.5 & 0 \\ 0 & 0 & -1 \end{bmatrix},$$

or $A = \Psi \Gamma \Psi^{-1}$,

$$A = \begin{bmatrix} 0 & -1 & 0 \\ 1 & 0 & 0 \\ 0.4 & -0.8 & 1 \end{bmatrix} \begin{bmatrix} -0.5 & 1 & 0 \\ -1 & -0.5 & 0 \\ 0 & 0 & -1 \end{bmatrix} \begin{bmatrix} 0 & -1 & 0 \\ 1 & 0 & 0 \\ 0.4 & -0.8 & 1 \end{bmatrix}^{-1}$$

$$= \begin{bmatrix} -0.5 & 1 & 0 \\ -1 & -0.5 & 0 \\ 0 & 1 & -1 \end{bmatrix}.$$

2.15   Nonhomogeneous First-Order Matrix Differential Equations

A nonhomogeneous first-order linear matrix differential equation has the form

$$\frac{dx}{dt} = Ax + f, \tag{2.53}$$

where $f = f(t)$ is some forcing (vector) function of time t. There are various ways to solve this problem. The following subsections will describe a few of them.

METHOD OF UNDETERMINED COEFFICIENTS

A method that is most similar to the scalar case is the method of undetermined coefficients. Here we find a particular solution that resembles the forcing function. The particular solution contains a number of coefficient vectors that are to be determined so that the particular solution satisfies the original matrix differential equation. The general solution is then the sum of the homogeneous solution and the particular solution. Finally, the coefficients in the homogeneous solution are determined when the general solution is made to satisfy the given initial conditions of the problem. It is best to illustrate this procedure by an example.

EXAMPLE 2.12

Consider the following problem,

$$\frac{dx}{dt} = \begin{bmatrix} -2 & 1 \\ 1 & -2 \end{bmatrix} x + \begin{bmatrix} 2e^{-t} \\ 3t \end{bmatrix},$$

subject to some initial condition $x(0) = x_0$. First, solving the eigenvalue problem

$$\begin{bmatrix} -2 - \lambda & 1 \\ 1 & -2 - \lambda \end{bmatrix} v = \begin{bmatrix} 0 \\ 0 \end{bmatrix}$$

yields two distinct eigenvalues and two eigenvectors, i.e.,

$$\lambda_1 = -3, \quad v^{(1)} = \begin{bmatrix} 1 \\ -1 \end{bmatrix} \quad \text{and} \quad \lambda_2 = -1, \quad v^{(2)} = \begin{bmatrix} 1 \\ 1 \end{bmatrix}.$$

Therefore, the homogeneous solution is

$$x_h(t) = c_1 \begin{bmatrix} 1 \\ -1 \end{bmatrix} e^{-3t} + c_2 \begin{bmatrix} 1 \\ 1 \end{bmatrix} e^{-t}.$$

Next, the forcing function

$$\begin{bmatrix} 2e^{-t} \\ 3t \end{bmatrix} = \begin{bmatrix} 2 \\ 0 \end{bmatrix} e^{-t} + \begin{bmatrix} 0 \\ 3 \end{bmatrix} t$$

motivates a candidate particular solution of the form

$$x_p(t) = \alpha_1 t e^{-t} + \alpha_2 e^{-t} + \alpha_3 t + \alpha_4.$$

At this point, it is perhaps useful to compare how the matrix version differs from the scalar version. First, $\alpha_1, \alpha_2, \alpha_3, \alpha_3$ are now 2×1 column vectors, not scalars. Because -1 is an eigenvalue and the term e^{-t} appears in the forcing function, the assumed solution has the term $\alpha_1 t e^{-t}$ in it. This is analogous to the scalar case.

However, unlike the scalar; case, it also includes the term $\alpha_2 e^{-t}$. This is because in the scalar case α_2 is a scalar; the term $\alpha_2 e^{-t}$ can be absorbed into the homogeneous solution of the differential equation (recall that the general solution is the sum of the homogeneous solution and the particular solution). Here, α_2 is a vector and $\alpha_2 e^{-t}$ is not a homogeneous solution of the matrix differential equation unless α_2 turns out to be a multiple of the eigenvector $v^{(2)}$. This possibility is very unlikely and should not be assumed at the start. The next step is substituting the candidate particular solution into the original nonhomogeneous differential equation. This step produces the equations that $\alpha_1, \alpha_2, \alpha_3$, and α_4 have to satisfy:

$$A\alpha_1 = -\alpha_1,$$

$$A\alpha_2 = \alpha_1 - \alpha_2 - \begin{bmatrix} 2 \\ 0 \end{bmatrix},$$

$$A\alpha_3 = \begin{bmatrix} 0 \\ -3 \end{bmatrix},$$

$$A\alpha_4 = \alpha_3.$$

There is freedom in choosing α_1 and α_2, For example, a particular combination of $\alpha_1, \alpha_2, \alpha_3, \alpha_4$ is

$$\alpha_1 = \begin{bmatrix} 1 \\ 1 \end{bmatrix}, \qquad \alpha_2 = \begin{bmatrix} 0 \\ -1 \end{bmatrix}, \qquad \alpha_3 = \begin{bmatrix} 1 \\ 2 \end{bmatrix}, \qquad \alpha_4 = -\frac{1}{3}\begin{bmatrix} 4 \\ 5 \end{bmatrix}.$$

Thus the general solution to the nonhomogeneous matrix differential equation is

$$x(t) = c_1 \begin{bmatrix} 1 \\ -1 \end{bmatrix} e^{-3t} + c_2 \begin{bmatrix} 1 \\ 1 \end{bmatrix} e^{-t} + \begin{bmatrix} 1 \\ 1 \end{bmatrix} t e^{-t} - \begin{bmatrix} 0 \\ 1 \end{bmatrix} e^{-t} + \begin{bmatrix} 1 \\ 2 \end{bmatrix} t - \frac{1}{3}\begin{bmatrix} 4 \\ 5 \end{bmatrix}.$$

Finally, the constants c_1 and c_2 are determined from the initial conditions specified for the problem, $x(0) = x_0$.

METHOD OF UNCOUPLING THE VARIABLES BY DIAGONALIZATION

A second method that is sometimes convenient to use is to transform the original set of equations into a new set of equations in which the (new) variables are uncoupled. We then obtain the solution for the new variables by solving one decoupled equation at a time. The results are then transformed back to the original variables. If the matrix A is diagonalizable, the matrix of eigenvectors T can be used to accomplish this procedure. Mathematically, starting with the set of equations, i.e., Eq. (2.53),

$$\frac{dx}{dt} = Ax + f,$$

we define a new variable y such that $x = Ty$ or $y = T^{-1}x$. The above equation can be expressed in terms of the new variable y as

$$\frac{dy}{dt} = (T^{-1}AT)y + T^{-1}f$$

$$= \Lambda y + g. \tag{2.54}$$

If $\mathcal{T}$ is chosen to be a matrix of the eigenvectors, the equations in y will become decoupled (because Λ is a diagonal matrix) and can be solved separately. Finally, the solution x is obtained from y by means of the relationship $x = \mathcal{T}y$.

METHOD OF MATRIX EXPONENTIAL

This method is considerably more simple to use. The derivation of this method is as follows. We first rearrange differential equation (2.53),

$$\frac{dx}{dt} - Ax = f,$$

and multiply both sides of the rearranged equation by e^{-At}:

$$e^{-At}\left(\frac{dx}{dt} - Ax\right) = e^{-At} f. \tag{2.55}$$

Recognize that the left-hand side is a perfect differential:

$$\frac{d}{dt}[e^{-At}x(t)] = e^{-At} f(t). \tag{2.56}$$

Integrating both sides from t_0 to t, using τ as a dummy variable, yields

$$\int_{t_0}^{t} \frac{d}{d\tau}[e^{-A\tau}x(\tau)]\, d\tau = \int_{t_0}^{t} e^{-A\tau} f(\tau)\, d\tau$$

$$= e^{-At}x(t) - e^{-At_0}x(t_0). \tag{2.57}$$

We can now solve for $x(t)$ from Eq. (2.57):

$$x(t) = e^{A(t-t_0)}x(t_0) + \int_{t_0}^{t} e^{A(t-\tau)} f(\tau)\, d\tau \tag{2.58}$$

This is a very convenient result that expresses the solution in term of its initial condition and the matrix exponential.

EXAMPLE 2.13

Solve the following matrix differential equation,

$$\frac{d}{dt}\begin{bmatrix} x_1 \\ x_2 \end{bmatrix} = \begin{bmatrix} 0 & 1 \\ -\omega^2 & 0 \end{bmatrix}\begin{bmatrix} x_1 \\ x_2 \end{bmatrix} + \begin{bmatrix} 0 \\ 1 \end{bmatrix},$$

subject to some initial condition

$$\begin{bmatrix} x_1(0) \\ x_2(0) \end{bmatrix}.$$

We must first find the matrix exponential e^{At} for

$$A = \begin{bmatrix} 0 & 1 \\ -\omega^2 & 0 \end{bmatrix}.$$

From Eq. (2.47), e^{At} can be expressed in terms of a power series as

$$e^{At} = I + At + \frac{1}{2!}A^2t^2 + \frac{1}{3!}A^3t^3 + \cdots$$

$$= \begin{bmatrix} 1 & 0 \\ 0 & 1 \end{bmatrix} + \begin{bmatrix} 0 & t \\ -\omega^2 t & 0 \end{bmatrix} + \frac{1}{2!}\begin{bmatrix} -\omega^2 t^2 & 0 \\ 0 & -\omega^2 t^2 \end{bmatrix}$$

$$+ \frac{1}{3!}\begin{bmatrix} 0 & -\omega^2 t^3 \\ \omega^4 t^3 & 0 \end{bmatrix} + \cdots$$

$$= \begin{bmatrix} \cos \omega t & \omega^{-1}\sin \omega t \\ -\omega \sin \omega t & \cos \omega t \end{bmatrix}$$

because

$$\cos \omega t = 1 - \frac{1}{2!}\omega^2 t^2 + \frac{1}{4!}\omega^4 t^4 + \cdots,$$

$$\sin \omega t = \omega t - \frac{1}{3!}\omega^3 t^3 + \frac{1}{5!}\omega^5 t^5 - \cdots.$$

The next step is simply carrying out the integration

$$\begin{bmatrix} x_1(t) \\ x_2(t) \end{bmatrix} = \begin{bmatrix} \cos \omega t & \omega^{-1}\sin \omega t \\ -\omega \sin \omega t & \cos \omega t \end{bmatrix}\begin{bmatrix} x_1(0) \\ x_2(0) \end{bmatrix}$$

$$+ \int_0^t \begin{bmatrix} \cos \omega(t-\tau) & \omega^{-1}\sin \omega(t-\tau) \\ -\omega \sin \omega(t-\tau) & \cos \omega(t-\tau) \end{bmatrix}\begin{bmatrix} 0 \\ 1 \end{bmatrix} d\tau$$

$$= \begin{bmatrix} x_1(0)\cos \omega t + x_2(0)\omega^{-1}\sin \omega t \\ -x_1(0)\omega \sin \omega t + x_2(0)\cos \omega t \end{bmatrix} + \begin{bmatrix} \int_0^t \omega^{-1}\sin \omega(t-\tau)\,d\tau \\ \int_0^t \cos \omega(t-\tau)\,d\tau \end{bmatrix},$$

which produces the solution

$$x_1(t) = x_1(0)\cos \omega t + x_2(0)\omega^{-1}\sin \omega t + \omega^{-2}(1 - \cos \omega t),$$

$$x_2(t) = -x_1(0)\omega \sin \omega t + x_2(0)\cos \omega t + \omega^{-1}\sin \omega t.$$

2.16 Concluding Remarks

This chapter reviews the basic results of linear algebra and matrix differential equations that are required for vibration analysis, system identification, and control of mechanical systems. Most types of mechanical systems can be characterized by vector equations with matrix coefficients. Similar to the case of scalar differential equations that describe single-degree-of-freedom systems, the matrix approach can be used to determine the dynamic response of multiple-degree-of-freedom systems. For data analysis and model determination in system identification, solving linear algebraic equations is a necessary step. The SVD is a common tool in determining the size of the system model and finding the basis vectors for forming the identified model.

2.17 Problems

2.1 Let A and B be nonsingular and similar matrices, i.e.,

$$A = T^{-1}BT.$$

Prove that

(a) A^k is similar to B^k for every integer k,
(b) A^{-1} is similar to B^{-1},
(c) A^T is similar to B^T,
(d) A and B have the same eigenvalues.

2.2 Let A be

$$A = \begin{bmatrix} -3 & 1 & 0 \\ 1 & -3 & 0 \\ 0 & 0 & 5 \end{bmatrix}.$$

(a) First derive the eigenvalues and eigenvectors of the symmetric matrices $A^T A$ and AA^T. Then use the eigensolution to solve for the SVD of the matrix A.
(b) Change A slightly to become

$$A = \begin{bmatrix} -3 & 1 & 0 \\ 0 & -3 & 0 \\ 0 & 0 & 5 \end{bmatrix}.$$

Repeat item (a) to determine the SVD of the new matrix A.

2.3 Let $A = [a_{ij}]$ as defined in Eq. (2.1) with $m = n$ and $x = \{x_i\}$ be a column vector of length n. Define

$$\|x\|_p = \left(\sum_{i=1}^{n} |x_i|^p \right)^{1/p} \quad \text{for} \quad p = 1, 2,$$

$$\|x\|_\infty = \max_{1 \le i \le n} \{|x_i|\},$$

$$\|A\|_p = \max_{\|x\|_p=1} \{\|Ax\|_p\} \quad \text{for} \quad p = 1, 2, \text{ or } \infty.$$

Prove that

$$\|A\|_1 = \max_{j=1}^{n} \sum_{i=1}^{n} |a_{ij}|,$$

$$\|A\|_2 = \sigma_{\max}(A),$$

$$\|A\|_\infty = \max_{i=1}^{n} \sum_{j=1}^{n} |a_{ij}|,$$

where $\sigma_{\max}(A)$ means the maximum singular value of A.

2.4 Consider the following two matrices:

$$A = \begin{bmatrix} 4 & 2 & -2 \\ 2 & 10 & 2 \\ -2 & 2 & 5 \end{bmatrix}, \quad B = \begin{bmatrix} -3 & 1 & 0 \\ 1 & -3 & 0 \\ 0 & 0 & 3 \end{bmatrix}.$$

Determine the following quantities:

(a) $\|A\|_1$, $\|B\|_1$, $\|A\|_\infty$, $\|B\|_\infty$, $\|A\|_2$, $\|B\|_2$.

(b) The condition numbers of $\kappa(A)$ and $\kappa(B)$, i.e.,

$$\kappa(A) = \frac{\sigma_{max}(A)}{\sigma_{min}(A)}, \qquad \kappa(B) = \frac{\sigma_{max}(B)}{\sigma_{min}(B)},$$

where $\sigma_{max}(A)$ and $\sigma_{min}(A)$ are the maximum and the minimum singular values of the matrix A, respectively. The same definitions are also applicable for the matrix B.

2.5 Let the matrix A be

$$A = \begin{bmatrix} 1 & -2 & 1 & 1 \\ -1 & 3 & 0 & 2 \\ 0 & 1 & 1 & 2 \end{bmatrix}.$$

(a) Find basis vectors for the column space, the row space, and the null space of A, respectively.

(b) For a linear system $Ax = y$ with column vectors x and y, suppose that x_0 is a solution to the system. Find the set of solutions to the system.

2.6 Given an $n \times n$ matrix A, its matrix potential can be computed by

$$e^A = I + \sum_{k=1}^{\infty} \frac{1}{k!} A^k.$$

This series can be simplified to become

$$e^A = \alpha_{n-1} A^{n-1} + \alpha_{n-2} A^{n-2} + \cdots + \alpha_1 A + \alpha_0 I,$$

where I is an $n \times n$ identity matrix. Derive a formula to solve for the constant scalars

$$\alpha_{n-1}, \ \alpha_{n-2}, \ \dots, \ \alpha_1, \ \alpha_0.$$

Hint: First prove that

$$e^\lambda = \alpha_{n-1} \lambda^{n-1} + \alpha_{n-2} \lambda^{n-2} + \cdots + \alpha_1 \lambda + \alpha_0,$$

where λ is an eigenvalue of A, i.e.,

$$|A - \lambda I| = 0.$$

2.7 Use the formulation derived in Problem 2.6 to determine e^A, where

$$A = \begin{bmatrix} 3 & -3 & 3 \\ -1 & 5 & -2 \\ -1 & 3 & 0 \end{bmatrix}.$$

2.8 Determine e^{At}, where t is a scalar and

$$A = \begin{bmatrix} 0 & 0 & 0 \\ 1 & 0 & 0 \\ 1 & 0 & 1 \end{bmatrix}.$$

2.9 Let

$$A = \begin{bmatrix} 0 & 1 \\ 0 & 0 \end{bmatrix}, \qquad B = \begin{bmatrix} 0 & 0 \\ -1 & 0 \end{bmatrix}.$$

Calculate

$$e^{At}e^{Bt}, \quad e^{(A+B)t}.$$

Are they equal?

2.10   Prove that $e^{At}e^{Bt} = e^{(A+B)t}$ if and only if A and B commute, i.e., $AB = BA$.

BIBLIOGRAPHY

[1]   Belman, R., *Introduction to Matrix Analysis*, 2nd ed., McGraw-Hill, New York, 1970.

[2]   Golub, G. H. and Van Voan, C. F., *Matrix Computation*, The Johns Hopkins University Press, Baltimore, MD, 1983.

[3]   Klema, V. C. and Laub, A. J., "The Singular Value Decomposition: Its Computation and Some Applications," *IEEE Trans. Autom. Control,* Vol. AC-25, No. 2, pp. 164–176, April 1980.

[4]   Stewart, G. W., "On the Early History of the Singular Value Decomposition," IMA Preprint Series, No. 952, Institute for Mathematics and its Applications, University of Minnesota, Minneapolis, MN, pp. 1–24.

3

Modeling Techniques

3.1 Introduction

In this chapter, methods for developing dynamic equations of motion are introduced. We start with Newton's laws, which are fundamental to dynamic systems. Direct applications of Newton's laws to develop dynamic equations of motion become difficult for dynamic systems with complex configurations such as aircraft, automobiles, etc. D'Alembert's principle is introduced to derive the principle of virtual work for a body composed of a number of particles acted on by a number of external forces. By the concept of virtual work, Hamilton's principle is shown. Hamilton's principle is then used to formulate Lagrange's equation of motion (Refs. [1–2]). In contrast to Newton's second law, which involves vector quantities such as forces and accelerations, Hamilton's principle and Lagrange's equation use scalar quantities, including kinetic energy, potential energy, and virtual work, to develop the system's equations of motion. Along a similar line, Gibbs–Appell equations of motion are introduced with the scalar formed by the squared amplitude of acceleration instead of kinetic and potential energies. Finally, Kane's equations are derived with the same approach used in formulating the Gibbs–Appell equations. Kane's equations introduce a generalized speed vector rather than a scalar quantity (Ref. [3]). All techniques are related. Selecting a method to derive dynamic equations of motion is problem dependent.

3.2 Newton's Three Fundamental Laws

These laws were formulated and published by Sir Isaac Newton (1642–1726) in 1687 in his famous *Principia* (*Philosophiae Naturalis Principia Mathematica*).

FIRST LAW

If the resultant force acting on a particle is zero, the particle at rest will remain at rest or the particle in motion will move with constant speed in a straight line.

SECOND LAW

If the resultant force acting on a particle is not zero, the particle will have an acceleration proportional to the magnitude of the resultant force and in the direction of this resultant force. This law may be symbolically stated as

$$\mathbf{f} = m\mathbf{a}, \tag{3.1}$$

where $\mathbf{f}$, m, and $\mathbf{a}$ represent the resultant force acting on the particle, the mass of the particle, and the acceleration of the particle, respectively. All the quantities $\mathbf{f}$, m, and $\mathbf{a}$ must be expressed in a consistent system of units. For example, a force of one Newton is equivalent to a one-kilogram mass with an acceleration of one meter per second squared, i.e., $1\,\mathrm{N} = (1\,\mathrm{kg})\,(1\,\mathrm{m/s^2})$.

The vector $m\mathbf{a}$ is commonly referred to as the effective force or the inertia force of the particle. Therefore Newton's second law states that the resultant force acting on a particle is equal to the inertia force of the particle.

The acceleration of a particle is related to the velocity $\mathbf{v}$ and the displacement $\mathbf{r}$ of the particle by

$$\mathbf{a} = \frac{d\mathbf{v}}{dt} = \frac{d^2\mathbf{r}}{dt^2},$$

or, in short,

$$\mathbf{a} = \dot{\mathbf{v}} = \ddot{\mathbf{r}} \tag{3.2}$$

Newton's second law can be written in a form slightly different from that of Eq. (3.1):

$$\mathbf{f} = m\ddot{\mathbf{r}} = m\frac{d\mathbf{v}}{dt} = \frac{d}{dt}(m\mathbf{v}), \tag{3.3}$$

where the particle mass m has been assumed constant. The quantity $m\mathbf{v}$ is called the linear momentum of the particle.

Now, taking the cross product of the vector $\mathbf{r}$ and Eq. (3.3) yields

$$\mathbf{r} \times \mathbf{f} = \mathbf{r} \times \frac{d}{dt}(m\mathbf{v}),$$

or

$$\mathbf{r} \times \mathbf{f} = \frac{d}{dt}[\mathbf{r} \times (m\mathbf{v})]. \tag{3.4}$$

Note that the equality $\mathbf{v} \times \mathbf{v} = 0$ has been used to derive Eq. (3.4). The quantity $\mathbf{r} \times (m\mathbf{v})$ is defined as the moment of momentum, or angular momentum of m with respect to the origin of $\mathbf{r}$, i.e., the cross product of the vector $\mathbf{r}$ and its linear momentum $m\mathbf{v}$. Equation (3.4) says that the rate of change of the angular momentum for constant m about a fixed point is equal to the moment of the resultant force about that point.

THIRD LAW

To every force of action, there always exists a counterforce called the reaction force. The forces of action and reaction caused by the contact of bodies have the same magnitude and same line of action but opposite directions. In other words, when two particles exert forces on one another, these forces will be equal in magnitude and opposite in direction.

3.3   D'Alembert's Principle

D'Albembert's principle is named after the French mathematician Jean le rond d'Alembert (1717–1783), even though d'Albembert's original statement was written in a somewhat different form. Consider a rigid body formed by a large number of particles dm and acted on by n forces $\mathbf{f}_1, \mathbf{f}_2, \ldots, \mathbf{f}_n$. D'Alembert's principle states that *the external forces acting on a rigid body are equivalent to the the inertia forces of the various particles forming the body.* D'Albembert's principle may be symbolically written as

$$\mathbf{f}_1 + \mathbf{f}_2 + \cdots + \mathbf{f}_n = \sum_{i=1}^{n} \mathbf{f}_i = \int_{\Omega} \ddot{\mathbf{r}} dm,$$

or, more generally,

$$\int_{\Omega} (d\mathbf{f} - \ddot{\mathbf{r}} dm) = 0, \qquad (3.5)$$

where $\int_{\Omega} d\mathbf{f}$ represents sum of all the external forces, regardless of whether they are continuous or discrete in nature, $\ddot{\mathbf{r}}$ is the acceleration of the particle with the mass dm, and Ω represents the total region of the rigid body

3.4 Principle of Virtual Work

Consider a particle acted on by n forces $\mathbf{f}_1, \mathbf{f}_2, \ldots, \mathbf{f}_n$. Assume that the particle undergoes a small displacement $\delta\mathbf{r}$ from a point to another point. The quantity $\delta\mathbf{r}$ is only a possible displacement, but will not necessarily take place. The particle may be at rest, or it may move in a direction different from that of $\delta\mathbf{r}$. Therefore, the displacement $\delta\mathbf{r}$ is purely imaginary, and it is commonly called a virtual displacement. The symbol $\delta\mathbf{r}$ represent a differential of the first order. However, it has somewhat different characteristics in differential operation from the displacement $d\mathbf{r}$ that would take place under actual motion. The difference in differential operation between $\delta\mathbf{r}$ and $d\mathbf{r}$ will be discussed later.

The work done by each of the forces $\mathbf{f}_1, \mathbf{f}_2, \ldots, \mathbf{f}_n$ during the virtual displacement $\delta\mathbf{r}$ is called virtual work. The virtual work of all the forces acting on the particle is

$$\delta\mathcal{W} = \mathbf{f}_1 \cdot \delta\mathbf{r} + \mathbf{f}_2 \cdot \delta\mathbf{r} + \cdots + \mathbf{f}_n \cdot \delta\mathbf{r}$$
$$= (\mathbf{f}_1 + \mathbf{f}_2 + \cdots + \mathbf{f}_n) \cdot \delta\mathbf{r},$$

or, equivalently,

$$\delta\mathcal{W} = \mathbf{f} \cdot \delta\mathbf{r}, \qquad (3.6)$$

where $\mathbf{f}$ is the vector resultant of the forces $\mathbf{f}_1, \mathbf{f}_2, \ldots, \mathbf{f}_n$. The total virtual work of the forces $\mathbf{f}_1, \mathbf{f}_2, \ldots, \mathbf{f}_n$ is equal to the virtual work of their resultant $\mathbf{f}$.

Any vector having three components can be expressed in matrix notation and operated according to the matrix operation rules. For example, let f_1, f_2, and f_3 represent the three components of $\mathbf{f}$ in a specific coordinate frame. Similarly, let δr_1, δr_2, and δr_3 represent the three components of $\delta\mathbf{r}$ in the same coordinate frame. The dot product of the two vectors $\mathbf{f}$ and $\delta\mathbf{r}$ becomes $\mathbf{f} \cdot \delta\mathbf{r} = f_1\delta r_1 + f_2\delta r_2 + f_3\delta r_3$. For simplicity, define

$$f = \begin{bmatrix} f_1 \\ f_2 \\ f_3 \end{bmatrix}, \qquad \delta r = \begin{bmatrix} \delta r_1 \\ \delta r_2 \\ \delta r_3 \end{bmatrix}$$

as the 3×1 column vectors in matrix notation. Then the dot product of the two vectors **f** and δ**r** becomes **f** $\cdot \delta$**r** $= f^T \delta r$, where the superscript T means the transpose. Equation (3.6) can thus be written as

$$\delta W = f^T \delta r.$$

Hereinafter, for simplicity, all vectors not in bold faced type will be treated as column vectors and operated according to the matrix operation rules, unless otherwise stated.

The principle of virtual work for a particle states that *if a particle is in static equilibrium, the total work of the forces acting on the particle is zero for any virtual displacement of the particle*, i.e.,

$$\delta W = f^T \delta r = 0. \tag{3.7}$$

This condition is obviously necessary because the resultant force f is zero for a particle in static equilibrium from Newton's second law, and it follows that the total virtual work δW is zero. The condition is also sufficient because zero virtual work, $\delta W = 0$, means that $f^T \delta r = 0$, which implies that $f = 0$ for any arbitrary δr.

Similarly, the principle of virtual work can be extended to the dynamic case. Newton's second law states that the resultant of the forces acting on a particle is equal to the inertia force of the particle or, equivalently,

$$f - m\ddot{r} = 0. \tag{3.8}$$

In this case, the particle is said to be in dynamic equilibrium. The equilibrium condition implies that

$$\{f - m\ddot{r}\}^T \delta r = 0. \tag{3.9}$$

The principle of virtual work for a particle in dynamic equilibrium states that *if a particle is in dynamic equilibrium, the total virtual work of the forces acting on the particle is equal to the virtual work done by the inertia force for any virtual displacement of the particle*.

For a rigid body formed by a number of particles with mass dm and acted on by a number of external forces df, the principle of virtual work, with the aid of Eq. (3.5), can be expressed by

$$\int_\Omega \delta r^T (df - \ddot{r}\, dm) = 0, \tag{3.10}$$

where Ω is the domain of the whole body. The principle of virtual work for a rigid body in dynamic equilibrium states that *if a rigid body is in dynamic equilibrium, the total virtual work of the forces acting on the body is equal to the total virtual work done by the inertia forces for any virtual displacement of the body*.

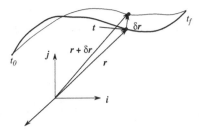

Figure 3.1. The true path and a virtual path of a particle.

3.5 Hamilton's Principle

One of the most basic and important principles in deriving equations of motion for physical systems is Hamilton's principle. For simplicity and clarity, consider a particle of mass m moving subject to a force f. Let r denote the position vector from the fixed origin of an inertia frame to the particle at time t. The path of the particle is governed by Newton's laws of motion,

$$m\ddot{r} = f. \tag{3.11}$$

Now consider any other path $r + \delta r$ as shown in Fig 3.1. Assume that the true path and the virtual path always coincide at two specific instants, t_o and t_f, which implies that the path variation δr vanishes at the two instants, i.e.,

$$\delta r|_{t=t_o} = \delta r|_{t=t_f} = 0. \tag{3.12}$$

At any other instant t between t_o and t_f, the path variation $\delta r(t)$ is arbitrary. Equation (3.12) *does not* imply that $\frac{d}{dt}\delta r|_{t=t_o} = 0$ or $\frac{d}{dt}\delta r|_{t=t_f} = 0$.

Take the vector product of Eq. (3.11) and the variation δr and integrate the result with respect to time over (t_o, t_f) to establish the relation

$$\int_{t_o}^{t_f} m\,\ddot{r}^T \delta r\,dt = \int_{t_o}^{t_f} f^T \delta r\,dt, \tag{3.13}$$

or, equivalently,

$$\int_{t_o}^{t_f} \{m\,\ddot{r} - f\}^T \delta r\,dt = 0.$$

This formula results from the principle of virtual work. The quantity $f^T \delta r$ represents the work done by the force f along the variation vector δr.

Integrate the left-hand side of Eq. (3.13) by parts[1] to obtain

$$m[\dot{r}^T \delta r]_{t_o}^{t_f} - \int_{t_o}^{t_f} m\,\dot{r}^T \delta\dot{r}\,dt = \int_{t_o}^{t_f} f^T \delta r\,dt. \tag{3.14}$$

The quantity $\delta\dot{r}$ represents the virtual velocity of the path.

Any quantity operated by δ corresponds to the virtual quantity. Conventionally, δ is referred to as the virtual operator, which is identical in operation to the differential

---

[1] Given any two vectors u and v, the formula for integration by parts is

$$\int_a^b u^T \frac{dv}{dt}\,dt = [u^T v]_a^b - \int_a^b \frac{du}{dt}^T v\,dt.$$

operator d except at the two points shown in Eq. (3.12). In addition, δ does not operate on the time variable, as the time variable is considered fixed. For example, let $T(q, \dot{q}, t)$ be an explicit function of q, $\dot{q}$, and t; its variation is

$$\delta T = \frac{\partial T}{\partial q} \delta q + \frac{\partial T}{\partial \dot{q}} \delta \dot{q},$$

whereas its differential is

$$dT = \frac{\partial T}{\partial q} dq + \frac{\partial T}{\partial \dot{q}} d\dot{q} + \frac{\partial T}{\partial t} dt.$$

When the function T does not depend on time explicitly, i.e., $T = T(q, \dot{q})$, both variation and differentiation are identical. The laws of variation of sums, products, ratios, and so forth are the same as those of differentiation. The two operators δ and d are commutable (i.e., interchangeable). For example, $\delta(dr/dt) = (d/dt)\,\delta r$, i.e., the virtual velocity of the path is equivalent to the velocity of the virtual path.

Now note the following equality:

$$m\,\dot{r}^T \delta \dot{r} = \frac{1}{2}m\,\delta(\dot{r}^T \dot{r}) = \delta T, \tag{3.15}$$

where T is the kinetic energy of the particle defined by

$$T = \frac{1}{2}m\,\dot{r}^T \dot{r}. \tag{3.16}$$

Substitution of Eqs. (3.12) and (3.15) into Eq. (3.14) produces

$$-\int_{t_o}^{t_f} \delta T\, dt = \int_{t_o}^{t_f} f^T \delta r\, dt$$

or

$$\int_{t_o}^{t_f} (\delta T + \delta W)\, dt = 0, \tag{3.17}$$

where $\delta W = f^T \delta r$ is the virtual work done by the external force f. This is Hamilton's principle in its most general form for the motion of a particle. It means that the integral of the sum of the varied kinetic energy δT and the virtual work δW over a time interval vanishes, under the assumption that the path variation δr vanishes at both ends of the same time interval.

EXAMPLE 3.1

Consider the mass–spring–dashpot system shown in Fig. 3.2. The quantity w is the displacement of the mass M moving subject to a resultant force f in one direction. As shown in Fig 3.2, there are three different forces simultaneously applied to the mass

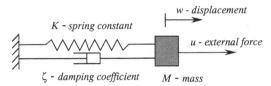

Figure 3.2. A simple mass–spring–dashpot system.

M, including the external force, the spring force, and the dashpot force. Let w be the displacement relative to the position of the mass when the spring force vanishes. The spring force is proportional to the negative displacement of the mass, whereas the dashpot force is proportional to the negative velocity of the mass. Therefore, the resultant force f is

$$f = u - Kw - \zeta \dot{w}, \tag{3.18}$$

where K is the spring constant and ζ is the damping coefficient of the dashpot. The equation of motion by Newton's laws is

$$M\ddot{w} = u - Kw - \zeta \dot{w},$$

or

$$M\ddot{w} + Kw + \zeta \dot{w} = u, \tag{3.19}$$

which is the second-order equation of motion for the simple mass–spring–dashpot system shown in Fig. 3.2.

Substituting Eq. (3.18) into Hamilton's principle, Eq. (3.17), with δr replaced with δw yields

$$\int_{t_o}^{t_f} [\delta T + (u - Kw - \zeta \dot{w})\delta w] \, dt = 0. \tag{3.20}$$

Let $\mathcal{V}$ denote the potential energy of the spring:

$$\mathcal{V} = \frac{1}{2} Kw^2. \tag{3.21}$$

The variation of the potential energy becomes

$$\delta \mathcal{V} = \frac{\partial \mathcal{V}}{\partial w} \delta w = Kw \delta w. \tag{3.22}$$

The spring force generated by the operation $f_c = -\frac{\partial \mathcal{V}}{\partial w} = -Kw$ is commonly referred to as the potential force or the conservative force. It is known that the spring force does not dissipate the energy. However, the dashpot force $-\zeta \dot{w}$ shown in Fig. 3.2 dissipates the energy and thus is called the dissipative force.

Similar to the conservative force $f_c = -kw$, the nonconservative damping force f_d of the type

$$f_d = -\zeta \dot{w} \tag{3.23}$$

can be derived from a function called Rayleigh's dissipation function, which is defined as

$$\mathcal{R} = \frac{1}{2}\zeta \dot{w}^2. \tag{3.24}$$

It is obvious that

$$f_d = -\frac{\partial \mathcal{R}}{\partial \dot{w}} = -\zeta \dot{w}. \tag{3.25}$$

From Eqs. (3.22) and (3.25), Eq. (3.20) can be rewritten as

$$\int_{t_o}^{t_f} \left[\delta(T - V) + \left(u - \frac{\partial \mathcal{R}}{\partial \dot{w}} \right) \delta w \right] dt = 0. \tag{3.26}$$

This is Hamilton's principle when a potential function exists. The energy difference defined by

$$\mathcal{L} = T - V \tag{3.27}$$

is called the Lagrangian. Equation (3.26) thus becomes

$$\int_{t_o}^{t_f} \left[\delta\mathcal{L} + \left(u - \frac{\partial \mathcal{R}}{\partial \dot{w}} \right) \delta w \right] dt = 0. \tag{3.28}$$

This equation simply states that the sum of the variation of the Lagrangian and the virtual work done by the nonconservative forces during any time interval t_o and t_f must vanish. This statement implies that the equation of motion can be derived from the variation of kinetic energy and potential energy and the virtual work of nonconservative forces including the damping force generated from Rayleigh's dissipative function.

From Example 3.1, it is clear that Hamilton's principle may be expressed in terms of the Lagrangian as defined in Eq. (3.27). Let $\mathcal{L} = T - V$ be the system Lagrangian, where T and V are the kinetic and the potential energy, respectively. The conservative force produced by the potential energy is

$$f_c = -\frac{\partial V}{\partial r} = \left\{ \begin{array}{c} -\dfrac{\partial V}{\partial r_1} \\[2mm] -\dfrac{\partial V}{\partial r_2} \\[2mm] -\dfrac{\partial V}{\partial r_3} \end{array} \right\}, \tag{3.29}$$

where r is a column vector having three components r_1, r_2, and r_3. Let the external force f in Eq. (3.17) be decomposed into two kind of forces, namely the conservative

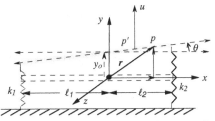

Figure 3.3. A rigid bar supported by two springs at both ends.

force f_c and the nonconservative force f_{nc}, such that

$$f = f_c + f_{nc}. \tag{3.30}$$

From Eqs. (3.29) and (3.30), Hamilton's principle shown in Eq. (3.17) becomes

$$\int_{t_o}^{t_f} (\delta\mathcal{L} + \delta W_{nc})\, dt = 0, \quad \delta W_{nc} = f_{nc}^T \delta r, \tag{3.31}$$

where δW_{nc} is the virtual work done by the nonconservative force f_{nc}. The nonconservative force may include the damping force generated from Rayleigh's dissipative function. Although Hamilton's principle has been derived here specifically for a particle (or a point mass) moving in response to external point forces, it is also applicable for multiple flexible bodies moving subject to external distributed forces and torques.

EXAMPLE 3.2

Consider a rigid bar supported by two springs at both ends, as shown in Fig. 3.3. Let x, y, and z denote the inertia coordinates with its origin located at the point • on the bar at rest (dashed bar in Fig. 3.3). When the system starts vibrating because of the external force u, the position of point p on the bar can be expressed by

$$r = \begin{Bmatrix} x \\ y \end{Bmatrix} = \begin{Bmatrix} x \\ y_o + x\theta \end{Bmatrix}, \tag{3.32}$$

where y_o is the displacement of the point • in the y direction and θ is the rotational angle relative to the z axis. For a small rotation, the change in height can be approximated by $y = y_o + x\theta$. Differentiating Eq. (3.32) relative to time yields

$$\dot{r} = \begin{Bmatrix} 0 \\ \dot{y}_o + x\dot{\theta} \end{Bmatrix}.$$

The kinetic energy of the system as a function of time is

$$T = \frac{1}{2} \int_{-\ell_1}^{\ell_2} \dot{r}^T \dot{r}\, \rho dx = \frac{1}{2} \int_{-\ell_1}^{\ell_2} (\dot{y}_o + x\dot{\theta})^2 \rho dx, \tag{3.33}$$

where ℓ_1 and ℓ_2 are, the left length and the right length of the bar divided by the point •, respectively, and ρ is the mass density of the bar per unit length. The potential energy of the system at time t that is due to the two springs at both ends is

$$V = \frac{1}{2} k_1 (y_o - \ell_1\theta)^2 + \frac{1}{2} k_2 (y_o + \ell_2\theta)^2, \tag{3.34}$$

where k_1 and k_2 are the left and the right spring constants, respectively.

The Lagrangian at time t can then be defined by

$$
\begin{aligned}
\mathcal{L} &= \mathcal{T} - \mathcal{V} \\
&= \frac{1}{2} \int_{-\ell_1}^{\ell_2} (\dot{y}_o + x\dot{\theta})^2 \rho dx - \frac{1}{2}k_1(y_o - \ell_1\theta)^2 - \frac{1}{2}k_2(y_o + \ell_2\theta)^2.
\end{aligned} \quad (3.35)
$$

Now integrate the variation of the Lagrangian with respect to time over (t_o, t_f) to obtain the formulation

$$
\begin{aligned}
\int_{t_o}^{t_f} \delta\mathcal{L}dt &= \int_{t_o}^{t_f} \delta[\mathcal{T} - \mathcal{V}]dt = \frac{1}{2}\int_{t_o}^{t_f} \left\{ \int_{-\ell_1}^{\ell_2} \delta[(\dot{y}_o + x\dot{\theta})^2]\rho dx \right. \\
&\qquad \left. - k_1\delta[(y_o - \ell_1\theta)^2] - k_2\delta[(y_o + \ell_2\theta)^2] \right\} dt \\
&= \int_{t_o}^{t_f} \left\{ \int_{-\ell_1}^{\ell_2} (\dot{y}_o + x\dot{\theta})(\delta\dot{y}_o + x\delta\dot{\theta})\rho dx \right\} dt \\
&\qquad - \int_{t_o}^{t_f} k_1(y_o - \ell_1\theta)(\delta y_o - \ell_1\delta\theta)\, dt \\
&\qquad - \int_{t_o}^{t_f} k_2(y_o + \ell_2\theta)(\delta y_o + \ell_2\delta\theta)\, dt \\
&= \int_{-\ell_1}^{\ell_2} [(\dot{y}_o + x\dot{\theta})(\delta y_o + x\delta\theta)]_{t_o}^{t_f} \rho dx \\
&\qquad - \int_{t_o}^{t_f} \left[\int_{-\ell_1}^{\ell_2} (\ddot{y}_o + x\ddot{\theta})(\delta y_o + x\delta\theta)\rho dx \right] dt \\
&\qquad - \int_{t_o}^{t_f} k_1(y_o - \ell_1\theta)(\delta y_o - \ell_1\delta\theta)\, dt \\
&\qquad - \int_{t_o}^{t_f} k_2(y_o + \ell_2\theta)(\delta y_o + \ell_2\delta\theta)\, dt,
\end{aligned} \quad (3.36)
$$

where the last equality results from the formula for integration by parts relative to time t. The quantities δy_o and $\delta\theta$ are called the virtual displacement and the virtual angle, respectively. The virtual operator δ is identical in operation to the differential operator dt except at two distinct instants, $t = t_o$ and $t = t_f$. The virtual displacement δy_o and the virtual angle $\delta\theta$ vanish at the two instants $t = t_o$ and $t = t_f$:

$$
\delta y_o(t_o) = \delta\theta(t_o) = \delta y_o(t_f) = \delta\theta(t_f) = 0. \quad (3.37)
$$

With Eq. (3.37), Eq. (3.36) becomes

$$
\begin{aligned}
\int_{t_o}^{t_f} \delta\mathcal{L}dt &= \int_{t_o}^{t_f} \delta[\mathcal{T} - \mathcal{V}]\, dt \\
&= -\int_{t_o}^{t_f} [m\ddot{y}_o + mx_c\ddot{\theta} + k_1(y_o - \ell_1\theta) + k_2(y_o + \ell_2\theta)]\delta y_o dt \\
&\qquad - \int_{t_o}^{t_f} [mx_c\ddot{y}_o + I\ddot{\theta} + k_1(-\ell_1 y_o + \ell_1^2\theta) + k_2(\ell_2 y_o + \ell_2^2\theta)]\,\delta\theta dt,
\end{aligned} \quad (3.38)
$$

where the total mass m, the total moment of inertia $\mathcal{I}$, and the center of mass x_c are defined by

$$m = \int_{-\ell_1}^{\ell_2} \rho dx, \qquad \mathcal{I} = \int_{-\ell_1}^{\ell_2} \rho x^2 dx, \qquad x_c = \frac{1}{m} \int_{-\ell_1}^{\ell_2} \rho x dx. \qquad (3.39)$$

Assume that there is a force u acting on the bar at point p' along the direction y. Let r' be the displacement vector from the origin of the inertia coordinate to point p'. Similar to the displacement vector r shown in Eq. (3.32), r' for a small displacement can be written as

$$r' = \begin{Bmatrix} x' \\ y' \end{Bmatrix} = \begin{Bmatrix} x' \\ y_o + x'\theta \end{Bmatrix}. \qquad (3.40)$$

The virtual work done by the force u relative to time over (t_o, t_f) becomes

$$\int_{t_o}^{t_f} \delta \mathcal{W}_{nc} dt = \int_{t_o}^{t_f} u(\delta y_o + x'\delta\theta) \, dt. \qquad (3.41)$$

Application of Hamilton's principle, i.e.,

$$\int_{t_o}^{t_f} (\delta \mathcal{L} + \delta \mathcal{W}_{nc}) \, dt = 0, \qquad (3.42)$$

with the help of Eq. (3.38) and (3.41), produces

$$- \int_{t_o}^{t_f} [m\ddot{y}_o + mx_c\ddot{\theta} + k_1(y_o - \ell_1\theta) + k_2(y_o + \ell_2\theta)]\delta y_o dt$$

$$- \int_{t_o}^{t_f} \left[mx_c\ddot{y}_o + \mathcal{I}\ddot{\theta} + k_1 \left(-\ell_1 y_o + \ell_1^2\theta \right) + k_2 \left(\ell_2 y_o + \ell_2^2\theta \right) \right] \delta\theta dt$$

$$+ \int_{t_o}^{t_f} u(\delta y_o + x'\delta\theta) \, dt = 0,$$

or, when the same virtual quantities are collected,

$$\int_{t_o}^{t_f} [m\ddot{y}_o + mx_c\ddot{\theta} + k_1(y_o - \ell_1\theta) + k_2(y_o + \ell_2\theta) - u]\delta y_o dt$$

$$+ \int_{t_o}^{t_f} \left[mx_c\ddot{y}_o + \mathcal{I}\ddot{\theta} + k_1 \left(-\ell_1 y_o + \ell_1^2\theta \right) + k_2 \left(\ell_2 y_o + \ell_2^2\theta \right) - x'u \right] \delta\theta dt = 0. \qquad (3.43)$$

Because the virtual quantities δy_o and $\delta\theta$ are arbitrary, the coefficients associated with these quantities must vanish to make Eq. (3.43) valid, i.e.,

$$m\ddot{y}_o + mx_c\ddot{\theta} + k_1(y_o - \ell_1\theta) + k_2(y_o + \ell_2\theta) = u,$$

$$mx_c\ddot{y}_o + \mathcal{I}\ddot{\theta} + k_1 \left(-\ell_1 y_o + \ell_1^2\theta \right) + k_2 \left(\ell_2 y_o + \ell_2^2\theta \right) = x'u, \qquad (3.44)$$

which can be written in matrix form:

$$\begin{bmatrix} m & mx_c \\ mx_c & \mathcal{I} \end{bmatrix} \begin{bmatrix} \ddot{y}_o \\ \ddot{\theta} \end{bmatrix} + \begin{bmatrix} k_1 + k_2 & -k_1\ell_1 + k_2\ell_2 \\ -k_1\ell_1 + k_2\ell_2 & k_1\ell_1^2 + k_2\ell_2^2 \end{bmatrix} \begin{bmatrix} y_o \\ \theta \end{bmatrix} = \begin{bmatrix} 1 \\ x' \end{bmatrix} u, \qquad (3.45)$$

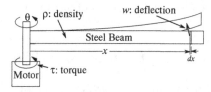

Figure 3.4. A rotating beam.

or, equivalently,

$$\mathcal{M}\ddot{w} + \mathcal{K}w = \mathcal{B}u. \tag{3.46}$$

$\mathcal{M}$ and $\mathcal{K}$ are generally called the mass matrix and the stiffness matrix, respectively, w is the state vector, and $\mathcal{B}$ is the force influence matrix, which are defined by

$$\mathcal{M} = \begin{bmatrix} m & mx_c \\ mx_c & \mathcal{I} \end{bmatrix}, \qquad w = \begin{bmatrix} y_o \\ \theta \end{bmatrix}, \qquad \mathcal{B} = \begin{bmatrix} 1 \\ x' \end{bmatrix},$$

$$\mathcal{K} = \begin{bmatrix} k_1 + k_2 & -k_1\ell_1 + k_2\ell_2 \\ -k_1\ell_1 + k_2\ell_2 & k_1\ell_1^2 + k_2\ell_2^2 \end{bmatrix}. \tag{3.47}$$

For the case in which $x_c = 0$ and $k_1\ell_1 = k_2\ell_2$, i.e., the point ● (see Fig. 3.3) coincides with the center of mass and the moment center of the two spring forces attached at the end of the bar, the mass and stiffness matrices become

$$\mathcal{M} = \begin{bmatrix} m & 0 \\ 0 & \mathcal{I} \end{bmatrix}, \qquad \mathcal{K} = \begin{bmatrix} k_1 + k_2 & 0 \\ 0 & k_1\ell_1^2 + k_2\ell_2^2 \end{bmatrix}.$$

Physically, this means that the translational and the rotational motions are completely decoupled.

EXAMPLE 3.3

Figure 3.4 shows a steel beam driven by a motor to rotate about a vertical axis. For simplicity, consider only an in-plane small bending motion. Let ρ be the mass density of the steel beam per unit length. When the steel beam starts rotating and vibrating, the velocity of the mass ρdx can be expressed by

$$\dot{r} = x\dot{\theta} + \dot{w}, \tag{3.48}$$

where x is the distance from the center of the rotating shaft to the mass ρdx, θ is the rotating angle, and w is the in-plane bending deflection.

The kinetic energy of the system at time t is

$$T = \frac{1}{2}\int_0^\ell (x\dot{\theta} + \dot{w})^2 \rho dx, \tag{3.49}$$

where ℓ is the total length of the steel beam. The potential energy at time t from the beam elasticity is

$$V = \frac{1}{2} \int_0^\ell EI \left(\frac{\partial^2 w}{\partial x^2} \right)^2 dx, \tag{3.50}$$

where EI is the bending rigidity of the steel beam. At this point, it is not important to understand how the potential energy is formulated. This example is chosen to demonstrate how we can use Hamilton's principle to model a continuous beam.

Now, integrate the kinetic energy with respect to time over (t_o, t_f) and take the variation of the result to obtain the formulation

$$\int_{t_o}^{t_f} \delta T \, dt = \frac{1}{2} \int_{t_o}^{t_f} \delta \int_0^\ell (x\dot\theta + \dot w)^2 \rho \, dx \, dt = \int_{t_o}^{t_f} \int_0^\ell (x\dot\theta + \dot w)(x\delta\dot\theta + \delta\dot w) \rho \, dx \, dt$$

$$= \int_0^\ell [(x\dot\theta + \dot w)(x\delta\theta + \delta w)]_{t_o}^{t_f} \rho \, dx - \int_{t_o}^{t_f} \int_0^\ell (x\ddot\theta + \ddot w)(x\delta\theta + \delta w) \rho \, dx \, dt$$

$$= - \int_{t_o}^{t_f} \int_0^\ell (x\ddot\theta + \ddot w)(x\delta\theta + \delta w) \rho \, dx \, dt, \tag{3.51}$$

where the third equality results from the formula for integration by parts relative to time t. The virtual bending deflection δw and the virtual angle $\delta\theta$ vanish at the two instants $t = t_o$ and $t = t_f$.

Similarly, integrate the potential energy with respect to time over (t_o, t_f) and take the variation of the result to obtain the formulation

$$\int_{t_o}^{t_f} \delta V \, dt = \frac{1}{2} \int_{t_o}^{t_f} \delta \int_0^\ell EI \left(\frac{\partial^2 w}{\partial x^2} \right)^2 dx \, dt$$

$$= \int_{t_o}^{t_f} \int_0^\ell EI \left(\frac{\partial^2 w}{\partial x^2} \right) \delta \left(\frac{\partial^2 w}{\partial x^2} \right) dx \, dt$$

$$= \int_{t_o}^{t_f} \left[EI \left(\frac{\partial^2 w}{\partial x^2} \right) \delta \left(\frac{\partial w}{\partial x} \right) \right]_0^\ell dx - \int_{t_o}^{t_f} \int_0^\ell EI \left(\frac{\partial^3 w}{\partial x^3} \right) \delta \left(\frac{\partial w}{\partial x} \right) dx \, dt$$

$$= \int_{t_o}^{t_f} \left[EI \left(\frac{\partial^2 w}{\partial x^2} \right) \delta \left(\frac{\partial w}{\partial x} \right) \right]_0^\ell dt - \int_{t_o}^{t_f} \left[EI \left(\frac{\partial^3 w}{\partial x^3} \right) \delta w \right]_0^\ell dt$$

$$+ \int_{t_o}^{t_f} \int_0^\ell EI \left(\frac{\partial^4 w}{\partial x^4} \right) (\delta w) \, dx \, dt, \tag{3.52}$$

where the last two equalities result from the formula for integration by parts relative to the spatial coordinate x. The virtual bending deflection δw and the virtual bending-deflection derivative $\delta(\partial w/\partial x)$ vanish at the root $x = 0$ because the steel beam is clamped at the root. In addition, the bending moment $EI(\partial^2 w/\partial x^2)$ and the bending shear force $EI(\partial^3 w/\partial x^3)$ are zero at the free end of the beam, i.e., $x = \ell$. Note that the quantities $EI(\partial^2 w/\partial x^2)$ and $EI(\partial^3 w/\partial x^3)$ are the bending moment and the bending shear force at location x, respectively, according to the flexural theory of flexible structures (Ref. [2, 4, 5]). As a result, if we write

$$\left[EI\left(\frac{\partial^2 w}{\partial x^2}\right) \delta\left(\frac{\partial w}{\partial x}\right) \right]_0^\ell = 0, \tag{3.53}$$

$$\left[EI\left(\frac{\partial^3 w}{\partial x^3}\right) \delta w \right]_0^\ell = 0, \tag{3.54}$$

we take into account the possibility that either $EI(\partial^2 w/\partial x^2)$ or $\delta(\partial w/\partial x)$, on the one hand, and either $EI(\partial^3 w/\partial x^3)$ or δw, on the other, vanishes at any of the ends $x = 0$ and $x = \ell$. From Eqs. (3.53) and (3.54), Eq. (3.52) becomes

$$\int_{t_0}^{t_f} \delta V dt = \int_{t_0}^{t_f} \int_0^\ell EI\left(\frac{\partial^4 w}{\partial x^4}\right)(\delta w)\,dx\,dt. \tag{3.55}$$

The virtual work from the motor torque is

$$\int_{t_0}^{t_f} \delta W_{nc} dt = \int_{t_0}^{t_f} \tau \delta\theta dt. \tag{3.56}$$

Application of Hamilton's principle with the aid of Eqs. (3.51)–(3.56) thus yields

$$\int_{t_0}^{t_f} \left\{ \left[\int_0^\ell (x\ddot\theta + \ddot w)x\rho dx \right] - \tau \right\} \delta\theta dt$$

$$+ \int_{t_0}^{t_f} \int_0^\ell \left[(x\ddot\theta + \ddot w)\rho + EI\left(\frac{\partial^4 w}{\partial x^4}\right) \right] \delta w\,dx\,dt = 0. \tag{3.57}$$

Because the variation quantities δw and $\delta\theta$ are arbitrary, the coefficients associated with these quantities must vanish to make Eq. (3.57) valid, i.e.,

$$\mathcal{I}\ddot\theta + \int_0^\ell x\rho\ddot w dx = \tau, \tag{3.58}$$

$$\rho\ddot w + EI\left(\frac{\partial^4 w}{\partial x^4}\right) + \rho x\ddot\theta = 0, \tag{3.59}$$

where $\mathcal{I} = \int_0^\ell \rho x^2 dx$ is the moment of inertia of the steel beam. Equations (3.58) and (3.59) are the differential equations of motion for the rotational beam shown in Fig. 3.4. These equations must be satisfied over the length of the steel beam. Equations (3.53) and (3.54) represent the boundary conditions.

3.6 Lagrange's Equations

Newton's second law involves vector quantities such as forces and accelerations. Similar to Hamilton's principle, Lagrange's equations use scalar quantities including the kinetic energy, potential energy, and virtual work to develop the system's equations of motion. Lagrange's equations can be derived directly from Hamilton's principle. Let

$q = (q_1, q_2, \ldots, q_n)^T$ be the generalized coordinate vector of dimension n, which may represent the physical coordinate vector. The displacement column vector r and its velocity vector $\dot{r}$ can then be expressed as

$$r = r(q, t), \tag{3.60}$$

$$\dot{r} = \frac{\partial r}{\partial q}\dot{q} + \frac{\partial r}{\partial t}, \tag{3.61}$$

or, in matrix form,

$$
\begin{Bmatrix} \dot{r}_1 \\ \dot{r}_2 \\ \dot{r}_3 \end{Bmatrix}
=
\begin{bmatrix}
\dfrac{\partial r_1}{\partial q_1} & \dfrac{\partial r_1}{\partial q_2} & \cdots & \dfrac{\partial r_1}{\partial q_n} \\[2mm]
\dfrac{\partial r_2}{\partial q_1} & \dfrac{\partial r_2}{\partial q_2} & \cdots & \dfrac{\partial r_2}{\partial q_n} \\[2mm]
\dfrac{\partial r_3}{\partial q_1} & \dfrac{\partial r_3}{\partial q_2} & \cdots & \dfrac{\partial r_3}{\partial q_n}
\end{bmatrix}
\begin{Bmatrix} \dot{q}_1 \\ \dot{q}_2 \\ \vdots \\ \dot{q}_n \end{Bmatrix}
+
\begin{Bmatrix} \dfrac{\partial r_1}{\partial t} \\[2mm] \dfrac{\partial r_2}{\partial t} \\[2mm] \dfrac{\partial r_3}{\partial t} \end{Bmatrix}.
\tag{3.62}
$$

The kinetic energy of a system is computed by

$$\mathcal{T} = \frac{1}{2}\int_\Omega \dot{r}^T \dot{r}\, dm, \tag{3.63}$$

where dm is the infinitesimal mass of a continuous system. Substitution of Eq. (3.61) for $\dot{r}$ into Eq. (3.63) clearly indicates that the kinetic energy of a system must be a function of $(q, \dot{q}, t)$. However, the potential energy of a system is a function of r only, and thus is a function of (q, t) only. Knowing that the kinetic energy $\mathcal{T} = \mathcal{T}(q, \dot{q}, t)$ and the potential energy $\mathcal{V} = \mathcal{T}(q, t)$, their variations are

$$\delta\mathcal{T} = \left(\frac{\partial\mathcal{T}}{\partial q}\right)^T \delta q + \left(\frac{\partial\mathcal{T}}{\partial \dot{q}}\right)^T \delta\dot{q} = \sum_{i=1}^{n}\frac{\partial\mathcal{T}}{\partial q_i}\delta q_i + \sum_{i=1}^{n}\frac{\partial\mathcal{T}}{\partial \dot{q}_i}\delta\dot{q}_i, \tag{3.64}$$

$$\delta\mathcal{V} = \left(\frac{\partial\mathcal{V}}{\partial q}\right)^T \delta q = \sum_{i=1}^{k}\frac{\partial\mathcal{V}}{\partial q_i}\delta q_i. \tag{3.65}$$

As discussed earlier, the time variation δt does not exist, and thus no such term is shown in Eqs. (3.64) and (3.65). The virtual work induced by the nonconservative force is

$$\delta\mathcal{W}_{\text{nc}} = f_{\text{nc}}^T \delta r = f_{\text{nc}}^T \left\{\frac{\partial r}{\partial q}\delta q\right\} = Q_{\text{nc}}^T \delta q = \sum_{i=1}^{k} Q_{\text{nc}_i}\delta q_i, \tag{3.66}$$

where Q_{nc} is the generalized force vector defined by

$$Q_{\text{nc}} = \left[\frac{\partial r}{\partial q}\right]^T f_{\text{nc}}, \tag{3.67}$$

and thus its elements $Q_{\text{nc}_1}, Q_{\text{nc}_2}, \ldots, Q_{\text{nc}_n}$ are computed by

$$Q_{\text{nc}_i} = \sum_{j=1}^{3}\frac{\partial r_j}{\partial q_i}f_{\text{nc}_j}, \quad i = 1, 2, \ldots, n. \tag{3.68}$$

Note that a subscript represents an element corresponding to the vector quantity. The partial derivative of a vector of dimension n with respect to another vector of dimension m results in an $n \times m$ matrix. For example, $\partial r / \partial q$ is a $3 \times n$ matrix.

Application of Hamilton's principle defined in Eq. (3.31) leads to

$$\int_{t_o}^{t_f} \left[\left(\frac{\partial \mathcal{T}}{\partial q} - \frac{\partial \mathcal{V}}{\partial q} + Q_{\text{nc}} \right)^T \delta q + \left(\frac{\partial \mathcal{T}}{\partial \dot{q}} \right)^T \delta \dot{q} \right] dt = 0. \tag{3.69}$$

The last term in the bracket can be integrated by parts to yield

$$\int_{t_o}^{t_f} \left(\frac{\partial \mathcal{T}}{\partial \dot{q}} \right)^T \delta \dot{q} \, dt = \left(\frac{\partial \mathcal{T}}{\partial \dot{q}} \right)^T \delta q \Bigg|_{t_o}^{t_f} - \int_{t_o}^{t_f} \frac{d}{dt} \left(\frac{\partial \mathcal{T}}{\partial \dot{q}} \right)^T \delta q \, dt. \tag{3.70}$$

The first term on the right-hand side vanishes because the motion is specified at the two ends t_o and t_f, i.e., $\delta q(t_o) = \delta q(t_f) = 0$. From Eq. (3.70), Eq. (3.69) becomes

$$\int_{t_o}^{t_f} \left\{ \left[-\frac{d}{dt} \left(\frac{\partial \mathcal{T}}{\partial \dot{q}} \right) + \frac{\partial \mathcal{T}}{\partial q} - \frac{\partial \mathcal{V}}{\partial q} + Q_{\text{nc}} \right]^T \delta q \right\} dt = 0. \tag{3.71}$$

Assume that the coordinates $q_1, q_2, \ldots, q_n$ are independent in the sense that any coordinate q_i cannot be obtained by any combination of the other coordinates. Because δq_i for $i = 1, 2, \ldots n$ are arbitrary, Eq. (3.71) leads to

$$\frac{d}{dt} \left(\frac{\partial \mathcal{T}}{\partial \dot{q}} \right) - \frac{\partial \mathcal{T}}{\partial q} + \frac{\partial \mathcal{V}}{\partial q} = Q_{\text{nc}}. \tag{3.72}$$

Noting that the potential energy $\mathcal{V}$ is a function of q only such that $(\partial \mathcal{V} / \partial \dot{q}) = 0$, we can rewrite Eq. (3.72) as

$$\frac{d}{dt} \left(\frac{\partial \mathcal{L}}{\partial \dot{q}} \right) - \frac{\partial \mathcal{L}}{\partial q} = Q_{\text{nc}}, \tag{3.73}$$

where $\mathcal{L} = \mathcal{T} - \mathcal{V}$ is the system Lagrangian, as discussed earlier in the chapter. Equation (3.73) is commonly referred to as Lagrange's equation of motion. It is convenient to write Lagrange's equation in a scalar form

$$\frac{d}{dt} \left(\frac{\partial \mathcal{L}}{\partial \dot{q}_i} \right) - \frac{\partial \mathcal{L}}{\partial q_i} = Q_{\text{nc}_i}, \quad i = 1, 2, \ldots, n. \tag{3.74}$$

EXAMPLE 3.4

Consider the two-link system shown in Fig. 3.5. The vectors $\mathbf{r}_1$ and $\boldsymbol{\ell}_1 + \mathbf{r}_2$ describe the positions of the infinitesimal masses dm_1 and dm_2 of the first and the second

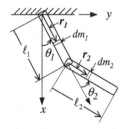

Figure 3.5. A planar two-link system.

links, respectively. Both vectors have two components in the Cartesian coordinates x and y, i.e.,

$$\mathbf{r}_1 = \left\{ \begin{array}{c} x_1 \\ y_1 \end{array} \right\} = \left\{ \begin{array}{c} r_1 \cos \theta_1 \\ r_1 \sin \theta_1 \end{array} \right\}, \tag{3.75}$$

$$\boldsymbol{\ell}_1 + \mathbf{r}_2 = \left\{ \begin{array}{c} x_2 \\ y_2 \end{array} \right\} = \left\{ \begin{array}{c} \ell_1 \cos \theta_1 + r_2 \cos(\theta_1 + \theta_2) \\ \ell_1 \sin \theta_1 + r_2 \sin(\theta_1 + \theta_2) \end{array} \right\}, \tag{3.76}$$

where θ_1 and θ_2 are the angles measured from the x axis and the first link, respectively. The quantities ℓ_1 and ℓ_2 represent the length of the first link and the second link, respectively. Let the generalized coordinate vector q be defined as

$$q = \left\{ \begin{array}{c} q_1 \\ q_2 \end{array} \right\} = \left\{ \begin{array}{c} \theta_1 \\ \theta_1 + \theta_2 \end{array} \right\}. \tag{3.77}$$

Equations (3.75) and (3.76) for position vectors $\mathbf{r}_1$ and $\boldsymbol{\ell}_1 + \mathbf{r}_2$ become

$$\mathbf{r}_1 = \left\{ \begin{array}{c} r_1 \cos q_1 \\ r_1 \sin q_1 \end{array} \right\}, \tag{3.78}$$

$$\boldsymbol{\ell}_1 + \mathbf{r}_2 = \left\{ \begin{array}{c} \ell_1 \cos q_1 + r_2 \cos q_2 \\ \ell_1 \sin q_1 + r_2 \sin q_2 \end{array} \right\}. \tag{3.79}$$

Differentiating Eqs. (3.78) and (3.79) with respect to time t yields

$$\dot{\mathbf{r}}_1 = r_1 \dot{q}_1 \left\{ \begin{array}{c} -\sin q_1 \\ \cos q_1 \end{array} \right\}, \tag{3.80}$$

$$\dot{\mathbf{r}}_2 + \dot{\boldsymbol{\ell}}_1 = \left\{ \begin{array}{c} -\ell_1 \dot{q}_1 \sin q_1 - r_2 \dot{q}_2 \sin q_2 \\ \ell_1 \dot{q}_1 \cos q_1 + r_2 \dot{q}_2 \cos q_2 \end{array} \right\}. \tag{3.81}$$

From Eqs. (3.80) and (3.81), the kinetic energy of the two-link system can now be calculated by

$$\begin{aligned} T &= \frac{1}{2} \int_0^{\ell_1} \dot{\mathbf{r}}_1^T \dot{\mathbf{r}}_1 dm_1 + \int_0^{\ell_2} (\dot{\boldsymbol{\ell}}_1 + \dot{\mathbf{r}}_2)^T (\dot{\boldsymbol{\ell}}_1 + \dot{\mathbf{r}}_2) dm_2 \\ &= \frac{1}{2} \left[\mathcal{I}_1 \dot{q}_1^2 + m_2 \ell_1^2 \dot{q}_1^2 + \mathcal{I}_2 \dot{q}_2^2 + 2 m_2 \ell_1 r_{2c} \dot{q}_1 \dot{q}_2 \cos(q_2 - q_1) \right], \end{aligned} \tag{3.82}$$

where m_1 and m_2 are the total masses of the first link and the second link, respectively, and

$$\mathcal{I}_1 = \int_0^{\ell_1} r_1^2 dm_1, \qquad \mathcal{I}_2 = \int_0^{\ell_2} r_2^2 dm_2, \qquad r_{2c} = \frac{1}{m_2} \int_0^{\ell_2} r_2 dm_2.$$

The quantity $\mathcal{I}_1$ is the moment of inertia of the first link relative to the origin of the Cartesian coordinates, and $\mathcal{I}_2$ represents the moment of inertia of the second link with respect to the joint of the two links. The quantity r_{2c} is the position of the center of mass of the second link measured from the joint of the two links. For uniform links, the infinitesimal masses dm_1 and dm_2 may be expressed by $dm_1 = \rho_1 dr_1$ and $dm_2 = \rho_2 dr_2$, with constants ρ_1 and ρ_2 representing the densities per unit length for first and second links, respectively. Quantities $\mathcal{I}_1, \mathcal{I}_2,$

and r_{2c} in the case of a slender rod can be easily shown to be

$$I_1 = \frac{1}{3}\ell_1^2 m_1, \quad I_2 = \frac{1}{3}\ell_2^2 m_2, \quad r_{2c} = \frac{1}{2}\ell_2.$$

The potential energy from the gravitational force is equivalent to the work done by the gravitational force when the subjected system moves from the current position back to the original position. Let the initial position of the two-link system be straight and along the y axis. The potential energy for the two-link system becomes

$$
\begin{aligned}
\mathcal{V} &= -\int_0^{\ell_1} g r_1 \cos\theta_1 \, dm_1 - \int_0^{\ell_2} g[\ell_1 \cos\theta_1 + r_2 \cos(\theta_1 + \theta_2)] \, dm_2 \\
&= -\int_0^{\ell_1} g r_1 \cos q_1 \, dm_1 - \int_0^{\ell_2} g(\ell_1 \cos q_1 + r_2 \cos q_2) \, dm_2 \\
&= -m_1 g r_{1c} \cos q_1 - m_2 g(\ell_1 \cos q_1 + r_{2c} \cos q_2),
\end{aligned}
\tag{3.83}
$$

where r_{1c} is the position of the center of mass for the first link from the origin of the Cartesian coordinates, i.e.,

$$r_{1c} = \frac{1}{m_1} \int_0^{\ell_1} r_1 \, dm_1,$$

which is similar to r_{2c} given earlier. For a uniform link, $r_{1c} = \frac{1}{2}\ell_1$.

With the Lagrangian function,

$$
\begin{aligned}
\mathcal{L} &= \mathcal{T} - \mathcal{V} \\
&= \frac{1}{2}\left[I_1 \dot{q}_1^2 + m_2 \ell_1^2 \dot{q}_1^2 + I_2 \dot{q}_2^2 + 2 m_2 \ell_1 r_{2c} \dot{q}_1 \dot{q}_2 \cos(q_2 - q_1) \right] \\
&\quad + (m_1 g r_{1c} + m_2 g \ell_1) \cos q_1 + m_2 g r_{2c} \cos q_2,
\end{aligned}
\tag{3.84}
$$

the equation of motion [see Eq. (3.74)] for q_1 is derived from

$$\frac{d}{dt}\left(\frac{\partial \mathcal{L}}{\partial \dot{q}_1}\right) - \frac{\partial \mathcal{L}}{\partial q_1} = 0. \tag{3.85}$$

Performing the differentiation in Eq. (3.85) yields

$$
\begin{aligned}
\left(I_1 + m_2 \ell_1^2\right) \ddot{q}_1 + m_2 \ell_1 r_{2c} \ddot{q}_2 \cos(q_2 - q_1) & \\
- m_2 \ell_1 r_{2c} \dot{q}_2 (\dot{q}_2 - \dot{q}_1) \sin(q_2 - q_1) + (m_1 g r_{1c} + m_2 g \ell_1) \sin q_1 &= 0.
\end{aligned}
\tag{3.86}
$$

Similarly, the equation of motion for q_2 is

$$\frac{d}{dt}\left(\frac{\partial \mathcal{L}}{\partial \dot{q}_2}\right) - \frac{\partial \mathcal{L}}{\partial q_2} = 0, \tag{3.87}$$

which yields

$$
\begin{aligned}
I_2 \ddot{q}_2 + m_2 \ell_1 r_{2c} \ddot{q}_1 \cos(q_2 - q_1) - m_2 \ell_1 r_{2c} \dot{q}_1 (\dot{q}_2 - \dot{q}_1) \sin(q_2 - q_1) & \\
+ m_2 g r_{2c} \sin q_2 &= 0.
\end{aligned}
\tag{3.88}
$$

Equations (3.86) and (3.88) constitute the equations of motion for the two-link system shown in Fig. (3.5).

3.6.1 Rayleigh's Dissipation Function

Assume that part of the virtual work shown in Eq. (3.66) is due to the damping force f_d that can be derived from Rayleigh's dissipation energy $\mathcal{R}$ as

$$f_d = -\frac{\partial \mathcal{R}}{\partial \dot{r}}. \tag{3.89}$$

It should be noted that, in general, this formulation is valid when $\dot{r}$ is the vector of relative velocity between two points (see Example 3.1). The virtual work done by the force f_d becomes

$$f_d^T \delta r = -\left\{\frac{\partial \mathcal{R}}{\partial \dot{r}}\right\}^T \left\{\frac{\partial r}{\partial q}\delta q\right\} = -\left\{\left(\frac{\partial \mathcal{R}}{\partial \dot{r}}\right)^T \frac{\partial \dot{r}}{\partial \dot{q}}\right\}\delta q$$

$$= -\left\{\frac{\partial \mathcal{R}}{\partial \dot{q}}\right\}^T \delta q = -Q_d^T \delta q = -\sum_{i=1}^k Q_{d_i}\delta q_i. \tag{3.90}$$

The equality $(\partial r/\partial q) = (\partial \dot{r}/\partial \dot{q})$ has been used to arrive at the second equality of Eq. (3.90), and the generalized damping force Q_d is defined as

$$Q_d = \frac{\partial \mathcal{R}}{\partial \dot{q}}, \tag{3.91}$$

where its elements $Q_{d_1}, Q_{d_2}, \ldots, Q_{d_n}$ are computed by

$$Q_{d_i} = \frac{\partial \mathcal{R}}{\partial \dot{q}_i}, \quad i = 1, 2, \ldots, n. \tag{3.92}$$

Because $\dot{r}$ is given by

$$\dot{r} = \frac{\partial r}{\partial q}\dot{q} + \frac{\partial r}{\partial t}, \tag{3.93}$$

taking the partial derivative of this equation with respect to $\dot{q}$ yields the following equality:

$$\frac{\partial r}{\partial q} = \frac{\partial \dot{r}}{\partial \dot{q}}. \tag{3.94}$$

The generalized force vector Q_{nc} shown in Eq. (3.67) can now be written as

$$Q_{\mathrm{nc}} = \left[\frac{\partial r}{\partial q}\right]^T f_{\mathrm{nc}} = \left[\frac{\partial r}{\partial q}\right]^T (f_{\mathrm{nd}} + f_d) = Q_{\mathrm{nd}} - Q_d,$$

or

$$Q_{\mathrm{nc}} = Q_{\mathrm{nd}} - \frac{\partial \mathcal{R}}{\partial \dot{q}}, \tag{3.95}$$

where f_{nd} and Q_{nd} represent the nonconservative physical force vector and the nonconservative generalized force vector, respectively, excluding the damping force vector derived from Rayleigh's dissipation function.

Substituting Eq. (3.95) into Eq. (3.73) yields

$$\frac{d}{dt}\left(\frac{\partial \mathcal{L}}{\partial \dot{q}}\right) - \frac{\partial \mathcal{L}}{\partial q} + \frac{\partial \mathcal{R}}{\partial \dot{q}} = Q_{nd}, \tag{3.96}$$

or, in terms of its components,

$$\frac{d}{dt}\left(\frac{\partial \mathcal{L}}{\partial \dot{q}_i}\right) - \frac{\partial \mathcal{L}}{\partial q_i} + \frac{\partial \mathcal{R}}{\partial \dot{q}_i} = Q_{nd_i}, \quad i = 1, 2, \ldots, n. \tag{3.97}$$

This is an alternative form of Lagrange's equation in terms of the Lagrangian and Rayleigh's dissipation function.

3.6.2 Constraint Equations

Lagrange's equations shown earlier are derived with the assumption that all coordinates are independent. There are many cases in which some coordinates are constrained by explicit algebraic or differential equations. In this section, Lagrange's multipliers are introduced to cope with the constraint equations.

Let the motion of a system be expressed by a set of n coordinates $q_1, q_2, \ldots, q_n$. Assume that the coordinates are not all independent but subject to m constraints

$$g_j(q_1, q_2, \ldots, q_n, t) = 0, \quad j = 1, 2, \ldots, m. \tag{3.98}$$

The variation in the function g_j corresponding to the virtual quantities δq_i ($i = 1, 2, \ldots, n$) is

$$\delta g_j = \frac{\partial g_j}{\partial q_1}\delta q_1 + \frac{\partial g_j}{\partial q_2}\delta q_2 + \cdots + \frac{\partial g_j}{\partial q_n}\delta q_n = 0, \quad j = 1, 2, \ldots, m, \tag{3.99}$$

or, in matrix form,

$$\delta g = \frac{\partial g}{\partial q}\delta q = 0, \tag{3.100}$$

where

$$\delta g = \left\{\begin{array}{c} \delta g_1 \\ \delta g_2 \\ \vdots \\ \delta g_m \end{array}\right\}, \qquad \delta q = \left\{\begin{array}{c} \delta q_1 \\ \delta q_2 \\ \vdots \\ \delta q_n \end{array}\right\},$$

$$\frac{\partial g}{\partial q} = \begin{bmatrix} \dfrac{\partial g_1}{\partial q_1} & \dfrac{\partial g_1}{\partial q_2} & \cdots & \dfrac{\partial g_1}{\partial q_n} \\[2ex] \dfrac{\partial g_2}{\partial q_1} & \dfrac{\partial g_2}{\partial q_2} & \cdots & \dfrac{\partial g_2}{\partial q_n} \\[2ex] \vdots & \vdots & \ddots & \vdots \\[2ex] \dfrac{\partial g_m}{\partial q_1} & \dfrac{\partial g_m}{\partial q_2} & \cdots & \dfrac{\partial g_m}{\partial q_n} \end{bmatrix}.$$

The operator δ represents the virtual character of the instantaneous variations at any time t, as opposed to the differential operator d that designates actual differentials taking place in the time interval dt during which the forces and constraints may change. Because the coordinates q_i are not independent, the variations δq_i are not arbitrary but must obey the relations shown in Eq. (3.99). We may solve Eq. (3.99) for m of δq_i ($i = n - m + 1, n - m + 2, \ldots, n$) and treat the remaining δq_i ($i = 1, 2, \ldots, n - m$) as independent quantities. With substitution of δq_i ($i = n - m + 1, n - m + 2, \ldots, n$) into Eq. (3.71) in terms of δq_i ($i = 1, 2, \ldots, n - m$), the number of Lagrange's equations, Eq. (3.97), reduces from n to $n - m$ for q_i ($i = 1, 2, \ldots, n - m$). The $n - m$ Lagrange's equations coupled with the m constraint equations represent a set of n equations to describe the motion of the system.

The method of Lagrange's multipliers provides an alternative approach. Multiplying Eq. (3.99) by an undermined multiplier λ_j and adding the resulting expressions to Eq. (3.71) yields

$$\int_{t_o}^{t_f} \left(\left\{ \left[-\frac{d}{dt}\left(\frac{\partial T}{\partial \dot{q}}\right) + \frac{\partial T}{\partial q} - \frac{\partial V}{\partial q} + Q_{nc} \right]^T + \lambda^T \frac{\partial g}{\partial q} \right\} \delta q \right) dt = 0, \qquad (3.101)$$

where λ is a column vector with components $\lambda_1, \lambda_2, \ldots, \lambda_m$. There are $n - m$ independent quantities δq_i ($i = 1, 2, \ldots, n - m$). The coefficients corresponding to these independent quantities must be zero in order to satisfy Eq. (3.101), i.e.,

$$\frac{d}{dt}\left(\frac{\partial T}{\partial \dot{q}_i}\right) - \frac{\partial T}{\partial q_i} + \frac{\partial V}{\partial q_i} = Q_{nc_i} + \lambda^T \frac{\partial g}{\partial q_i}, \qquad i = 1, 2, \ldots, n - m. \qquad (3.102)$$

The remaining m quantities δq_i ($i = n - m + 1, n - m + 2, \ldots, n$) are not arbitrary. Let us choose $\lambda_1, \lambda_2, \ldots, \lambda_m$ such that

$$\frac{d}{dt}\left(\frac{\partial T}{\partial \dot{q}_i}\right) - \frac{\partial T}{\partial q_i} + \frac{\partial V}{\partial q_i} = Q_{nc_i} + \lambda^T \frac{\partial g}{\partial q_i},$$

$$i = n - m + 1, n - m + 2, \ldots, n. \qquad (3.103)$$

Equation (3.101) is clearly satisfied by the insertion of Eqs. (3.102) and (3.103). Equations (3.102) and (3.103) are identical in form with different subscripts. As a result, Lagrange's equations with its multipliers and constraint equations are

$$g_j(q_1, q_2, \ldots, q_n, t) = 0, \qquad j = 1, 2, \ldots, m,$$

$$\frac{d}{dt}\left(\frac{\partial T}{\partial \dot{q}_i}\right) - \frac{\partial T}{\partial q_i} + \frac{\partial V}{\partial q_i} = Q_{nc_i} + \sum_{j=1}^{m} \lambda_j \frac{\partial g_j}{\partial q_i}, \qquad i = 1, 2, \ldots, n. \qquad (3.104)$$

There are a total of $n + m$ equations for $n + m$ unknowns to be determined, including q_i ($i = 1, 2, \ldots, n$) and λ_j ($j = 1, 2, \ldots, m$).

EXAMPLE 3.5

A two-dimensional weightless and rigid pendulum link is pinned frictionlessly at the origin, as shown in Fig. 3.6. On the link there is an electronic mouse with mass m. The gravitational force is acting in the positive x direction with gravitational acceleration constant g. The mouse is programmed to climb up and down the rigid link in such a

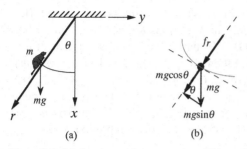

Figure 3.6. (a) An electronic mouse on a rigid pendulum link, (b) free-body diagram.

way that the following nonholonomic constraint is satisfied exactly:

$$\Psi(\dot{x}, \dot{y}, t) = \frac{dx}{dt} - \eta(x, y, t) = 0,$$

where (x, y) is the Cartesian coordinate of the mouse and $(\dot{x}, \dot{y})$ is its Cartesian velocity. The $\eta(x, y, t)$ function is given and is differentiable with respect to all its arguments.

Derive two equations of motion for the two configuration variables r and θ for the case $\eta(x, y, t) = 0$.

(1) Do this problem by using the Lagrangian with $q_1 = r$ and $q_2 = \theta$, where r is the length along the link from the pinned point and θ is measured clockwise from the downward vertical.
(2) Redo the problem by using Lagrangian and its multipliers associated with the constraint equation.
(3) Do this problem by drawing a free-body diagram and doing it the simple way in a coordinate system chosen as desired. Obviously, the answers should agree with that derived with the Lagrangian.

The solutions for the above three items are shown as follows.

(1) The kinematic energy T is

$$T = \frac{1}{2}(mr^2\dot{\theta}^2 + m\dot{r}^2),$$

and the potential energy from the gravitational force V is

$$V = -mgr \cos\theta.$$

The Lagrangian $\mathcal{L}$ can then be written as

$$\mathcal{L} = T - V = \frac{1}{2}(mr^2\dot{\theta}^2 + m\dot{r}^2) + mgr \cos\theta, \tag{3.105}$$

which gives

$$\frac{d}{dt}\left(\frac{\partial\mathcal{L}}{\partial\dot{\theta}}\right) - \frac{\partial\mathcal{L}}{\partial\theta} = mr^2\ddot{\theta} + 2mr\dot{\theta}\dot{r} + mgr \sin\theta, \tag{3.106}$$

$$\frac{d}{dt}\left(\frac{\partial\mathcal{L}}{\partial\dot{r}}\right) - \frac{\partial\mathcal{L}}{\partial r} = m\ddot{r} - mr\dot{\theta}^2 - mg \cos\theta. \tag{3.107}$$

From the constrained equation,

$$\frac{dx}{dt} = \frac{d}{dt} r \cos\theta = 0,$$

we obtain the following two equations:

$$\dot{r}\cos\theta - r\dot{\theta}\sin\theta = 0, \qquad (3.108)$$

$$x = r\cos\theta = r_0\cos\theta_0, \qquad (3.109)$$

where r_0 and θ_0 is an initial position of the mouse. Solving for $\dot{r}$ from Eq. (3.108) and r from Eq. (3.109) yields

$$\frac{\dot{r}}{r} = \frac{\dot{\theta}\sin\theta}{\cos\theta}, \qquad (3.110)$$

$$r = \frac{r_0\cos\theta_0}{\cos\theta}. \qquad (3.111)$$

With the help of Eq. (3.110), Eq. (3.106) becomes

$$\frac{d}{dt}\left(\frac{\partial\mathcal{L}}{\partial\dot{\theta}}\right) - \frac{\partial\mathcal{L}}{\partial\theta} = r^2\left(m\ddot{\theta} + 2m\frac{\sin\theta}{\cos\theta}\dot{\theta}^2 + \frac{mg}{r_0\cos\theta_0}\sin\theta\cos\theta\right). \quad (3.112)$$

Because there is no other external force applied along the direction of θ, i.e., perpendicular to the rigid link, application of Lagrange's equation of motion for θ yields

$$\frac{d}{d}\left(\frac{\partial\mathcal{L}}{\partial\dot{\theta}}\right) - \frac{\partial\mathcal{L}}{\partial\theta} = 0,$$

or

$$m\ddot{\theta} + 2m\frac{\sin\theta}{\cos\theta}\dot{\theta}^2 + \frac{mg}{r_0\cos\theta_0}\sin\theta\cos\theta = 0. \qquad (3.113)$$

Equation (3.113) provides the time history of θ for any given initial conditions r_0 and θ_0. As soon as the time history of θ is known, the time history of r can be solved from Eq. (3.111).

At this moment, we may determine that the problem is solved. Where does Lagrange's equation of motion for r go? How can the mouse move up and down along the link? Obviously, there must exist an external force along the rigid link for the mouse to move along the link. Let the external force along the link be denoted by f_r. Lagrange's equation of motion for r is

$$\frac{d}{dt}\left(\frac{\partial\mathcal{L}}{\partial\dot{r}}\right) - \frac{\partial\mathcal{L}}{\partial r} = f_r,$$

from which we find

$$m\ddot{r} - mr\dot{\theta}^2 - mg\cos\theta = f_r. \qquad (3.114)$$

The force f_r is computed from the time histories of r and θ that are solved from Eqs. (3.111) and (3.113). Lagrange's equation of motion for r is used to determine the external force f_r required for the mouse to climb along the link. If the link is assumed to be frictionless, then there is no friction force for the mouse to climb up and down. That means that Eq. (3.114) would not be satisfied. In other words, the problem has no physical solution, even though Eqs. (3.111) and (3.113) are mathematically correct.

(2) Now, taking the variation of Eq. (3.109) yields

$$\delta x = \cos\theta \, \delta r - r \sin\theta \, \delta\theta = 0.$$

The variation is valid when the mouse is allowed to move freely in the coordinate (r, θ). Because the mouse is allowed to climb only up and down the rigid link, the variation $\delta\theta$ must vanish at any instant time t, i.e., $\delta\theta = 0$. Therefore the variation of the constrained equation becomes

$$\delta x = \cos\theta \, \delta r = 0.$$

Application of Lagrange's equation of motion with Lagrangian multipliers thus yields

$$\frac{d}{dt}\left(\frac{\partial \mathcal{L}}{\partial \dot\theta}\right) - \frac{\partial \mathcal{L}}{\partial \theta} = 0,$$

or

$$mr^2\ddot\theta + 2mr\dot\theta\dot r + mgr\sin\theta = 0, \tag{3.115}$$

and

$$\frac{d}{dt}\left(\frac{\partial \mathcal{L}}{\partial \dot r}\right) - \frac{\partial \mathcal{L}}{\partial r} = \lambda\cos\theta,$$

or

$$m\ddot r - mr\dot\theta^2 - mg\cos\theta = \lambda\cos\theta, \tag{3.116}$$

where λ is the Lagrangian multiplier.

Equations (3.108) and (3.115) are used to solve for the motion of the mouse in the coordinate (r, θ). Equation (3.116) can then be used to calculate the force that is required for moving the mouse up and down the rigid link to satisfy the constrained equation $(dx/dt) = 0$.

Combination of Eqs. (3.108) and (3.115) yields

$$\ddot\theta + 2\frac{\sin\theta}{\cos\theta}\dot\theta^2 + \frac{g}{r_0\cos\theta_0}\sin\theta\cos\theta = 0, \tag{3.117}$$

which is identical to Eq. (3.113). This equation does not have r involved except in the initial position r_0. As long as the initial conditions θ_0, $\dot\theta_0$, and r_0 are given, Eq. (3.117) can be solved alone to obtain the time history of θ. However, there is a singularity at the position where $r_0 \neq 0$ and $\theta_0 = 90°$. The mouse must have an infinite speed along the rigid link to keep the position of $x = r_0\cos\theta_0 = 0$. For the case in which the initial conditions $\theta_0 = 0°$ and $r_0 \neq 0$, the trivial solution for θ is $\theta = \dot\theta = 0$.

(3) From the free-body diagram of Fig. 3.6, only the gravitational force has the component in the direction of θ. Using the angular momentum equation of

motion in the direction of θ thus yields

$$\frac{d}{dt}mr^2\dot\theta = -mgr\sin\theta \implies mr^2\ddot\theta + 2mr\dot r\dot\theta + mgr\sin\theta = 0,$$

or, with the substitution of r from Eq. (3.111),

$$\ddot\theta + 2\frac{\sin\theta}{\cos\theta}\dot\theta^2 + \frac{g}{r_0\cos\theta_0}\sin\theta\cos\theta = 0 \tag{3.118}$$

Equations (3.113), (3.117), and (3.118) are identical. On the other hand, application of the linear momentum equation in the direction of r gives

$$\frac{d}{dt}m\dot r = mr\dot\theta^2 + mg\cos\theta + f_r,$$

or

$$m\ddot r - mr\dot\theta^2 - mg\cos\theta = f_r, \tag{3.119}$$

where $mr\dot\theta^2$ is the centrifugal force generated by the rotational motion θ because of the gravitational force. Equation (3.119) is identical to Eq. (3.116) if λ is chosen such that $f_r = \lambda\cos\theta$.

3.7 Gibbs–Appell Equations

Gibbs–Appell equations were first discovered by Gibbs in 1879 and independently discovered and studied in detail by Appell 20 years later. There are many ways of deriving these equations. It is our intention in this section to introduce the Gibbs–Appell equations and later correlate them with the other equations of motion already introduced in this chapter. It seems to be a common practice to derive the Gibbs–Appell equations by using the principle of virtual work in conjunction with the generalized coordinates.

Consider a system consisting a number of particles with mass dm. Let $q = (q_1, q_2, \ldots, q_n)^T$ be the generalized coordinate vector of dimension n that may represent physical coordinates. The displacement vector $r = (r_1, r_2, r_3)^T$ of a particle in the system can then be expressed as

$$r = r(q, t). \tag{3.120}$$

The maximal dimension of the displacement vector r is only 3 for the three directions in translational motion. Nevertheless, for a point mass, the number of generalized coordinates does not exceed three, i.e., $n \le 3$. For a rigid body, n may be as large as 6, composed of three translational and three rotational motions. For a flexible body, n may be unlimited in theory.

The virtual displacement δr becomes

$$\delta r = \frac{\partial r}{\partial q}\delta q = \begin{bmatrix} \dfrac{\partial r_1}{\partial q_1} & \dfrac{\partial r_1}{\partial q_2} & \cdots & \dfrac{\partial r_1}{\partial q_n} \\[2mm] \dfrac{\partial r_2}{\partial q_1} & \dfrac{\partial r_2}{\partial q_2} & \cdots & \dfrac{\partial r_2}{\partial q_n} \\[2mm] \dfrac{\partial r_3}{\partial q_1} & \dfrac{\partial r_3}{\partial q_2} & \cdots & \dfrac{\partial r_3}{\partial q_n} \end{bmatrix} \begin{Bmatrix} \delta q_1 \\ \delta q_2 \\ \vdots \\ \delta q_n \end{Bmatrix}, \tag{3.121}$$

where $\partial r / \partial q$ is a $3 \times n$ matrix and δq is an $n \times 1$ vector. As discussed earlier, the time t is assumed fixed in performing the variation of a quantity. Therefore, δt does not exist.

Application of the principle of virtual work leads to [see Eq. (3.10)]

$$\int_{\Omega} \delta r^T (df - \ddot{r} dm) = \{\delta q\}^T \int_{\Omega} \left[\frac{\partial r}{\partial q} \right]^T \{df - \ddot{r} dm\} = 0, \tag{3.122}$$

where Ω means the domain of the whole body. Assume that coordinates $q_1, q_2, \ldots, q_n$ are linearly independent. Because the components $\delta q_1, \delta q_2, \ldots, \delta q_n$ of the variation vector δq are linearly independent and arbitrary, the following equation must be satisfied:

$$\int_{\Omega} \left[\frac{\partial r}{\partial q} \right]^T \{df - \ddot{r} dm\} = 0, \tag{3.123}$$

which is the principle of virtual work in terms of generalized coordinates.

From Eq. (3.120), we can write

$$dr = \frac{\partial r}{\partial q} dq + \frac{\partial r}{\partial t} dt,$$

which gives the velocity of each component,

$$\dot{r}_i = \left\{ \frac{\partial r_i}{\partial q} \right\}^T \dot{q} + \frac{\partial r_i}{\partial t}, \quad i = 1, 2, 3, \tag{3.124}$$

and the acceleration of each component

$$\ddot{r}_i = \dot{q}^T \frac{\partial^2 r_i}{\partial q \partial q^T} \dot{q} + \left\{ \frac{\partial r_i}{\partial q} \right\}^T \ddot{q} + \left\{ \frac{\partial^2 r_i}{\partial t \partial q} \right\}^T \dot{q} + \frac{\partial^2 r_i}{\partial t^2}, \quad i = 1, 2, 3, \tag{3.125}$$

where

$$\frac{\partial^2 r_i}{\partial q \partial q^T} = \begin{bmatrix} \dfrac{\partial^2 r_i}{\partial q_1 \partial q_1} & \dfrac{\partial r_i}{\partial q_1 \partial q_2} & \cdots & \dfrac{\partial^2 r_i}{\partial q_1 \partial q_n} \\[2mm] \dfrac{\partial^2 r_i}{\partial q_2 \partial q_1} & \dfrac{\partial r_i}{\partial q_2 \partial q_2} & \cdots & \dfrac{\partial^2 r_i}{\partial q_2 \partial q_n} \\[2mm] \vdots & \vdots & \ddots & \vdots \\[2mm] \dfrac{\partial^2 r_i}{\partial q_n \partial q_1} & \dfrac{\partial r_i}{\partial q_n \partial q_2} & \cdots & \dfrac{\partial^2 r_i}{\partial q_n \partial q_n} \end{bmatrix},$$

$$\frac{\partial r_i}{\partial q} = \left\{ \begin{array}{c} \dfrac{\partial r_i}{\partial q_1} \\[2mm] \dfrac{\partial r_i}{\partial q_2} \\[1mm] \vdots \\[1mm] \dfrac{\partial r_i}{\partial q_n} \end{array} \right\}, \quad \frac{\partial^2 r_i}{\partial t \partial q} = \left\{ \begin{array}{c} \dfrac{\partial^2 r_i}{\partial t \partial q_1} \\[2mm] \dfrac{\partial^2 r_i}{\partial t \partial q_2} \\[1mm] \vdots \\[1mm] \dfrac{\partial^2 r_i}{\partial t \partial q_n} \end{array} \right\}.$$

Quantities $\partial r_i / \partial q$, $\partial r_i / \partial t$, $\partial^2 r_i / \partial q^T \partial q$, and $\partial^2 r_i / \partial t \partial q$ are functions of q and t only.

Therefore, taking the partial derivative of Eq. (3.124) with respect to $\dot{q}$ and Eq. (3.125) with respect to $\ddot{q}$ yields

$$\frac{\partial r}{\partial q} = \frac{\partial \dot{r}}{\partial \dot{q}} = \frac{\partial \ddot{r}}{\partial \ddot{q}}, \tag{3.126}$$

where $\partial \dot{r}/\partial \dot{q}$ and $\partial \ddot{r}/\partial \ddot{q}$ are $3 \times n$ matrices similar to $\partial r/\partial q$ defined in Eq. (3.121). From Eq. (3.126), Eq. (3.123) gives

$$\int_\Omega \left[\frac{\partial r}{\partial q} \right]^T df = \int_\Omega \left[\frac{\partial \ddot{r}}{\partial \ddot{q}} \right]^T \ddot{r}\, dm$$

$$= \int_\Omega \left[\frac{\partial \dot{r}}{\partial \dot{q}} \right]^T \ddot{r}\, dm. \tag{3.127}$$

The first equality provides the basis for deriving the Gibbs–Appell equations, whereas the second equality is used for the development of Kane's equations, which will be introduced in the following section.

Now, note the following equality:

$$\left[\frac{\partial \ddot{r}}{\partial \ddot{q}} \right]^T \ddot{r} = \frac{1}{2} \frac{\partial}{\partial \ddot{q}} \left(\ddot{r}^T \ddot{r} \right).$$

Next, define the generalized force f_q as

$$f_q = \int_\Omega \left[\frac{\partial r}{\partial q} \right]^T df, \tag{3.128}$$

and the Gibbs–Appell function G as

$$G = \frac{1}{2} \int_\Omega \left(\ddot{r}^T \ddot{r} \right) dm. \tag{3.129}$$

Substituting Eqs. (3.128) and (3.129) into Eq. (3.127) produces

$$f_q = \frac{\partial G}{\partial \ddot{q}}. \tag{3.130}$$

Equation (3.130) is the Gibbs–Appell equation of motion for a system consisting of a number of particles with mass dm.

Similar to Lagrange's equations, Gibbs–Appell equations also use a scalar function G defined in Eq. (3.129) to derive equations of motion. The Lagrangian function is the difference between kinetic and potential energies whereas the Gibbs–Appell function is one half of the sum of the squared amplitude of the acceleration multiplied by its corresponding mass.

EXAMPLE 3.6

Consider the simple pendulum system shown in Fig. 3.7 with the position vector r expressed by

$$r = \begin{Bmatrix} x \\ y \end{Bmatrix} = \begin{Bmatrix} r \cos \theta \\ r \sin \theta \end{Bmatrix}, \tag{3.131}$$

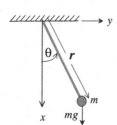

Figure 3.7. A simple pendulum system.

where θ is the angle measured from the x axis. Differentiating Eq. (3.131) with respect to time t yields

$$\dot{r} = \frac{dr}{dt} = \begin{Bmatrix} -r\sin\theta \\ r\cos\theta \end{Bmatrix} \dot{\theta}, \tag{3.132}$$

$$\ddot{r} = \frac{d}{dt}\left(\frac{dr}{dt}\right) = \begin{Bmatrix} -r\sin\theta \\ r\cos\theta \end{Bmatrix} \ddot{\theta} + \begin{Bmatrix} -r\cos\theta \\ -r\sin\theta \end{Bmatrix} \dot{\theta}^2. \tag{3.133}$$

Therefore the Gibbs–Appell function defined in Eq. (3.129) becomes

$$
\begin{aligned}
G &= \frac{1}{2}m(\ddot{r}^T \ddot{r}) \\
&= \frac{1}{2}mr^2 \{-\ddot{\theta}\sin\theta - \dot{\theta}^2\cos\theta \quad \ddot{\theta}\cos\theta - \dot{\theta}^2\sin\theta\} \begin{Bmatrix} -\ddot{\theta}\sin\theta - \dot{\theta}^2\cos\theta \\ \ddot{\theta}\cos\theta - \dot{\theta}^2\sin\theta \end{Bmatrix} \\
&= \frac{1}{2}mr^2(\dot{\theta}^4 + \ddot{\theta}^2).
\end{aligned}
\tag{3.134}$$

The position of the mass m described by the vector r has two components in the Cartesian coordinates x and y, which can be written in vector form.

The gravitational force mg is always in the direction of x, as shown in Fig. 3.7, and thus the external force f is expressed as

$$f = \begin{Bmatrix} mg \\ 0 \end{Bmatrix}. \tag{3.135}$$

It is quite clear from Eqs. (3.134) (Gibbs–Appell function) and (3.135) (gravitational force) that there is only one independent variable, i.e., the angle θ. Consider the angle θ as the generalized coordinate, i.e., $q_1 = \theta$. The generalized force f_q defined in Eq. (3.128) is computed by

$$f_q = \left\{\frac{\partial r}{\partial \theta}\right\}^T f = [-r\sin\theta \quad r\cos\theta]\begin{bmatrix} mg \\ 0 \end{bmatrix}$$

$$= -mgr\sin\theta. \tag{3.136}$$

Application of Gibbs–Appell equation (3.130) produces

$$f_q = -mgr\sin\theta = \frac{\partial G}{\partial \ddot{\theta}} = mr^2\ddot{\theta},$$

or

$$\ddot{\theta} = -\frac{g}{r}\sin\theta, \tag{3.137}$$

which is the familiar equation of motion for the pendulum shown in Fig. 3.7.

3.8 Kane's Equations

Deriving Kane's equations requires introducing a generalized vector p of dimension n that is a function of q and $\dot{q}$ such that

$$p = A(q,t)\dot{q} + b(q,t), \tag{3.138}$$

where A is an $n \times n$ square matrix and b is an $n \times 1$ vector. The generalized vector p has a linear relationship with $\dot{q}$ but may have a nonlinear relationship with q. Taking a partial derivative of Eq. (3.138) with respect to $\dot{q}$ gives

$$\frac{\partial p}{\partial \dot{q}} = A(q,t). \tag{3.139}$$

Because the velocity vector $\dot{r}$ is a function of q and $\dot{q}$, it must also function of q and p. Using the chain rule of differentiation thus gives

$$\frac{\partial \dot{r}}{\partial \dot{q}} = \frac{\partial \dot{r}}{\partial p}\frac{\partial p}{\partial \dot{q}} = \frac{\partial \dot{r}}{\partial p}A(q,t). \tag{3.140}$$

From Eqs. (3.126) and (3.140), Eq. (3.127) can be rewritten as

$$\int_{\Omega}\left[\frac{\partial \dot{r}}{\partial \dot{q}}\right]^{T} df = \int_{\Omega}\left[\frac{\partial \dot{r}}{\partial \dot{q}}\right]^{T} \ddot{r}\, dm,$$

and thus

$$A^{T}(q,t)\left(\int_{\Omega}\left[\frac{\partial \dot{r}}{\partial p}\right]^{T} df - \int_{\Omega}\left[\frac{\partial \dot{r}}{\partial p}\right]^{T} \ddot{r}\, dm\right) = 0 \tag{3.141}$$

For a nonsingular matrix $A(q,t)$, Eq. (3.141) is equivalent to

$$\int_{\Omega}\left[\frac{\partial \dot{r}}{\partial p}\right]^{T} df = \int_{\Omega}\left[\frac{\partial \dot{r}}{\partial p}\right]^{T} \ddot{r}\, dm. \tag{3.142}$$

Equation (3.142) is known as Kane's equation of motion for dynamic systems.

EXAMPLE 3.7

Consider the same simple pendulum system as that described in Example 3.6. Let the generalized speed p be defined as

$$p = \dot{\theta}. \tag{3.143}$$

The velocity vector $\dot{r}$ shown in Eq. (3.132) can then be written as

$$\dot{r} = \begin{bmatrix} -r\sin\theta \\ r\cos\theta \end{bmatrix} p.$$
(3.144)

Similarly, from Eq. (3.133), we obtain

$$\ddot{r} = \begin{bmatrix} -r\sin\theta \\ r\cos\theta \end{bmatrix} \dot{p} + \begin{bmatrix} -r\cos\theta \\ -r\sin\theta \end{bmatrix} p^2.$$
(3.145)

The gravitational force mg is always in the direction of x and thus the external force f can be written as shown in Eq. (3.135). Taking the partial derivative of Eq. (3.144) with respect to p and substituting the result in combination with Eqs. (3.135) and (3.145) into Eq. (3.142) produces

$$\begin{bmatrix} -r\sin\theta & r\cos\theta \end{bmatrix} \left(\begin{bmatrix} mg \\ 0 \end{bmatrix} - m \begin{bmatrix} -r\sin\theta \\ r\cos\theta \end{bmatrix} \dot{p} - m \begin{bmatrix} -r\cos\theta \\ -r\sin\theta \end{bmatrix} p^2 \right) = 0,$$
(3.146)

which yields

$$\dot{p} = -\frac{g}{r}\sin\theta.$$
(3.147)

Equations (3.143) and (3.147) together constitute the equations of motion for the pendulum shown in Fig. 3.7. Indeed, Eqs. (3.143) and (3.147) may be combined to become the second-order equation of motion

$$\ddot{\theta} = -\frac{g}{r}\sin\theta,$$
(3.148)

which is the familiar pendulum equation of motion.

The way of deriving equations of motion for the simple pendulum problem shown in Example 3.7 seems tedious and unnecessary, because it may be easier to obtain simply by use of the second-order equation of motion. Nevertheless, Kane's equation has been used with good success in deriving equations of motion for multibody dynamic systems.

3.9 Concluding Remarks

Several methods for deriving dynamic equations of motion are presented. Each method has its own advantages and disadvantages. For example, the Newtonian method involves physical vector quantities such as forces and accelerations to formulate equations of motion and thus provides physical insights to characterize the system motion easily. For a system composed of multiple bodies, other methods such as Hamilton's principle and Lagrange's equation are used instead that involve scalar quantities such as kinetic energy and potential energy. The use of scalar quantities avoids the possibility of making directional errors associated with vector quantities. However, considerable care must be taken for dependent coordinates associated with constraint equations. Gibbs–Appell equations use, instead of kinetic energy and potential energy, the scalar formed by the squared amplitude of acceleration. Gibbs–Appell approach seems very simple in concept, but the explicit expression for acceleration is generally very complex in form for multibody dynamic systems. On the other hand, Kane's equation uses a generalized speed vector that is an arbitrary function of generalized displacement coordinates but linearly related to generalized velocity coordinates. The generalized speed vector may be defined so as to reduce considerably the complexity of the explicit expression for acceleration. Some dynamicists have successfully applied Kane's equation to derive formulations for complex multibody systems for numerical simulations with speed and accuracy.

3.10 Problems

3.1 Two masses of magnitude m_1 and m_2, as shown in Fig. 3.8, are attached to springs of stiffness k_1 and k_2. Let the horizontal displacements of the masses, measured from the equilibrium position, be w_1 and w_2. By using Hamilton's principle, derive the equations of motion for the system subject to an external force u.

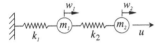

Figure 3.8. Two-mass–spring system.

3.2 Consider the motion of a rigid body that is supported on two springs, as shown in Fig. 3.9. Only vertical motion is involved. The body of mass M has moment of inertia $3Ma^2/4$ about its center of gravity G, where a is the distance of G from the centerline of each spring. The stiffnesses of the springs are k_1 and k_2. The vertical displacements w_1 and w_2 of the top of the springs are measured from its equilibrium position. Derive the equations of motion for the system.

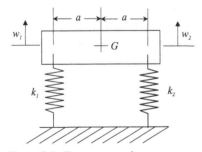

Figure 3.9. Two-mass–spring system.

3.3 Figure 3.10 shows an automobile simulated by a simplified two-degree-of-freedom system including a rigid bar supported by two springs with spring constants k_1 and k_2 and two dashpots with damping coefficients ζ_1 and ζ_2. The displacement $w_1(t)$ and the rotation angle $\theta(t)$ define the motions of the two-degree-of-freedom system. Note that w_1 is the displacement of the center of mass. The force $u(t)$ provides the excitation source to the system for the purpose of structural tests. Prove that the second-order differential equations for this system are (see Ref. [4])

$$M\ddot{w} + \zeta\dot{w} + Kw = B_2 u(t),$$

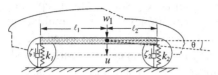

Figure 3.10. A simulated automobile system.

where

$$M = \begin{bmatrix} m & 0 \\ 0 & \mathcal{I} \end{bmatrix},$$

$$K = \begin{bmatrix} k_1 + k_2 & -(k_1\ell_1 - k_2\ell_2) \\ -(k_1\ell_1 - k_2\ell_2) & k_1\ell_1^2 + k_2\ell_2^2 \end{bmatrix},$$

$$\zeta = \begin{bmatrix} \zeta_1 + \zeta_2 & -(\zeta_1\ell_1 - \zeta_2\ell_2) \\ -(\zeta_1\ell_1 - \zeta_2\ell_2) & \zeta_1\ell_1^2 + \zeta_2\ell_2^2 \end{bmatrix},$$

$$w = \begin{bmatrix} w_1 \\ \theta \end{bmatrix},$$

$$B_2 = \begin{bmatrix} 1 \\ 0 \end{bmatrix}.$$

The quantity m is the total mass of the system, and $\mathcal{I}$ is the moment of inertia relative to the center of mass.

3.4   An inverted pendulum system shown in Fig. 3.11 includes a cart of mass M and a massless rigid rod attached with a tip mass m subject to the gravity acceleration g. The motion of the cart is controlled horizontally by the force $u(t)$. The displacement $x(t)$ and the rotation angle $\theta(t)$ define the motions of the two-degree-of-freedom system. Derive the equations of motion for the system by using Hamilton's principle and Lagrange's method.

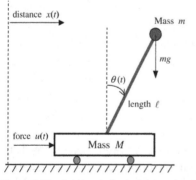

Figure 3.11. An inverted pendulum.

3.5   Figure 3.12 shows two pendulums of mass m and length ℓ coupled by a spring of stiffness k are controlled by two forces u_1 and u_2. The length h is the distance from the ceiling to the attachment of the spring to the pendulum. The rotation angles θ_1 and θ_2 define the motions of the two-degree-of-freedom system. Derive (1) the equations

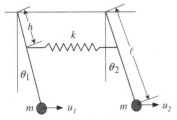

Figure 3.12. Two pendulums coupled by a spring.

of motion for the system by using Hamilton's principle and Lagrange's method and (2) the approximate linear equations of motion for small angles (Ref. [5]).

3.6 Figure 3.13 shows a small mass that is restrained by four orthogonal springs. Each of the springs has a stiffness constant k and an unstressed length ℓ. Derive (1) the equations of motion for the system by using Hamilton's principle and Lagrange's method and (2) the approximate linear equations of motion for small angles (Ref. [5]).

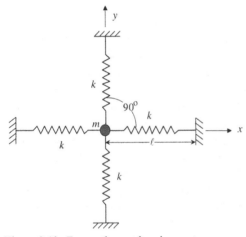

Figure 3.13. Four-orthogonal-spring system.

3.7 Take the plane of a rectangular membrane shown in Fig. 3.14 as the x–y plane. Let a and b be the lengths of the sides of the rectangular membrane. Denote w as the displacement of any point on the membrane in the direction perpendicular to the x–y plane, T as the uniform tension per unit length of the boundary, and ρ as the mass density per unit area.

For small deflections, the potential energy may be approximated by

$$V = \frac{T}{2} \iint \left[\left(\frac{\partial w}{\partial x} \right)^2 + \left(\frac{\partial w}{\partial y} \right)^2 \right] dx\,dy.$$

The kinetic energy of the membrane during vibration is

$$T = \frac{\rho}{2} \iint \left(\frac{dw}{dt} \right)^2 dx\,dy.$$

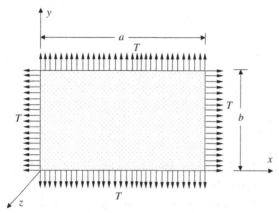

Figure 3.14. Rectangular membrane.

Assume that the displacement function w can be represented within the limits of the rectangle by the double series

$$w = \sum_{m=1}^{\infty} \sum_{n=1}^{\infty} \phi_{mn}(t) \sin \frac{m\pi x}{a} \sin \frac{n\pi y}{b},$$

where the coefficients $\phi_{mn}(t)$ are functions of time t. Each term of the above series satisfies the boundary conditions, i.e., $w = 0$ for $x = 0$ and $x = a$, and $w = 0$ for $y = 0$ and $y = b$. Derive the equations of motion for the membrane subject to a point force $f(t)$ at the center of the membrane.

3.8   Consider a rectangular plate with a uniform thickness h that is assumed to be small compared with other dimensions. Assume that the plate is simply supported. Take the middle plane of the plate as the x–y plane. Let a and b be the lengths of the sides of the plate. The shape of the plate and its definition of coordinates are identical to the membrane shown in Fig. 3.14. Assume that the deflection w in the z direction is small compared with the thickness h. Denote D as the flexural rigidity of the plate, v as Poisson's ratio, and ρh as the mass density per unit area. The potential energy of bending is

$$V = \frac{D}{2} \iint \left\{ \left[\left(\frac{\partial^2 w}{\partial x^2} \right)^2 + \left(\frac{\partial^2 w}{\partial y^2} \right)^2 \right. \right.$$
$$\left. \left. + 2v \frac{\partial^2 w}{\partial x^2} \frac{\partial^2 w}{\partial y^2} + 2(1 - v) \frac{\partial^2 w}{\partial x \partial y} \right] \right\} dx dy.$$

The kinetic energy of the vibrating plate is

$$T = \frac{\rho h}{2} \iint \left(\frac{dw}{dt} \right)^2 dx dy.$$

Take the displacement function w as the double series

$$w = \sum_{m=1}^{\infty} \sum_{n=1}^{\infty} \phi_{mn}(t) \sin \frac{m\pi x}{a} \sin \frac{n\pi y}{b},$$

where the coefficients $\phi_{mn}(t)$ are functions of time t. Each term of the series satisfies the boundary conditions (simply supported edges), i.e., $w = \partial^2 w / \partial x^2 = 0$

for $x = 0$ and $x = a$, and $w = \partial^2 w/\partial y^2 = 0$ for $y = 0$ and $y = b$. Derive the equations of motion for the plate with a point force $f(t)$ applied at its center.

BIBLIOGRAPHY

[1] Meirovitch, L., *Methods of Analytical Dynamics*, McGraw-Hill, New York, 1970.

[2] Meirovitch, L., *Analytical Methods in Vibrations*, Macmillan, New York, 1960.

[3] Kane, T. R. and Levinson, D. A., *Dynamics: Theory and Applications*, McGraw-Hill, New York, 1985.

[4] Timoshenko, S., Young, D. H., and Weaver, W., Jr., *Vibration Problems in Engineering*, 4th ed., Wiley, New York, 1974.

[5] Bishop, R. E. D. and Johnson, D. C., *The Mechanics of Vibration*, Cambridge University Press, London, U.K., 1960.

4

Finite-Element Method

4.1 Introduction

In the analysis of continuous systems, the formulations describing the system response are governed by partial differential equations, as presented in the last chapter. The exact solution of the partial differential equations satisfying all boundary conditions is possible for only relatively simple systems such as a uniform beam. Numerical procedures must be used to approximate the partial differential equations and predict the system response.

The finite-element method is a very popular technique for the numerical solution of complex problems in engineering. It is a technique for solving partial differential equations that represent a physical system by discretizing them in their space dimensions. The discretization is performed locally over small regions of simple but arbitrary shape, i.e., finite elements. For example, in structural engineering, a structure is typically represented as an assemblage of discrete truss and beam elements. The discretization process converts the partial differential equations into matrix equations relating the input at specified points in the elements to the output at these same points. To solve equations over large regions, the matrix equations for the smaller subregions are summed node by node to yield global matrix equations.

Our objective in this chapter is to present the fundamental principles of the finite-element method. It is not our goal to summarize all the finite-element formulations available, but rather to establish only the basic and general principles that provide the foundation for a preliminary understanding of the finite-element method. There are a large number of publications on the finite-element method (see, for example, Refs. [1–6]) that may be used for the readers who are interested in further studies.

In this chapter, we restrict ourselves to the analysis of structural problems. We shall first concentrate on the beam element. Interpolation functions, matrix equation of motion, and boundary conditions will be described. The element assembly will then be presented, including the definition of constraint equations, transformation of coordinates, and the reduced matrix equation of motion. The final two sections present the general formulation for truss structures. The basic truss element will be discussed, including longitudinal motion and rigid-body motion. The energy method will be introduced for truss-element assembly.

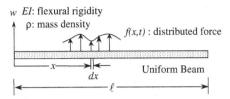

Figure 4.1. A simple uniform beam.

4.2 Uniform Beam Element

In this section, we deal with a simple uniform beam element. Beam elements are commonly used in structural engineering for building frames and bridges. The equation of motion for a uniform beam element, as shown in Fig. 4.1, can be expressed by

$$\rho \frac{\partial^2 w(x,t)}{\partial t^2} + EI \frac{\partial^4 w(x,t)}{\partial x^4} = f(x,t), \tag{4.1}$$

where ρ is the mass density per unit length, E is Young's modulus of the material, I is the second moment of area of the cross section about the neutral axis through its centroid, and f is the applied lateral distributed force. It is known that the homogeneous part of the partial differential equation can be solved by the separation principle, i.e., the displacement $w(x,t)$ is the product of two separate functions; one is space dependent and the other is time dependent. Equation (4.1) is quite simple, and a closed solution is obtainable by the separation principle. However, there are many other partial differential equations in practice that represent real physical systems that have no closed-form solution. The variable separation approach is still useful for obtaining an approximate solution.

4.2.1 Interpolation Functions

Let us assume that the displacement $w(x,t)$ can be approximated by

$$w(x,t) \approx [\phi_1(x) \quad \varphi_1(x) \quad \phi_2(x) \quad \varphi_2(x)] \begin{bmatrix} w_1(t) \\ \theta_1(t) \\ w_2(t) \\ \theta_2(t) \end{bmatrix}, \tag{4.2}$$

where $w_1(t)$ and $w_2(t)$, as shown in Fig. 4.2, are the displacements at point $x = 0$ and $x = \ell$ and the variables $\theta_1(t)$ and $\theta_2(t)$ represent rotational angles at $x = 0$ and $x = \ell$,

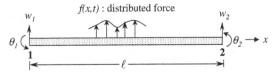

Figure 4.2. A uniform beam element.

respectively. The rotational angles can be computed by

$$\theta_1(t) = \left.\frac{\partial w(x,t)}{\partial x}\right|_{x=0}, \qquad \theta_2(t) = \left.\frac{\partial w(x,t)}{\partial x}\right|_{x=\ell}. \tag{4.3}$$

The four quantities $\phi_1(x)$, $\varphi_1(x)$, $\phi_2(x)$, and $\varphi_2(x)$ are commonly called interpolation functions of the variable x, depending on boundary conditions at $x = 0$ and $x = \ell$. The interpolation functions are not unique in general. They allow the displacement to be specified, or interpolated, at points along the structure that lie between nodes. The term interpolation must satisfy certain boundary conditions. The fundamental requirement is to choose the four functions that satisfy the following conditions:

$$w(0, t) = w_1(t),$$

$$\left.\frac{\partial w(x,t)}{\partial x}\right|_{x=0} = \theta_1(t),$$

$$w(\ell, t) = w_2(t),$$

$$\left.\frac{\partial w(x,t)}{\partial x}\right|_{x=\ell} = \theta_2(t). \tag{4.4}$$

Equations (4.4) imply that the displacement is exactly represented by approximation (4.2) at both end points $x = 0$ and $x = \ell$. In other words, the approximation sign should be replaced with the equal sign in approximation (4.2) at $x = 0$ and $x = \ell$. Approximation (4.2) will eliminate the space dependence for a partial differential equation such as Eq. (4.1) to become an ordinary differential equation.

Substituting Eqs. (4.4) into Eq. (4.2) thus produces the following boundary conditions:

$$\phi_1(0) = 1, \quad \left.\frac{d\phi_1(x)}{dx}\right|_{x=0} = 0, \quad \phi_1(\ell) = 0, \quad \left.\frac{d\phi_1(x)}{dx}\right|_{x=\ell} = 0, \tag{4.5}$$

$$\varphi_1(0) = 0, \quad \left.\frac{d\varphi_1(x)}{dx}\right|_{x=0} = 1, \quad \varphi_1(\ell) = 0, \quad \left.\frac{d\varphi_1(x)}{dx}\right|_{x=\ell} = 0, \tag{4.6}$$

$$\phi_2(0) = 0, \quad \left.\frac{d\phi_2(x)}{dx}\right|_{x=0} = 0, \quad \phi_1(\ell) = 1, \quad \left.\frac{d\phi_2(x)}{dx}\right|_{x=\ell} = 0, \tag{4.7}$$

$$\varphi_2(0) = 0, \quad \left.\frac{d\varphi_2(x)}{dx}\right|_{x=0} = 0, \quad \varphi_2(\ell) = 0, \quad \left.\frac{d\varphi_2(x)}{dx}\right|_{x=\ell} = 1. \tag{4.8}$$

EXAMPLE 4.1

Let us choose a third-order polynomial for ϕ_1 such that

$$\phi_1(x) = c_0 + c_1 x + c_2 x^2 + c_3 x^3, \tag{4.9}$$

where c_0, c_1, c_2, and c_3 are constant coefficients to be determined from boundary conditions (4.5). The third-order polynomial is the most intuitive and simple function that is commonly used in practice for the finite-element method. Inserting Eq. (4.9) into conditions (4.5) yields

$$\phi_1(0) = 1 \Longrightarrow c_0 = 1,$$

$$\frac{d\phi_1}{dx}\bigg|_{x=0} = 0 \Longrightarrow c_1 = 0,$$

$$\phi_1(\ell) = 0 \Longrightarrow 1 + c_2\ell^2 + c_3\ell^3 = 0,$$

$$\frac{d\phi_1}{dx}\bigg|_{x=\ell} = 0 \Longrightarrow 2c_2\ell + 3c_3\ell^2 = 0. \tag{4.10}$$

Solving the last two equations of Eqs. (4.10) results in

$$c_2 = -\frac{3}{\ell^2}, \qquad c_3 = \frac{2}{\ell^3}. \tag{4.11}$$

The function from Eq. (4.9) thus becomes

$$\phi_1(x) = 1 - 3\frac{x^2}{\ell^2} + 2\frac{x^3}{\ell^3}. \tag{4.12}$$

It is easy to verify that Eq. (4.12) satisfies conditions (4.5).

A similar process can be used to determine the other functions $\varphi_1(x)$, $\phi_2(x)$, and $\varphi_2(x)$. Let us use the same third-order polynomial for $\varphi_1(x)$, $\phi_2(x)$, and $\varphi_2(x)$. Using boundary conditions (4.6), (4.7), and (4.8) should produce

$$\varphi_1(x) = x - 2\frac{x^2}{\ell} + \frac{x^3}{\ell^2}, \tag{4.13}$$

$$\phi_2(x) = 3\frac{x^2}{\ell^2} - 2\frac{x^3}{\ell^3}, \tag{4.14}$$

$$\varphi_2(x) = -\frac{x^2}{\ell} + \frac{x^3}{\ell^2}. \tag{4.15}$$

As an illustration, Fig. 4.3 shows the shapes of interpolation functions $\phi_1(x)$, $\varphi_1(x)$, $\phi_2(x)$, and $\varphi_2(x)$ with $\ell = 1$, for simplicity, and their derivatives. With the space functions $\phi_1(x)$, $\varphi_1(x)$, $\phi_2(x)$, and $\varphi_2(x)$ determined, the next step is to find a formulation to solve for the time-dependent functions $w_1(t)$, $\theta_1(t)$, $w_2(t)$, and $\theta_2(t)$.

4.2.2   Matrix Equation of Motion for an Element

Let us first define the interpolation function $\chi(x)$ as

$$\chi(x) \equiv \begin{bmatrix} \chi_1(x) \\ \chi_2(x) \\ \chi_3(x) \\ \chi_4(x) \end{bmatrix} = \begin{bmatrix} \phi_1(x) \\ \varphi_1(x) \\ \phi_2(x) \\ \varphi_2(x) \end{bmatrix}, \tag{4.16}$$

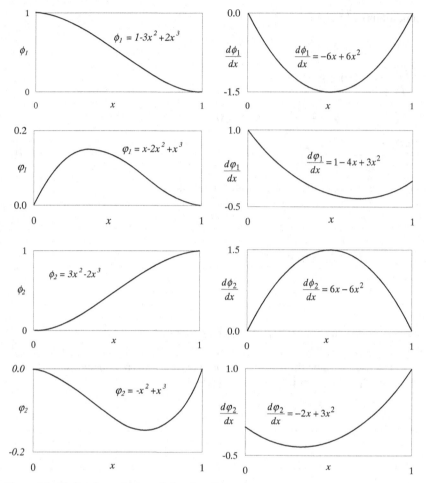

Figure 4.3. Finite-element interpolation functions.

$$q(t) \equiv \begin{bmatrix} q_1(t) \\ q_2(t) \\ q_3(t) \\ q_4(t) \end{bmatrix} = \begin{bmatrix} w_1(t) \\ \theta_1(t) \\ w_2(t) \\ \theta_2(t) \end{bmatrix}. \qquad (4.17)$$

Substituting approximation (4.2) into Eq. (4.1) with the aid of Eqs. (4.16) and (4.17) yields

$$\rho \chi^T(x) \frac{d^2 q(t)}{dt^2} + EI \frac{d^4 \chi^T(x)}{dx^4} q(t) = f(x, t). \qquad (4.18)$$

The displacement approximation defined in approximation (4.2) transforms partial differential equation (4.1) into an ordinary differential equation in the discretized space variables $\phi_1(x)$, $\varphi_1(x)$, $\phi_2(x)$, and $\varphi_2(x)$. The problem now reduces to finding values for $q(t)$, i.e., $w_1(t)$, $\theta_1(t)$, $w_2(t)$, and $\theta_2(t)$. Among the many methods that could be used to solve for $q(t)$, Galerkin's method is most widely used in finite-element work.

The method consists of premultiplying Eq. (4.18) by the interpolation function $\chi(x)$ and integrating over the element, i.e.,

$$\int_0^\ell \chi(x) \left[\rho \chi^T(x) \frac{d^2 q(t)}{dt^2} + EI \frac{d^4 \chi^T(x)}{dx^4} q(t) \right] dx = \int_0^\ell \chi(x) f(x,t) \, dx. \quad (4.19)$$

If the interpolation functions in $\chi(x)$ are approximated by the third-order polynomial, as shown in Example 4.1, differentiating the interpolation function four times would cause it to vanish, and yet the interpolation function defined in Eq. (4.9) cannot be higher than a third-order polynomial because there are only four boundary conditions to solve for four unknown coefficients. The way to escape this dilemma is by changing the form of Eq. (4.19).

Premultiplying Eq. (4.1) by $\chi(x)$ and integrating the resulting equation over the element produces

$$\int_0^\ell \chi(x) \left[\rho \frac{\partial^2 w(x,t)}{\partial t^2} + EI \frac{\partial^4 w(x,t)}{\partial x^4} - f(x,t) \right] dx = 0. \quad (4.20)$$

By applying integration by parts (i.e., Green's theorem) to the integral expression of the second term on the left-hand side, we obtain

$$\int_0^\ell \chi(x) EI \frac{\partial^4 w(x)}{\partial x^4} dx = \chi(x) EI \frac{\partial^3 w(x,t)}{\partial x^3} \Big|_0^\ell - \int_0^\ell \frac{d\chi(x)}{dx} EI \frac{\partial^3 w(x,t)}{\partial x^3} dx$$

$$= \chi(x) EI \frac{\partial^3 w(x,t)}{\partial x^3} \Big|_0^\ell - \frac{d\chi(x)}{dx} EI \frac{\partial^2 w(x,t)}{\partial x^2} \Big|_0^\ell$$

$$+ \int_0^\ell \frac{d^2 \chi(x)}{dx^2} EI \frac{\partial^2 w(x,t)}{\partial x^2} dx. \quad (4.21)$$

From Eqs. (4.5)–(4.8) and (4.16), we have

$$\chi(\ell) = \begin{bmatrix} 0 \\ 0 \\ 1 \\ 0 \end{bmatrix}, \qquad \chi(0) = \begin{bmatrix} 1 \\ 0 \\ 0 \\ 0 \end{bmatrix}, \qquad (4.22)$$

$$\frac{d\chi(x)}{dx} \Big|_{x=\ell} = \begin{bmatrix} 0 \\ 0 \\ 0 \\ 1 \end{bmatrix}, \qquad \frac{d\chi(x)}{dx} \Big|_{x=0} = \begin{bmatrix} 0 \\ 1 \\ 0 \\ 0 \end{bmatrix}. \qquad (4.23)$$

Substituting Eqs. (4.22) and (4.23) into Eq. (4.21) yields

$$
\int_0^\ell \chi(x) EI \frac{\partial^4 w(x,t)}{\partial x^4}\, dx =
\begin{bmatrix}
-EI \left. \dfrac{\partial^3 w(x,t)}{\partial x^3} \right|_{x=0} \\[2mm]
EI \left. \dfrac{\partial^2 w(x,t)}{\partial x^2} \right|_{x=0} \\[2mm]
EI \left. \dfrac{\partial^3 w(x,t)}{\partial x^3} \right|_{x=\ell} \\[2mm]
-EI \left. \dfrac{\partial^2 w(x,t)}{\partial x^2} \right|_{x=\ell}
\end{bmatrix}
$$

$$
+ \int_0^\ell \frac{d^2 \chi(x)}{dx^2} EI \frac{\partial^2 w(x,t)}{\partial x^2}\, dx. \tag{4.24}
$$

We do not use the interpolation functions in the boundary conditions, i.e., the first term on the right-hand side of Eq. (4.24), because they are related to the natural boundary conditions and the nodal forces, which will be discussed later. Substituting Eq. (4.2) with the aid of Eqs. (4.16), (4.17) and (4.24) into Eq. (4.20) yields

$$
\left[\int_0^\ell \rho \chi(x) \chi^T(x)\, dx \right] \frac{d^2 q(t)}{dt^2} + \left[\int_0^\ell EI \frac{d^2 \chi(x)}{dx^2} \frac{d^2 \chi^T(x)}{dx^2}\, dx \right] q(t)
$$

$$
= \int_0^\ell \chi(x) f(x,t)\, dx + F, \tag{4.25}
$$

or, equivalently,

$$
M\ddot{q} + Kq = F_d + F, \tag{4.26}
$$

where

$$
\ddot{q} = \frac{d^2 q(t)}{dt^2}, \tag{4.27}
$$

$$
M = \int_0^\ell \rho \chi(x) \chi^T(x)\, dx, \tag{4.28}
$$

$$
K = \int_0^\ell EI \frac{d^2 \chi(x)}{dx^2} \frac{d^2 \chi^T(x)}{dx^2}\, dx, \tag{4.29}
$$

$$
F_d = \int_0^\ell \chi(x) f(x,t)\, dx, \tag{4.30}
$$

$$
F \equiv \begin{bmatrix} f_1 \\ m_1 \\ f_2 \\ m_2 \end{bmatrix} = \begin{bmatrix} EI \dfrac{\partial^3 w(x,t)}{\partial x^3}\bigg|_{x=0} \\[2mm] -EI \dfrac{\partial^2 w(x,t)}{\partial x^2}\bigg|_{x=0} \\[2mm] -EI \dfrac{\partial^3 w(x,t)}{\partial x^3}\bigg|_{x=\ell} \\[2mm] EI \dfrac{\partial^2 w(x,t)}{\partial x^2}\bigg|_{x=\ell} \end{bmatrix}. \tag{4.31}
$$

The matrix M is called the mass matrix and K is referred to as the stiffness matrix. Note that integration by parts brings into effect the boundary information and lowers to two the fourth-order derivative appearing in the differential equation of motion, Eq. (4.1).

Two forces, F_d and F, act on the beam element. Force F_d results from the distributed force $f(x,t)$, whereas F comes from the shear forces and moments at both ends. The vector F_d is commonly called the generalized load vector or consistent nodal load vector. It is known from the beam theory that

$$
f_1 = EI \frac{\partial^3 w(x,t)}{\partial x^3}\bigg|_{x=0}, \qquad m_1 = -EI \frac{\partial^2 w(x,t)}{\partial x^2}\bigg|_{x=0} \tag{4.32}
$$

are shear force and bending moment, respectively, at node 1, shown in Fig. 4.2, and

$$
f_2 = -EI \frac{\partial^3 w(x,t)}{\partial x^3}\bigg|_{x=\ell}, \qquad m_2 = EI \frac{\partial^2 w(x,t)}{\partial x^2}\bigg|_{x=\ell} \tag{4.33}
$$

are shear force and bending moment, respectively, at node 2. Therefore, f_1 and f_2 are called the nodal forces whereas m_1 and m_2 are referred to as the nodal moments.

Equation (4.26) is the matrix equation for a two-node line element. The natural boundary conditions are taken into account when we assemble the element matrices. During the assembly of the element matrices, the boundary-condition terms F may cancel at all interior nodes, leaving only the natural boundary conditions to be evaluated at external end points.

EXAMPLE 4.2

Let us use the third-order polynomial derived in Example 4.1 as the interpolation functions, i.e.,

$$
\chi(x) \equiv \begin{bmatrix} \chi_1(x) \\ \chi_2(x) \\ \chi_3(x) \\ \chi_4(x) \end{bmatrix} = \begin{bmatrix} \phi_1(x) \\ \varphi_1(x) \\ \phi_2(x) \\ \varphi_2(x) \end{bmatrix} = \begin{bmatrix} 1 - 3\dfrac{x^2}{\ell^2} + 2\dfrac{x^3}{\ell^3} \\[2mm] x - 2\dfrac{x^2}{\ell} + \dfrac{x^3}{\ell^2} \\[2mm] 3\dfrac{x^2}{\ell^2} - 2\dfrac{x^3}{\ell^3} \\[2mm] -\dfrac{x^2}{\ell} + \dfrac{x^3}{\ell^2} \end{bmatrix}.
$$

Note that we may use a negative interpolation function instead, for example, $-\varphi_2(x)$ rather than $\varphi_2(x)$. The negative interpolation function will not alter the characteristics of the finite-element model such as the system eigenvalues and mode shapes. Substituting the above results into Eqs. (4.28)–(4.30) thus yields

$$M = \int_0^\ell \rho \chi(x) \chi^T(x)\, dx$$

$$= \rho \begin{bmatrix} \dfrac{13\ell}{35} & \dfrac{11\ell^2}{210} & \dfrac{9\ell}{70} & -\dfrac{13\ell^2}{420} \\[2ex] \dfrac{11\ell^2}{210} & \dfrac{\ell^3}{105} & \dfrac{13\ell^2}{420} & -\dfrac{\ell^3}{140} \\[2ex] \dfrac{9\ell}{70} & \dfrac{13\ell^2}{420} & \dfrac{13\ell}{35} & -\dfrac{11\ell^2}{210} \\[2ex] -\dfrac{13\ell^2}{420} & -\dfrac{\ell^3}{140} & -\dfrac{11\ell^2}{210} & \dfrac{\ell^3}{105} \end{bmatrix}, \tag{4.34}$$

$$K = \int_0^\ell EI \frac{d^2\chi(x)}{dx^2}\frac{d^2\chi^T(x)}{dx^2}\, dx$$

$$= EI \begin{bmatrix} \dfrac{12}{\ell^3} & \dfrac{6}{\ell^2} & -\dfrac{12}{\ell^3} & \dfrac{6}{\ell^2} \\[2ex] \dfrac{6}{\ell^2} & \dfrac{4}{\ell} & -\dfrac{6}{\ell^2} & \dfrac{2}{\ell} \\[2ex] -\dfrac{12}{\ell^3} & -\dfrac{6}{\ell^2} & \dfrac{12}{\ell^3} & -\dfrac{6}{\ell^2} \\[2ex] \dfrac{6}{\ell^2} & \dfrac{2}{\ell} & -\dfrac{6}{\ell^2} & \dfrac{4}{\ell} \end{bmatrix}, \tag{4.35}$$

$$F_d = \int_0^\ell \chi(x) f(x,t)\, dx = f \begin{bmatrix} \dfrac{\ell}{2} \\[2ex] \dfrac{\ell^2}{12} \\[2ex] \dfrac{\ell}{2} \\[2ex] -\dfrac{\ell^2}{12} \end{bmatrix}, \tag{4.36}$$

where the external forcing function f is assumed uniformly constant. One may argue at this moment that all the elements in the mass matrix M should be positive to be consistent with the natural mass property. Indeed, if we switch the sign of $\varphi_2(x)$ to become $-\varphi_2(x)$ used as the interpolation function, the last column and the last row of the mass matrix M will all be positive. Nevertheless, the mass matrix M in its present form is mathematically accurate as long as it is positive definite.

4.2.3 Boundary Conditions

To find the solution of the matrix equation of motion, it is necessary to specify the natural boundary conditions (i.e., the end points for the beam). The geometric boundary conditions relate to the deflection w and the angle (slope) $\partial w/\partial x$. The force boundary conditions relate to the bending moment $EI\partial^2 w/\partial x^2$ and the shear force $EI\partial^3 w/\partial x^3$. These boundary conditions are given by the following:

$$\text{clamped end,} \quad w(x,t) = \frac{\partial w(x,t)}{\partial x} = 0;$$

$$\text{pinned end,} \quad w(x,t) = \frac{\partial^2 w(x,t)}{\partial x^2} = 0;$$

$$\text{sliding end,} \quad \frac{\partial w(x,t)}{\partial x} = \frac{\partial^3 w(x,t)}{\partial x^3} = 0;$$

$$\text{free end,} \quad \frac{\partial^2 w(x,t)}{\partial x^2} = \frac{\partial^3 w(x,t)}{\partial x^3} = 0. \tag{4.37}$$

If a harmonic nodal force $f \sin(\omega t + \psi)$ of amplitude f, frequency ω, and phase angle ψ acts at an otherwise free end of a beam, then

$$f_2 = -EI\frac{\partial^3 w(x,t)}{\partial x^3}\bigg|_{x=\ell} = f \sin(\omega t + \psi). \tag{4.38}$$

On the other hand, if a harmonic nodal bending moment $h \sin(\omega t + \psi)$ of amplitude h, frequency ω, and phase angle ψ acts at an otherwise free end, then

$$m_2 = EI\frac{\partial^2 w(x,t)}{\partial x^2}\bigg|_{x=\ell} = h \sin(\omega t + \psi). \tag{4.39}$$

The following example illustrates the procedure for finding the matrix equation of motion after boundary conditions are specified.

EXAMPLE 4.3
The clamped–free uniform beam shown in Fig. 4.4 is applied by a static constant force f that acts downward at the end. The appropriate boundary conditions are the following:

$$\text{when } x = 0, \quad w(x,t) = \frac{\partial w(x,t)}{\partial x} = 0;$$

$$\text{when } x = \ell, \quad m_2 = \frac{\partial^2 w(x,t)}{\partial x^2} = 0, \quad f_2 = -EI\frac{\partial^3 w(x,t)}{\partial x^3} = -f. \tag{4.40}$$

The first boundary condition in Eqs. (4.40) implies that both w_1 and θ_1 defined in Eqs. (4.4) are zero. Note that w_1 and θ_1 are the first and the second elements of the vector q defined in Eq. (4.17). Assume that we are interested in only the static deflection of the beam, i.e., $\ddot{q} = 0$. With the aid of Eqs. (4.31) and (4.35), Eq. (4.26)

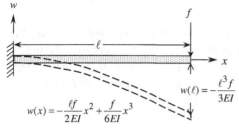

Figure 4.4. A uniform beam with static force applied at the end.

becomes

$$
EI
\begin{bmatrix}
\dfrac{12}{\ell^3} & \dfrac{6}{\ell^2} & -\dfrac{12}{\ell^3} & \dfrac{6}{\ell^2} \\[2mm]
\dfrac{6}{\ell^2} & \dfrac{4}{\ell} & -\dfrac{6}{\ell^2} & \dfrac{2}{\ell} \\[2mm]
-\dfrac{12}{\ell^3} & -\dfrac{6}{\ell^2} & \dfrac{12}{\ell^3} & -\dfrac{6}{\ell^2} \\[2mm]
\dfrac{6}{\ell^2} & \dfrac{2}{\ell} & -\dfrac{6}{\ell^2} & \dfrac{4}{\ell}
\end{bmatrix}
\begin{bmatrix}
0 \\ 0 \\ w_2 \\ \theta_2
\end{bmatrix}
=
\begin{bmatrix}
EI\dfrac{\partial^3 w(x,t)}{\partial x^3}\Big|_{x=0} \\[2mm]
-EI\dfrac{\partial^2 w(x,t)}{\partial x^2}\Big|_{x=0} \\[2mm]
-f \\[2mm]
0
\end{bmatrix}.
\tag{4.41}
$$

The quantities w_2 and θ_2 can be easily solved as

$$
\begin{bmatrix}
w_2 \\ \theta_2
\end{bmatrix}
=
\frac{1}{EI}
\begin{bmatrix}
\dfrac{12}{\ell^3} & -\dfrac{6}{\ell^2} \\[2mm]
-\dfrac{6}{\ell^2} & \dfrac{4}{\ell}
\end{bmatrix}^{-1}
\begin{bmatrix}
-f \\ 0
\end{bmatrix}
\tag{4.42}
$$

or

$$
\begin{bmatrix}
w_2 \\ \theta_2
\end{bmatrix}
=
\frac{1}{EI}
\begin{bmatrix}
\dfrac{\ell^3}{3} & \dfrac{\ell^2}{2} \\[2mm]
\dfrac{\ell^2}{2} & \ell
\end{bmatrix}
\begin{bmatrix}
-f \\ 0
\end{bmatrix}
$$

$$
=
\begin{bmatrix}
-\dfrac{\ell^3 f}{3EI} \\[2mm]
-\dfrac{\ell^2 f}{2EI}
\end{bmatrix}.
\tag{4.43}
$$

Substituting Eq. (4.43) into Eq. (4.41) thus yields

$$
\begin{bmatrix}
EI\dfrac{\partial^3 w(x,t)}{\partial x^3}\Big|_{x=0} \\[2mm]
EI\dfrac{\partial^2 w(x,t)}{\partial x^2}\Big|_{x=0}
\end{bmatrix}
=
\begin{bmatrix}
f \\ -f\ell
\end{bmatrix},
\tag{4.44}
$$

which gives the force and moment at $x = 0$. With the aid of Eq. (4.16), the displacement $w(x, t)$ can thus be obtained by

$$
\begin{aligned}
w(x) &= \chi^T(x)q = \phi_2(x)w_2 + \varphi_2(x)\theta_2 \\
&= -\left(3\frac{x^2}{\ell^2} - 2\frac{x^3}{\ell^3}\right)\frac{\ell^3 f}{3EI} - \left(-\frac{x^2}{\ell} + \frac{x^3}{\ell^2}\right)\frac{\ell^2 f}{2EI} \\
&= \left(\frac{\ell^3 f}{6EI}\right)\left(-3\frac{x^2}{\ell^2} + \frac{x^3}{\ell^3}\right) = -\frac{\ell f}{2EI}x^2 + \frac{f}{6EI}x^3.
\end{aligned}
\tag{4.45}
$$

The solution should be correct because it satisfies the balance law of force and moment. The deflection at the end of the beam is

$$
w(\ell, t) = -\frac{\ell^3 f}{3EI} = w_2,
\tag{4.46}
$$

and the slope is

$$
\left.\frac{\partial w(x, t)}{\partial x}\right|_{x=\ell} = -\frac{\ell^2 f}{2EI} = \theta_2.
\tag{4.47}
$$

From this example, it is seen that the interpolation function defined in Eq. (4.16) is the static beam-deflection function. Other functions may be used as long as they satisfy the boundary conditions defined in Eqs. (4.5)–(4.8).

EXAMPLE 4.4

Let us look at another case in which a harmonic nodal force $f \sin(\omega t + \psi)$ acts at the end, as shown in Fig. 4.5. The appropriate boundary conditions are the following:

$$
\text{when } x = 0, \quad w(x, t) = \frac{\partial w(x, t)}{\partial x} = 0;
$$

$$
\text{when } x = \ell, \quad m_2 = \frac{\partial^2 w(x, t)}{\partial x^2} = 0,
$$

$$
f_2 = -EI\frac{\partial^3 w(x, t)}{\partial x^3} = -f \sin(\omega t + \psi).
\tag{4.48}
$$

The first boundary condition in Eqs. (4.48) implies that both w_1 and θ_1 defined in Eqs. (4.4) are zero. With the aid of Eqs. (4.31), (4.34), and (4.35), Eq. (4.26) produces

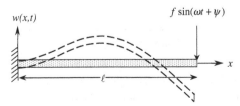

Figure 4.5. A uniform beam with a harmonic force applied at the end.

two matrix equations:

$$\rho \begin{bmatrix} \dfrac{13\ell}{35} & -\dfrac{11\ell^2}{210} \\ -\dfrac{11\ell^2}{210} & \dfrac{\ell^3}{105} \end{bmatrix} \begin{bmatrix} \ddot{w}_2 \\ \ddot{\theta}_2 \end{bmatrix} + EI \begin{bmatrix} \dfrac{12}{\ell^3} & -\dfrac{6}{\ell^2} \\ -\dfrac{6}{\ell^2} & \dfrac{4}{\ell} \end{bmatrix} \begin{bmatrix} w_2 \\ \theta_2 \end{bmatrix}$$

$$= \begin{bmatrix} -f\sin(\omega t + \psi) \\ 0 \end{bmatrix}, \tag{4.49}$$

$$\rho \begin{bmatrix} \dfrac{9\ell}{70} & -\dfrac{13\ell^2}{420} \\ \dfrac{13\ell^2}{420} & -\dfrac{\ell^3}{140} \end{bmatrix} \begin{bmatrix} \ddot{w}_2 \\ \ddot{\theta}_2 \end{bmatrix} + EI \begin{bmatrix} -\dfrac{12}{\ell^3} & \dfrac{6}{\ell^2} \\ \dfrac{6}{\ell^2} & \dfrac{2}{\ell} \end{bmatrix} \begin{bmatrix} w_2 \\ \theta_2 \end{bmatrix}$$

$$= \begin{bmatrix} EI \dfrac{\partial^3 w(x,t)}{\partial x^3}\Big|_{x=0} \\ -EI \dfrac{\partial^2 w(x,t)}{\partial x^2}\Big|_{x=0} \end{bmatrix}. \tag{4.50}$$

From Eq. (4.49), the quantities w_2 and θ_2 can be solved with the initial conditions $w_2(0)$ and $\theta_2(0)$ given. Once w_2 and the θ_2 are determined, the shear force and the moment at $x = 0$ can be computed from Eq. (4.50).

It will be shown in Chapter 5 that beam equation (4.1) possesses an infinite number of natural frequencies, whereas Eq. (4.49) has only one natural frequency in it. An example deflection is shown in Fig. 4.5 (dashed line), which corresponds to the second mode of a clamped–free (cantilevered) beam. If the forcing frequency ω is close to the frequency of the second mode, the beam would be deflected as shown in Fig. 4.5. Therefore, the solution produced by Eq. (4.49) may be considerably inaccurate, in particular when the forcing frequency ω is close to any beam natural frequency. In other words, the interpolation function used in this example does not approximate the displacement $w(x, t)$ accurately enough. One way to solve this problem is to change the interpolation function. The other way is to increase the number of elements and assemble the elements together to form a bigger matrix equation to produce more natural frequencies.

4.3 Element Assembly

The fundamental assumption used in element assembly is that the discrete elements that have simplified elastic properties are so interconnected as to represent the actual continuous structure. The boundary conditions are compatible at the node points where the elements are joined. The resulting assembly of the discrete elements with interpolation functions specified satisfactorily represents the structural properties such as natural frequencies and mode shapes.

4.3.1 Combined Matrix Equations of Motion

Matrix equations of motion for beam elements in isolation have been shown to be given by Eq. (4.26), i.e.,

$$M^{(i)}\ddot{q}^{(i)} + K^{(i)}q^{(i)} = F_d^{(i)} + F^{(i)}, \tag{4.51}$$

where the superscript signifies the element number. Each individual element has its own matrix elements defined as

$$q^{(i)} \equiv \begin{bmatrix} q_1^{(i)} \\ q_2^{(i)} \\ q_3^{(i)} \\ q_4^{(i)} \end{bmatrix} = \begin{bmatrix} w_1^{(i)} \\ \theta_1^{(i)} \\ w_2^{(i)} \\ \theta_2^{(i)} \end{bmatrix}, \tag{4.52}$$

where $q^{(i)}$ is commonly called the element generalized coordinate vector. Similarly, the element mass matrix is defined as

$$M^{(i)} \equiv \begin{bmatrix} m_{11}^{(i)} & m_{12}^{(i)} & m_{13}^{(i)} & m_{14}^{(i)} \\ m_{12}^{(i)} & m_{22}^{(i)} & m_{23}^{(i)} & m_{24}^{(i)} \\ m_{13}^{(i)} & m_{23}^{(i)} & m_{33}^{(i)} & m_{34}^{(i)} \\ m_{14}^{(i)} & m_{24}^{(i)} & m_{34}^{(i)} & m_{44}^{(i)} \end{bmatrix}, \tag{4.53}$$

and the element stiffness matrix is represented by

$$K^{(i)} \equiv \begin{bmatrix} k_{11}^{(i)} & k_{12}^{(i)} & k_{13}^{(i)} & k_{14}^{(i)} \\ k_{12}^{(i)} & k_{22}^{(i)} & k_{23}^{(i)} & k_{24}^{(i)} \\ k_{13}^{(i)} & k_{23}^{(i)} & k_{33}^{(i)} & k_{34}^{(i)} \\ k_{14}^{(i)} & k_{24}^{(i)} & k_{34}^{(i)} & k_{44}^{(i)} \end{bmatrix}. \tag{4.54}$$

The corresponding distributed-force vector and the nodal-force vector become

$$F_d^{(i)} \equiv \begin{bmatrix} f_{d1}^{(i)} \\ f_{d2}^{(i)} \\ f_{d3}^{(i)} \\ f_{d4}^{(i)} \end{bmatrix}, \qquad F^{(i)} \equiv \begin{bmatrix} f_1^{(i)} \\ m_1^{(i)} \\ f_2^{(i)} \\ m_2^{(i)} \end{bmatrix}. \tag{4.55}$$

The next problem is to assemble the elements and so derive the matrix equation of motion for a multiple-element system. For illustration, let us assume that we have only two elements, as shown in Fig. 4.6, to be assembled. Each beam element has two nodes, denoted by 1 and 2. The assembled beam has three nodes denoted as ①, ②, and ③ (the numbers within circles).

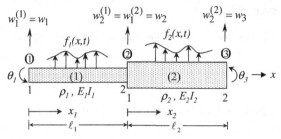

Figure 4.6. Two beam elements.

The combined matrix equation of motion for the two-element system is

$$M'\ddot{q}' + K'q' = F_d' + F', \tag{4.56}$$

where

$$q' = \begin{bmatrix} q^{(1)} \\ q^{(2)} \end{bmatrix}, \tag{4.57}$$

$$M' = \begin{bmatrix} M^{(1)} & 0_{2\times2} \\ 0_{2\times2} & M^{(2)} \end{bmatrix}, \tag{4.58}$$

$$K' = \begin{bmatrix} K^{(1)} & 0_{2\times2} \\ 0_{2\times2} & K^{(2)} \end{bmatrix}, \tag{4.59}$$

$$F_d' = \begin{bmatrix} F_d^{(1)} \\ F_d^{(2)} \end{bmatrix}, \qquad F' = \begin{bmatrix} F^{(1)} \\ F^{(2)} \end{bmatrix}, \tag{4.60}$$

in which $0_{2\times2}$ is a zero matrix of 2×2. There is a total of four nodes with two equations for each node, implying that eight equations are involved in Eq. (4.56). One question arises as to whether all eight equations are independent. If not, some constraint conditions must be imposed.

4.3.2   Constraint Equations

The second node of the first element is identical to the first node of the second element. In addition, assume that the two elements are rigidly linked at the middle node, i.e., the displacement and the slope at this node are identical. Then two constraint equations can be written as

$$w_2^{(1)} = w_1^{(2)} \implies q_3^{(1)} = q_1^{(2)}, \tag{4.61}$$

$$\theta_2^{(1)} = \theta_1^{(2)} \implies q_4^{(1)} = q_2^{(2)}. \tag{4.62}$$

Equations (4.61) and (4.62) can be combined into a matrix equation:

$$
\begin{bmatrix} 1 & 0 & -1 & 0 \\ 0 & 1 & 0 & -1 \end{bmatrix}
\begin{bmatrix} q_3^{(1)} \\ q_4^{(1)} \\ q_1^{(2)} \\ q_2^{(2)} \end{bmatrix} = 0,
\tag{4.63}
$$

or, equivalently,

$$
\Gamma q' = 0,
\tag{4.64}
$$

where Γ is defined as

$$
\Gamma = \begin{bmatrix} 0 & 0 & 1 & 0 & -1 & 0 & 0 & 0 \\ 0 & 0 & 0 & 1 & 0 & -1 & 0 & 0 \end{bmatrix}.
\tag{4.65}
$$

This is the constraint equation to determine the independent nodes for the assembled system. To derive the matrix equation of motion in terms of independent coordinates for system equation (4.56), subject to constraint equation (4.64), a coordinate transformation must be derived.

Now examine the following matrix:

$$
\Gamma_0 = \begin{bmatrix}
1 & 0 & 0 & 0 & 0 & 0 & 0 & 0 \\
0 & 1 & 0 & 0 & 0 & 0 & 0 & 0 \\
0 & 0 & 1 & 0 & 1 & 0 & 0 & 0 \\
0 & 0 & 0 & 1 & 0 & 1 & 0 & 0 \\
0 & 0 & 0 & 0 & 0 & 0 & 1 & 0 \\
0 & 0 & 0 & 0 & 0 & 0 & 0 & 1
\end{bmatrix}.
\tag{4.66}
$$

This matrix has two very important properties: All rows of Γ_0 are linearly independent and it is orthogonal to the constraint matrix Γ, i.e.,

$$
\Gamma \Gamma_0^T = 0_{6 \times 6}.
\tag{4.67}
$$

Note that there are an infinite number of matrices that are orthogonal to Γ. Any combination of rows in Γ_0 is orthogonal to the rows in Γ. In other words, the rows in Γ_0 can be used as the basis vectors to generate a row space that is orthogonal to the space generated by the rows of Γ. The matrix Γ_0 shown in Eq. (4.66) is the simplest and the most intuitive one that is commonly used. We may obtain other orthogonal matrices by taking the singular value decomposition of Γ and by using the orthonormal basis vectors for the subspaces with zero singular values (See Chap. 2 for details).

4.3.3   Assembled Matrix Equations of Motion

A coordinate transformation may now be defined as

$$q' = \Gamma_0^T q, \tag{4.68}$$

where q is a 6×1 vector of independent coordinates, defined as

$$q \equiv \begin{bmatrix} q_1 \\ q_2 \\ q_3 \\ q_4 \\ q_5 \\ q_6 \end{bmatrix} = \begin{bmatrix} w_1^{(1)} \\ \theta_1^{(1)} \\ w_2^{(1)} \\ \theta_2^{(1)} \\ w_2^{(2)} \\ \theta_2^{(2)} \end{bmatrix} = \begin{bmatrix} w_1 \\ \theta_1 \\ w_2 \\ \theta_2 \\ w_3 \\ \theta_3 \end{bmatrix}. \tag{4.69}$$

The numbers in parentheses in Eq. (4.69) correspond to the beam elements shown in Fig. 4.6. The subscripts for the variables without superscripts in the last column vector in Eq. (4.69) signify the node number of the assembled beam. Substituting Eq. (4.68) into Eq. (4.64) produces

$$\Gamma \Gamma_0^T q = 0_{6 \times 6}\, q = 0_{6 \times 1}. \tag{4.70}$$

This implies that the new coordinate vector q automatically satisfies the constraint equation.

The next step is to transform combined equation (4.56) into a reduced order of the matrix equation of motion. Substituting transformation equation (4.68) into Eq. (4.56) and premultiplying the resulting equation by Γ_0 yield

$$M\ddot{q} + Kq = F_d + F, \tag{4.71}$$

where

$$M = \Gamma_0 M' \Gamma_0^T, \tag{4.72}$$

$$K = \Gamma_0 K' \Gamma_0^T, \tag{4.73}$$

$$F_d = \Gamma_0 F_d', \tag{4.74}$$

$$F = \Gamma_0 F'. \tag{4.75}$$

Equation (4.71) is the matrix equation of motion for the assembled beam.

From Eqs. (4.53), (4.58), and (4.66), the assembled mass matrix becomes

$$M = \Gamma_0 M' \Gamma_0^T$$

$$= \begin{bmatrix}
m_{11}^{(1)} & m_{12}^{(1)} & m_{13}^{(1)} & m_{14}^{(1)} & 0 & 0 \\
m_{12}^{(1)} & m_{22}^{(1)} & m_{23}^{(1)} & m_{24}^{(1)} & 0 & 0 \\
m_{13}^{(1)} & m_{23}^{(1)} & m_{33}^{(1)} + m_{11}^{(2)} & m_{34}^{(1)} + m_{12}^{(2)} & m_{13}^{(2)} & m_{14}^{(2)} \\
m_{14}^{(1)} & m_{24}^{(1)} & m_{34}^{(1)} + m_{12}^{(2)} & m_{44}^{(1)} + m_{22}^{(2)} & m_{23}^{(2)} & m_{24}^{(2)} \\
0 & 0 & m_{13}^{(2)} & m_{23}^{(2)} & m_{33}^{(2)} & m_{34}^{(2)} \\
0 & 0 & m_{14}^{(2)} & m_{24}^{(2)} & m_{34}^{(2)} & m_{44}^{(2)}
\end{bmatrix}. \tag{4.76}$$

It is seen from Fig. 4.6 that the assembled-beam node ② corresponds to element node 2 in the first element (1) and element node 1 in the second element (2). The term $m_{33}^{(1)}$ from element (1) and the term $m_{11}^{(2)}$ from element (2) would be added together and would appear in location 3, 3 in the assembled mass matrix M. Similarly, $m_{44}^{(1)}$ from element (1) and $m_{22}^{(2)}$ from element (2) corresponding to the rotational angles should also be added together and so on.

Similarly, from Eqs. (4.54), (4.59), and (4.66), the assembled stiffness matrix becomes

$$K = \Gamma_0 K' \Gamma_0^T$$

$$= \begin{bmatrix}
k_{11}^{(1)} & k_{12}^{(1)} & k_{13}^{(1)} & k_{14}^{(1)} & 0 & 0 \\
k_{12}^{(1)} & k_{22}^{(1)} & k_{23}^{(1)} & k_{24}^{(1)} & 0 & 0 \\
k_{13}^{(1)} & k_{23}^{(1)} & k_{33}^{(1)} + k_{11}^{(2)} & k_{34}^{(1)} + k_{12}^{(2)} & k_{13}^{(2)} & k_{14}^{(2)} \\
k_{14}^{(1)} & k_{24}^{(1)} & k_{34}^{(1)} + k_{12}^{(2)} & k_{44}^{(1)} + k_{22}^{(2)} & k_{23}^{(2)} & k_{24}^{(2)} \\
0 & 0 & k_{13}^{(2)} & k_{23}^{(2)} & k_{33}^{(2)} & k_{34}^{(2)} \\
0 & 0 & k_{14}^{(2)} & k_{24}^{(2)} & k_{34}^{(2)} & k_{44}^{(2)}
\end{bmatrix}. \tag{4.77}$$

Both mass matrix M and stiffness matrix K are symmetric and possess the useful property of bandedness. The terms are concentrated around the main diagonal that stretches from the upper left to the lower right. No term in any row can be more than three locations away from the main diagonal so that the system is said to have a semibandwidth of 3. There are only 17 unique terms, including the main diagonal terms plus a maximum of 3 terms to the left or right of the diagonal in each row. Often, with a slight decrease of efficiency, the symmetrical half of a band matrix is stored as a rectangular array with a size equal to the number of systems times the semibandwidth plus 1, i.e., 6×4 for Eq. (4.76).

Again, from Eqs. (4.55), (4.60), and (4.66), the assembled force vectors become

$$F_d = \Gamma_0 F_d' = \begin{bmatrix} f_{d1}^{(1)} \\ f_{d2}^i \\ f_{d3}^{(1)} + f_{d1}^{(2)} \\ f_{d4}^{(1)} + f_{d2}^{(2)} \\ f_{d3}^{(1)} \\ f_{d4}^{(2)} \end{bmatrix}, \tag{4.78}$$

$$F \equiv \begin{bmatrix} f_1 \\ m_1 \\ f_2 \\ m_2 \\ f_3 \\ m_3 \end{bmatrix} = \Gamma_0 F' = \begin{bmatrix} f_1^{(1)} \\ m_1^{(1)} \\ f_2^{(1)} + f_1^{(2)} \\ m_2^{(1)} + m_1^{(2)} \\ f_2^{(2)} \\ m_2^{(2)} \end{bmatrix}$$

$$= \begin{bmatrix} E_1 I_1 \dfrac{\partial^3 w^{(1)}(x_1, t)}{\partial x_1^3}\bigg|_{x_1=0} \\[2ex] -E_1 I_1 \dfrac{\partial^2 w^{(1)}(x_1, t)}{\partial x_1^2}\bigg|_{x_1=0} \\[2ex] -E_1 I_1 \dfrac{\partial^3 w^{(1)}(x_1, t)}{\partial x_1^3}\bigg|_{x_1=\ell_1} + E_2 I_2 \dfrac{\partial^3 w^{(2)}(x_2, t)}{\partial x_2^3}\bigg|_{x_2=0} \\[2ex] E_1 I_1 \dfrac{\partial^2 w^{(1)}(x_1, t)}{\partial x_1^2}\bigg|_{x_1=\ell_1} - E_2 I_2 \dfrac{\partial^2 w^{(2)}(x_2, t)}{\partial x_2^2}\bigg|_{x_2=0} \\[2ex] -E_2 I_2 \dfrac{\partial^3 w^{(2)}(x_2, t)}{\partial x_2^3}\bigg|_{x_2=\ell_2} \\[2ex] E_2 I_2 \dfrac{\partial^2 w^{(2)}(x_2, t)}{\partial x_2^2}\bigg|_{x_2=\ell_2} \end{bmatrix}, \tag{4.79}$$

where f_i and m_i without superscripts represent nodal force and bending moment applied at the ith node of the assembled beam. Here the quantity $w^{(1)}(x_1, t)$ represents the displacement for the first beam element and $w^{(2)}(x_2, t)$ for the second beam element. The third entry of the boundary vector F represents the sum of the internal shear forces at the end of the first element and the beginning of the second element. Similarly, the fourth entry represents the sum of the internal moments at the end of the first element and the beginning of the second element. Both sums of internal shear forces and moments must be zero to guarantee the continuity of the assembled beam if there are no external force and bending moment applied at the corresponding node. In other words, the boundary terms must cancel at the interior node if $f_2 = m_2 = 0$ (no external force and moment act on the node), i.e.,

$$E_1 I_1 \frac{\partial^3 w^{(1)}(x_1, t)}{\partial x_1^3}\bigg|_{x_1=\ell_1} = E_2 I_2 \frac{\partial^3 w^{(2)}(x_2, t)}{\partial x_2^3}\bigg|_{x_2=0}, \tag{4.80}$$

$$E_1 I_1 \frac{\partial^2 w^{(1)}(x_1, t)}{\partial x_1^2}\bigg|_{x_1=\ell_1} = E_2 I_2 \frac{\partial^2 w^{(2)}(x_2, t)}{\partial x_2^2}\bigg|_{x_2=0}. \tag{4.81}$$

EXAMPLE 4.5

As shown in Fig. 4.6, assume that the flexural rigidity and the mass density of the first element are $E_1 I_1$ and ρ_1, respectively, whereas the flexile rigidity and mass density of the second element are $E_2 I_2$ and ρ_2, respectively. If we use the interpolation functions derived in Example 4.1, the substitution of the mass matrix, stiffness matrix, and force vector derived in Example 4.2 for an individual element into Eqs. (4.76) and (4.77) produces the following assembled mass matrix M,

$$\begin{bmatrix}
\dfrac{13\rho_1\ell_1}{35} & \dfrac{11\rho_1\ell_1^2}{210} & \dfrac{9\rho_1\ell_1}{70} & \dfrac{-13\rho_1\ell_1^2}{420} & 0 & 0 \\[2ex]
\dfrac{11\rho_1\ell_1^2}{210} & \dfrac{\rho_1\ell_1^3}{105} & \dfrac{13\rho_1\ell_1^2}{420} & \dfrac{-\rho_1\ell_1^3}{140} & 0 & 0 \\[2ex]
\dfrac{9\rho_1\ell_1}{70} & \dfrac{13\rho_1\ell_1^2}{420} & \dfrac{13(\rho_1\ell_1 + \rho_2\ell_2)}{35} & \dfrac{11(-\rho_1\ell_1^2 + \rho_2\ell_2^2)}{210} & \dfrac{9\rho_2\ell_2}{70} & \dfrac{-13\rho_2\ell_2^2}{420} \\[2ex]
\dfrac{-13\rho_1\ell_1^2}{420} & \dfrac{-\rho_1\ell_1^3}{140} & \dfrac{11(-\rho_1\ell_1^2 + \rho_2\ell_2^2)}{210} & \dfrac{(\rho_1\ell_1^3 + \rho_2\ell_2^3)}{105} & \dfrac{13\rho_2\ell_2^2}{420} & \dfrac{-\rho_2\ell_2^3}{140} \\[2ex]
0 & 0 & \dfrac{9\rho_2\ell_2}{70} & \dfrac{13\rho_2\ell_2^2}{420} & \dfrac{13\rho_2\ell_2}{35} & \dfrac{-11\rho_2\ell_2^2}{210} \\[2ex]
0 & 0 & \dfrac{-13\rho_2\ell_2^2}{420} & \dfrac{-\rho_2\ell_2^3}{140} & \dfrac{-11\rho_2\ell_2^2}{210} & \dfrac{\rho_2\ell_2^3}{105}
\end{bmatrix},$$

the stiffness matrix K,

$$\begin{bmatrix}
\dfrac{12E_1 I_1}{\ell_1^3} & \dfrac{6E_1 I_1}{\ell_1^2} & \dfrac{-12E_1 I_1}{\ell_1^3} & \dfrac{6E_1 I_1}{\ell_1^2} & 0 & 0 \\[2ex]
\dfrac{6E_1 I_1}{\ell_1^2} & \dfrac{4E_1 I_1}{\ell_1} & \dfrac{-6E_1 I_1}{\ell_1^2} & \dfrac{2E_1 I_1}{\ell_1} & 0 & 0 \\[2ex]
\dfrac{-12E_1 I_1}{\ell_1^3} & \dfrac{-6E_1 I_1}{\ell_1^2} & \dfrac{12E_1 I_1}{\ell_1^3} + \dfrac{12E_2 I_2}{\ell_2^3} & \dfrac{-6E_1 I_1}{\ell_1^2} + \dfrac{6E_2 I_2}{\ell_2^2} & \dfrac{-12E_2 I_2}{\ell_2^3} & \dfrac{6E_2 I_2}{\ell_2^2} \\[2ex]
\dfrac{-6E_1 I_1}{\ell_1^2} & \dfrac{2E_1 I_1}{\ell_1} & \dfrac{-6E_1 I_1}{\ell_1^2} + \dfrac{6E_2 I_2}{\ell_2^2} & \dfrac{4E_1 I_1}{\ell_1} + \dfrac{4E_2 I_2}{\ell_2} & \dfrac{-6E_2 I_2}{\ell_2^2} & \dfrac{2E_2 I_2}{\ell_2} \\[2ex]
0 & 0 & \dfrac{-12E_2 I_2}{\ell_2^3} & \dfrac{-6E_2 I_2}{\ell_2^2} & \dfrac{12E_2 I_2}{\ell_2^3} & \dfrac{-6E_2 I_2}{\ell_2^2} \\[2ex]
0 & 0 & \dfrac{6E_2 I_2}{\ell_2^2} & \dfrac{2E_2 I_2}{\ell_2} & \dfrac{-6E_2 I_2}{\ell_2^2} & \dfrac{4E_2 I_2}{\ell_2}
\end{bmatrix},$$

and the forcing vector,

$$
F_d = \begin{bmatrix}
\dfrac{\ell_1 f_1}{2} \\[2ex]
\dfrac{\ell_1^2 f_1}{12} \\[2ex]
\dfrac{\ell_1}{2} f_1 + \dfrac{\ell_2}{2} f_2 \\[2ex]
-\dfrac{\ell_1^2}{12} f_1 + \dfrac{\ell_2^2}{12} f_2 \\[2ex]
\dfrac{\ell_2 f_2}{2} \\[2ex]
-\dfrac{\ell_2^2 f_2}{12}
\end{bmatrix}.
$$

Before we finalize the matrix equation of motion, boundary conditions must be specified. For a free–free beam without any external force and moment acting on any node, the appropriate boundary conditions are the following:

$$
\text{when } x_1 = 0, \quad E_1 I_1 \frac{\partial^2 w^{(1)}(x_1, t)}{\partial x_1^2}\bigg|_{x_1=0} = E_1 I_1 \frac{\partial^3 w^{(1)}(x_1, t)}{\partial x_1^3}\bigg|_{x_1=0} = 0,
$$

$$
\text{when } x_2 = \ell_2, \quad E_2 I_2 \frac{\partial^2 w^{(2)}(x_2, t)}{\partial x_2^2}\bigg|_{x_2=\ell_2} = E_2 I_2 \frac{\partial^3 w^{(2)}(x_2, t)}{\partial x_2^3}\bigg|_{x_2=\ell_2} = 0,
$$

which implies that

$$
F = \begin{bmatrix}
0 \\
0 \\
0 \\
0 \\
0 \\
0
\end{bmatrix}.
$$

The matrix equation of motion will then have six independent equations to be solved for six independent coordinates, i.e.,

$$
M\ddot{q} + Kq = F_d,
$$

where M and K are 6×6 symmetric matrices and F_d is a 6×1 vector.

4.4   Truss Structures

Truss and beam elements are very widely used in structural engineering to model building frames and bridges. We will show how to derive a finite-element model for a truss structure. Instead of using the constraint method presented in the last section, we will use the energy method in conjunction with Lagrange's equations to derive the equation of motion. With the application of Lagrange's equations, the equations of motion of a truss structure modeled by one or more finite elements can be derived.

Three steps are involved. First, model the truss structure as several finite truss elements. Second, perform coordinate transformation from local coordinates of each element to global coordinates. Third, add the total energy of each element to produce the total energy of the complete structure and apply Lagrange's equations to produce the dynamic equation for the structure.

4.4.1 Truss Element: Longitudinal Motion

The simplest type of truss element is a pin-connected uniform bar. Consider a uniform bar and its coordinate system, shown in Fig. 4.7. Because the nodes are pinned, no moments may exist.

Define each end of the element as a node. The element has length ℓ, x is the local coordinate along the bar, and u denotes the longitudinal displacement of points on the bar. In the absence of any longitudinal body force, the differential equation governing the longitudinal motion is

$$\rho \frac{\partial^2 u}{\partial t^2} - EA \frac{\partial^2 u}{\partial x^2} = 0, \tag{4.82}$$

where ρ is the mass per unit length, E is Young's modulus, and A is the cross-sectional area. All three quantities, ρ, E, and A, are assumed constant for the uniform bar. In the finite-element technique, the continuous variable u is approximated in terms of nodal values u_1 and u_2 through simple functions of the space variable x, i.e., interpolation functions.

Let the longitudinal displacement $u(x, t)$ be approximated by

$$u(x, t) \approx \phi_1(x)u_1(t) + \phi_2(x)u_2(t),$$

or, in matrix form,

$$u(x, t) \approx [\phi_1(x) \quad \phi_2(x)] \begin{bmatrix} u_1(t) \\ u_2(t) \end{bmatrix}$$

$$= \chi^T(x)q_u(t), \tag{4.83}$$

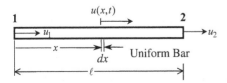

Figure 4.7. A uniform bar.

where the interpolation vector $\chi(x)$ is

$$\chi(x) = \begin{bmatrix} \phi_1(x) \\ \phi_2(x) \end{bmatrix}, \tag{4.84}$$

and the displacement vector $q_u(t)$ is

$$q_u(t) = \begin{bmatrix} u_1(t) \\ u_2(t) \end{bmatrix}. \tag{4.85}$$

The interpolation functions must satisfy the following boundary conditions:

$$\phi_1(0) = 1, \quad \phi_1(\ell) = 0, \tag{4.86}$$

$$\phi_2(0) = 0, \quad \phi_2(\ell) = 1. \tag{4.87}$$

EXAMPLE 4.6

Assume that the interpolation functions are chosen to be the first-order polynomial,

$$\phi_1(x) = c_0 + c_1 x.$$

It is easier to prove that

$$\phi_1(x) = 1 - \frac{x}{\ell}, \tag{4.88}$$

and

$$\phi_2(x) = \frac{x}{\ell} \tag{4.89}$$

satisfy Eqs. (4.86) and (4.87), respectively.

Premultiplying Eq. (4.82) by $\chi(x)$ and integrating the resulting equation over the element produces

$$\int_0^\ell \chi(x) \left(\rho \frac{\partial^2 u}{\partial t^2} - EA \frac{\partial^2 u}{\partial x^2} \right) dx = 0. \tag{4.90}$$

Applying integration by parts to the integral expression of the second term on the left-hand side yields

$$\int_0^\ell \chi(x) EA \frac{\partial^2 u(x)}{\partial x^2} dx = \chi(x) EA \frac{\partial u(x, t)}{\partial x} \Big|_0^\ell$$

$$- \int_0^\ell \frac{d\chi(x)}{dx} EA \frac{\partial u(x, t)}{\partial x} dx. \tag{4.91}$$

From Eqs. (4.86) and (4.87), we have

$$\chi(\ell) = \begin{bmatrix} 0 \\ 1 \end{bmatrix}, \qquad \chi(0) = \begin{bmatrix} 1 \\ 0 \end{bmatrix}. \tag{4.92}$$

Substituting Eqs. (4.92) into Eq. (4.91) yields

$$\int_0^\ell \chi(x) EA \frac{\partial^2 u(x,t)}{\partial x^2} dx = \begin{bmatrix} -EA \dfrac{\partial u(x,t)}{\partial x} \Big|_{x=0} \\ EA \dfrac{\partial u(x,t)}{\partial x} \Big|_{x=\ell} \end{bmatrix}$$

$$-EA \int_0^\ell \frac{d\chi(x)}{dx} \frac{\partial u(x,t)}{\partial x} dx. \tag{4.93}$$

With the insertion of Eq. (4.83) into Eq. (4.93), differential equation (4.90) for the uniform bar element becomes

$$\left[\rho \int_0^\ell \chi(x) \chi^T(x) dx \right] \frac{d^2 q_u(t)}{dt^2}$$

$$+ \left[EA \int_0^\ell \frac{d\chi(x)}{dx} \frac{d\chi^T(x)}{dx} dx \right] q_u(t) = F_u, \tag{4.94}$$

or, equivalently,

$$M_u \ddot{q}_u + K_u q_u = F_u, \tag{4.95}$$

where

$$M_u = \rho \int_0^\ell \chi(x) \chi^T(x) \, dx, \tag{4.96}$$

$$K_u = EA \int_0^\ell \frac{d\chi(x)}{dx} \frac{d\chi^T(x)}{dx} dx, \tag{4.97}$$

$$F_u \equiv \begin{bmatrix} f_{u1} \\ f_{u2} \end{bmatrix} = \begin{bmatrix} -EA \dfrac{\partial u(x,t)}{\partial x} \Big|_{x=0} \\ EA \dfrac{\partial u(x,t)}{\partial x} \Big|_{x=\ell} \end{bmatrix}. \tag{4.98}$$

Both the mass matrix and the stiffness matrix are determined from the interpolation functions chosen to describe the nodal displacements. The quantities f_{u1} and f_{u2} represent the longitudinal nodal forces acting on both ends of the bar.

EXAMPLE 4.7

Substitution of Eqs. (4.88) and (4.89) from Example 4.6 into Eq. (4.84), and then Eqs. (4.96) and (4.97) yield the following mass and stiffness matrices:

$$
M_u = \rho \int_0^\ell \chi(x) \chi^T(x) \, dx
$$

$$
= \rho \int_0^\ell \begin{bmatrix} 1 - \dfrac{x}{\ell} \\[2mm] \dfrac{x}{\ell} \end{bmatrix} \begin{bmatrix} 1 - \dfrac{x}{\ell} & \dfrac{x}{\ell} \end{bmatrix} dx \tag{4.99}
$$

$$
= \rho\ell \begin{bmatrix} \dfrac{1}{3} & \dfrac{1}{6} \\[2mm] \dfrac{1}{6} & \dfrac{1}{3} \end{bmatrix},
$$

$$
K_u = EA \int_0^\ell \frac{d\chi(x)}{dx} \frac{d\chi^T(x)}{dx} \, dx
$$

$$
= EA \int_0^\ell \begin{bmatrix} -\dfrac{1}{\ell} \\[2mm] \dfrac{1}{\ell} \end{bmatrix} \begin{bmatrix} -\dfrac{1}{\ell} & \dfrac{1}{\ell} \end{bmatrix} dx
$$

$$
= EA \begin{bmatrix} \dfrac{1}{\ell} & -\dfrac{1}{\ell} \\[2mm] -\dfrac{1}{\ell} & \dfrac{1}{\ell} \end{bmatrix}. \tag{4.100}
$$

The mass matrix M_u is symmetric and positive definite because its eigenvalues are $\rho\ell/6 > 0$ and $\rho\ell/2 > 0$ for positive ρ and ℓ. However, the stiffness matrix K_u is symmetric and positive semidefinite because its eigenvalues are $2EA/\ell$ and 0, i.e., one is positive for positive EA and another is zero. The zero eigenvalue signifies that a rigid-body motion exists. For a free–free bar, there exist indeed a rigid-body motion and an elastic motion in the longitudinal axis. If one end of the bar is clamped, then the rigid-body motion will be taken away.

4.4.2 Truss Element: Rigid-Body Motion

There are two vertical motions orthogonal to the longitudinal motion. One of the two vertical motions is shown in Fig. 4.8. The other motion is orthogonal to both x and y directions. Both vertical motions will have the same form of equations of motion if boundary conditions for both directions are identical

Because the bar is pin connected, no moment exists and thus the motion in the y direction is considered as rigid-body motion, i.e., no elastic motion is involved. When

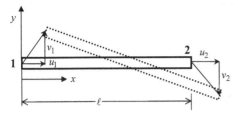

Figure 4.8. Rigid-body and elastic motions of a uniform bar.

the bar is moving, the dynamic equilibrium condition for inertial and external forces in the y direction is that the integral sum of the inertial forces must be equal to the sum of external forces, i.e.,

$$\int_0^\ell \frac{\partial^2 v}{\partial t^2} \rho \, dx = f_{v1} + f_{v2}, \tag{4.101}$$

where $v(x, t)$ is the displacement in the y direction, and f_{v1} and f_{v2} are the external forces applied at the ends of $x = 0$ and $x = \ell$, respectively. Given that $v_1 = v(0, t)$ at $x = 0$ and $v_2 = v(\ell, t)$ at $x = \ell$, the displacement $v(x, t)$ can be interpolated as

$$v(x, t) = v_1 + \frac{v_2 - v_1}{\ell} x = \left(1 - \frac{x}{\ell}\right) v_1 + \frac{x}{\ell} v_2. \tag{4.102}$$

This interpolation is exact when no elastic motion exists in the y direction. Substituting Eq. (4.102) into Eq. (4.101) produces

$$\frac{1}{2} \rho \ell (\ddot{v}_1 + \ddot{v}_2) = f_{v1} + f_{v2}. \tag{4.103}$$

This equation describes the translation motion of the bar in the y direction.

The moment equilibrium condition requires that the integral sum of the inertial moments relative to a specific point must be equal to the sum of the external torque relative to the same point. Let us choose $x = 0$ as the origin for computing the moments and torques. The moment equilibrium condition thus gives

$$\int_0^\ell x \frac{\partial^2 v}{\partial t^2} \rho \, dx = f_{v2} \ell. \tag{4.104}$$

The only external torque relative to the origin $x = 0$ is the product of the force f_{v2} applied at the end $x = \ell$ and the arm ℓ. Small displacements v_1 and v_2 are assumed in comparison with the bar length ℓ. Substitution of v from Eq. (4.102) into Eq. (4.104) produces

$$\rho \ell \left(\frac{1}{6} \ddot{v}_1 + \frac{1}{3} \ddot{v}_2\right) = f_{v2}. \tag{4.105}$$

Inserting f_{v2} from Eq. (4.105) into Eq. (4.103) and solving for f_{v1} yields

$$\rho \ell \left(\frac{1}{3} \ddot{v}_1 + \frac{1}{6} \ddot{v}_2 \right) = f_{v1}. \tag{4.106}$$

Combining Eqs. (4.105) and (4.106), we obtain

$$\rho \ell \begin{bmatrix} \dfrac{1}{3} & \dfrac{1}{6} \\ \dfrac{1}{6} & \dfrac{1}{3} \end{bmatrix} \begin{bmatrix} \ddot{v}_1 \\ \ddot{v}_2 \end{bmatrix} = \begin{bmatrix} f_{v1} \\ f_{v2} \end{bmatrix}, \tag{4.107}$$

or, in a compact form,

$$M_v \ddot{q}_v = F_v, \tag{4.108}$$

where the mass matrix M_v is defined as

$$M_v = \rho \ell \begin{bmatrix} \dfrac{1}{3} & \dfrac{1}{6} \\ \dfrac{1}{6} & \dfrac{1}{3} \end{bmatrix}, \tag{4.109}$$

and the displacement vector q_v and the force vector F_v are

$$q_v = \begin{bmatrix} v_1 \\ v_2 \end{bmatrix}, \quad F_v = \begin{bmatrix} f_{v1} \\ f_{v2} \end{bmatrix}. \tag{4.110}$$

The mass matrix M_v is symmetric and positive definite because its eigenvalues are positive for positive ρ and ℓ. It is interesting to note that the mass matrix M_v is identical to M_u, shown in Example 4.7. The interpolation function used in Example 4.7 is a linear polynomial of first order that happens to be identical to the interpolation function shown in Eq. (4.102). Equation (4.108) is the matrix equation describing the rigid-body motion. It is clear that the stiffness term is missing. No stiffness matrix is needed to describe the rigid-body motion.

To describe the complete planar motion of a pin-connected truss element, both Eqs. (4.95) and (4.108) are needed for two orthogonal directions. Combining Eqs. (4.95) and (4.108) produces

$$M \ddot{q} + K q = F, \tag{4.111}$$

where

$$q = \begin{bmatrix} q_u \\ q_v \end{bmatrix} \tag{4.112}$$

is the displacement vector for the end nodes of the truss element. The mass matrix is

$$M = \begin{bmatrix} M_u & 0_{2\times2} \\ 0_{2\times2} & M_v \end{bmatrix}, \tag{4.113}$$

where $0_{2\times2}$ is a 2×2 zero matrix. The stiffness matrix is

$$K = \begin{bmatrix} K_u & 0_{2\times2} \\ 0_{2\times2} & 0_{2\times2} \end{bmatrix}, \tag{4.114}$$

with zeros in most of the places except the upper left corner because of the elastic longitudinal motion. The force vector representing external forces applied at both nodes of the truss element is

$$F = \begin{bmatrix} F_u \\ F_v \end{bmatrix}. \tag{4.115}$$

Equation (4.111) is the matrix equation of motion for a pin-connected truss element. The motion in the longitudinal direction is completely decoupled from its vertical direction. In practice, motion coupling between these two directions may take place because of different boundary conditions. Note that the mass matrix M is symmetric and positive definite whereas the stiffness matrix is symmetric and positive semidefinite because it has at least two zero eigenvalues.

In practice, the truss element would have motion in a three-dimensional space generated by three orthogonal axes, x, y, and z. The direction x is the element longitudinal axis, whereas y and z are vertical axes orthogonal to x. If only rigid-body motion is allowed in both the y and the z directions, then Eq. (4.108) derived for the y direction is also applicable for the z direction. In other words, we will have another state vector q_w, mass matrix M_w, and force vector F_w to form another Eq. (4.108) for the z direction. As a result, Eq. (4.111) must be augmented to include the motion in the z direction. This is easily done when M_w is augmented on the lower main diagonal of the mass matrix M defined in Eq. (4.113), $0_{2\times2}$ on the lower main diagonal of the stiffness matrix K shown in Eq. (4.114), and F_w on the lower part of F in Eq. (4.115).

EXAMPLE 4.8
When the mass matrix M_u from Example 4.7 is used for the longitudinal motion and M_v from Eq. (4.109) is used for the rigid-body motion, the overall mass matrix M for the truss element becomes

$$M = \begin{bmatrix} M_u & 0_{2\times2} \\ 0_{2\times2} & M_v \end{bmatrix} = \frac{\rho\ell}{6} \begin{bmatrix} 2 & 1 & 0 & 0 \\ 1 & 2 & 0 & 0 \\ 0 & 0 & 2 & 1 \\ 0 & 0 & 1 & 2 \end{bmatrix}.$$

The mass matrix M is obviously symmetric and positive definite because its eigenvalues are $\rho\ell/6 > 0$, $\rho\ell/6 > 0$, $\rho\ell/2 > 0$, and $\rho\ell/2 > 0$ for positive ρ and ℓ.

Similarly, when the stiffness matrix K_u from Example 4.7 is used for the longitudinal motion, the overall stiffness matrix for the truss element becomes

$$K = \begin{bmatrix} K_u & 0_{2\times2} \\ 0_{2\times2} & 0_{2\times2} \end{bmatrix} = \frac{EA}{\ell} \begin{bmatrix} 1 & -1 & 0 & 0 \\ -1 & 1 & 0 & 0 \\ 0 & 0 & 0 & 0 \\ 0 & 0 & 0 & 0 \end{bmatrix}.$$

The stiffness matrix K_u is symmetric and positive semidefinite because its eigenvalues are $2EA/\ell, 0, 0$, and 0. For a free–free truss element, it has a translational motion and an elastic motion in the longitudinal axis and translational and rotational motion in the vertical axis. A total of three rigid-body motions exists, corresponding to three zero eigenvalues.

4.4.3 Coordinate Transformation

In the evaluation of the element matrices M, K, and F, we have established these matrices corresponding to the local coordinates. To determine the global assemblage matrices of the complete truss structure, a common global coordinate system must be established first for all unassembled structural elements. The choice of the global coordinate system is arbitrary.

Consider the pin-connected truss member and coordinate system shown in Fig. 4.9. Define each end of the element to be a node where two components of force and displacement in the global $(\bar{x}, \bar{y})$ coordinate may act. In the global coordinate, it has two degrees of freedom at each end instead of one. As a result, the element has four degrees of freedom. Let us assume that the change of displacement in length at both ends is small compared with its initial length and its angle of inclination remains the same.

From Fig. 4.9, the relationship between the local and the global displacements for the first node can be expressed by

$$u_1 = \bar{u}_1 \cos\alpha + \bar{v}_1 \sin\alpha, \tag{4.116a}$$

$$v_1 = -\bar{u}_1 \sin\alpha + \bar{v}_1 \cos\alpha, \tag{4.116b}$$

and, similarly, for the second node,

$$u_2 = \bar{u}_2 \cos\alpha + \bar{v}_2 \sin\alpha, \tag{4.117a}$$

$$v_2 = -\bar{u}_2 \sin\alpha + \bar{v}_2 \cos\alpha, \tag{4.117b}$$

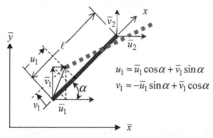

Figure 4.9. A truss member and its coordinate systems.

where α is the angle between the local axis x and the global axis $\bar{x}$, u_1 and v_1 are displacements along the local axes x and y, and $\bar{u}_1$ and $\bar{v}_1$ are displacements along the global axes $\bar{x}$ and $\bar{y}$, respectively. We obtain these relationships by resolving global displacements in the direction of the local coordinate. Equations (4.116) and (4.117) can be combined into the matrix equation

$$
\begin{bmatrix} u_1 \\ u_2 \\ v_1 \\ v_2 \end{bmatrix} = \begin{bmatrix} \cos\alpha & \sin\alpha & 0 & 0 \\ 0 & 0 & \cos\alpha & \sin\alpha \\ -\sin\alpha & \cos\alpha & 0 & 0 \\ 0 & 0 & -\sin\alpha & \cos\alpha \end{bmatrix} \begin{bmatrix} \bar{u}_1 \\ \bar{v}_1 \\ \bar{u}_2 \\ \bar{v}_2 \end{bmatrix},
\tag{4.118}
$$

or, in a compact form,

$$
q = \Gamma \bar{q}.
\tag{4.119}
$$

The transformation matrix,

$$
\Gamma = \begin{bmatrix} \cos\alpha & \sin\alpha & 0 & 0 \\ 0 & 0 & \cos\alpha & \sin\alpha \\ -\sin\alpha & \cos\alpha & 0 & 0 \\ 0 & 0 & -\sin\alpha & \cos\alpha \end{bmatrix},
\tag{4.120}
$$

is a matrix of coefficients actually obtained from the direct cosines of angles between the local coordinate and the global coordinate, i.e.,

$$
q = \begin{bmatrix} u_1 \\ u_2 \\ v_1 \\ v_2 \end{bmatrix}, \qquad \bar{q} = \begin{bmatrix} \bar{u}_1 \\ \bar{v}_1 \\ \bar{u}_2 \\ \bar{v}_2 \end{bmatrix}.
\tag{4.121}
$$

We have rearranged the positions of the global coordinates to be convenient for further analysis in the following sections. Note that the transformation matirx Γ is a unitary matrix having the following properties:

$$
\Gamma^T \Gamma = I_{4\times4}, \qquad \Gamma^T = \Gamma^{-1},
\tag{4.122}
$$

where $I_{4\times4}$ is a 4 × 4 identity matrix.

With Eq. (4.119), the matrix equation of motion shown in Eq. (4.111) can be transformed to become

$$
\bar{M}\ddot{\bar{q}} + \bar{K}\bar{q} = \bar{F},
\tag{4.123}
$$

where

$$
\bar{M} = \Gamma^T M \Gamma,
\tag{4.124}
$$

$$
\bar{K} = \Gamma^T K \Gamma,
\tag{4.125}
$$

$$
\bar{F} = \Gamma^T F.
\tag{4.126}
$$

Because Γ is unitary, this transformation does not alter the characteristics (eigenvalues) of the original matrices that are transformed.

EXAMPLE 4.9

When the system mass and stiffness matrices from Example 4.8 are used for the truss element in the local coordinates, the matrix Γ shown in Eq. (4.120) will transform them into the global coordinates to become

$$\bar{M} = \Gamma^T M \Gamma$$

$$= \frac{\rho\ell}{6} \begin{bmatrix} 2 & 0 & 1 & 0 \\ 0 & 2 & 0 & 1 \\ 1 & 0 & 2 & 0 \\ 0 & 1 & 0 & 2 \end{bmatrix},$$

$$\bar{K} = \Gamma^T K \Gamma$$

$$= \frac{EA}{\ell} \begin{bmatrix} \cos^2\alpha & \cos\alpha\sin\alpha & -\cos^2\alpha & -\cos\alpha\sin\alpha \\ \cos\alpha\sin\alpha & \sin^2\alpha & -\cos\alpha\sin\alpha & -\sin^2\alpha \\ -\cos^2\alpha & -\cos\alpha\sin\alpha & \cos^2\alpha & \cos\alpha\sin\alpha \\ -\cos\alpha\sin\alpha & -\sin^2\alpha & \cos\alpha\sin\alpha & \sin^2\alpha \end{bmatrix}.$$

Examination of the stiffness matrix $\bar{K}$ reveals that the first column distinguishes itself from the third column by only a negative sign and also the second column equals the minus of the fourth column. Furthermore, the second column can be obtained by multiplication of the first column by $\sin\alpha/\cos\alpha$. As a result, there is only one independent column in the stiffness matrix $\bar{K}$, implying that the stiffness matrix has a rank of one. Thus $\bar{K}$ possesses three zero eigenvalues.

4.5 Energy Method for Element Assembly

The energy method is based on the fact that kinetic energy and potential energy are scalar quantities. The total energy of a structure is the sum of the total energy contributions of all of its elements. For simplicity, we choose a simple plane truss structure to illustrate the concept.

The simplest and probably the most widely used truss structure is the three-node triangular one, as shown in Fig. 4.10. In the following, we will show how to determine the total energy of the truss structure. Instead of using the constraint method presented in the last section, Lagrange's equations will be used to derive the equation of motion.

The kinetic energy of the ith truss element for $i = 1, 2, 3$, denoted by $T^{(i)}$, is known to be

$$T^{(i)} = \frac{1}{2}\dot{q}^{(i)T} M^{(i)} \dot{q}^{(i)}, \tag{4.127}$$

where $q^{(i)}$ and $M^{(i)}$ are defined in Eqs. (4.112) and (4.113) for the ith truss element. As discussed in the above section, the local coordinate vector $q^{(i)}$ should be transformed

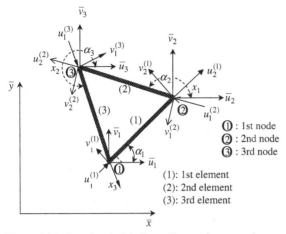

Figure 4.10. Local and global coordinates for truss-element assembly.

into the global coordinate. Substituting the following transformation equation

$$\dot{q}^{(i)} = \Gamma_i \dot{\bar{q}}^{(i)} \tag{4.128}$$

into the kinetic-energy equation yields

$$T^{(i)} = \frac{1}{2} \dot{\bar{q}}^{(i)T} \Gamma_i^T M^{(i)} \Gamma_i \dot{\bar{q}}^{(i)}, \tag{4.129}$$

where

$$\Gamma_i = \begin{bmatrix} \cos\alpha_i & \sin\alpha_i & 0 & 0 \\ 0 & 0 & \cos\alpha_i & \sin\alpha_i \\ -\sin\alpha_i & \cos\alpha_i & 0 & 0 \\ 0 & 0 & -\sin\alpha_i & \cos\alpha_i \end{bmatrix}. \tag{4.130}$$

The angles α_i for each element cannot be arbitrarily specified. They must follow a special pattern for all elements. In this development, each angle α_i is defined counterclockwise, starting from the global coordinate $\bar{x}$ to the local coordinate x_i. Note that the global coordinate vector $\bar{q}^{(i)}$ is different for a different truss element. For example, the vectors $\bar{q}^{(1)}$, $\bar{q}^{(2)}$, and $\bar{q}^{(3)}$ for the truss structure shown in Fig. 4.10 are

$$\bar{q}^{(1)} = \begin{bmatrix} \bar{u}_1 \\ \bar{v}_1 \\ \bar{u}_2 \\ \bar{v}_2 \end{bmatrix}, \qquad \bar{q}^{(2)} = \begin{bmatrix} \bar{u}_2 \\ \bar{v}_2 \\ \bar{u}_3 \\ \bar{v}_3 \end{bmatrix}, \qquad \bar{q}^{(3)} = \begin{bmatrix} \bar{u}_3 \\ \bar{v}_3 \\ \bar{u}_1 \\ \bar{v}_1 \end{bmatrix}. \tag{4.131}$$

To avoid the differences among the global vectors $\bar{q}^{(1)}$, $\bar{q}^{(2)}$, and $\bar{q}^{(3)}$, a system global vector containing all the global coordinates is defined as

$$\bar{q} = \begin{bmatrix} \bar{u}_1 \\ \bar{v}_1 \\ \bar{u}_2 \\ \bar{v}_2 \\ \bar{u}_3 \\ \bar{v}_3 \end{bmatrix}. \tag{4.132}$$

The relationship between $\bar{q}^{(1)}$ and $\bar{q}$ is obviously

$$\bar{q}^{(1)} = \begin{bmatrix} \bar{u}_1 \\ \bar{v}_1 \\ \bar{u}_2 \\ \bar{v}_2 \end{bmatrix} = \begin{bmatrix} 1 & 0 & 0 & 0 & 0 & 0 \\ 0 & 1 & 0 & 0 & 0 & 0 \\ 0 & 0 & 1 & 0 & 0 & 0 \\ 0 & 0 & 0 & 1 & 0 & 0 \end{bmatrix} \begin{bmatrix} \bar{u}_1 \\ \bar{v}_1 \\ \bar{u}_2 \\ \bar{v}_2 \\ \bar{u}_3 \\ \bar{v}_3 \end{bmatrix}, \tag{4.133}$$

or

$$\bar{q}^{(1)} = H_1 \bar{q}, \tag{4.134}$$

where

$$H_1 = \begin{bmatrix} 1 & 0 & 0 & 0 & 0 & 0 \\ 0 & 1 & 0 & 0 & 0 & 0 \\ 0 & 0 & 1 & 0 & 0 & 0 \\ 0 & 0 & 0 & 1 & 0 & 0 \end{bmatrix}. \tag{4.135}$$

Note that H_1 is a 4×6 matrix in which the number of rows is equal to the number of rows of $\bar{q}^{(1)}$ and the number of columns is equal to the number of rows of $\bar{q}$. The matrix H_1 consists of only ones and zeros. It is a common practice in numerical computation to store the locator information as a locator vector that lists the system global coordinates corresponding to the respective local global coordinates.

Similar to Eq. (4.134), two other transformations can be defined as

$$\bar{q}^{(2)} = H_2 \bar{q}, \qquad \bar{q}^{(3)} = H_3 \bar{q}, \tag{4.136}$$

with

$$H_2 = \begin{bmatrix} 0 & 0 & 1 & 0 & 0 & 0 \\ 0 & 0 & 0 & 1 & 0 & 0 \\ 0 & 0 & 0 & 0 & 1 & 0 \\ 0 & 0 & 0 & 0 & 0 & 1 \end{bmatrix}, \qquad H_3 = \begin{bmatrix} 0 & 0 & 0 & 0 & 1 & 0 \\ 0 & 0 & 0 & 0 & 0 & 1 \\ 1 & 0 & 0 & 0 & 0 & 0 \\ 0 & 1 & 0 & 0 & 0 & 0 \end{bmatrix}. \tag{4.137}$$

Substitution of transformation equation (4.134) into kinetic-energy equation (4.129) produces

$$T^{(i)} = \frac{1}{2}\dot{q}^T \left[H_i^T \Gamma_i^T M^{(i)} \Gamma_i H_i \right] \dot{q}. \tag{4.138}$$

In Eq. (4.138), only the quantity in the bracket is element dependent whereas the coordinate vector contains all the independent global coordinates. The total kinetic energy for the complete truss structure becomes

$$T = \frac{1}{2} \sum_{i=1}^{3} T^{(i)} = \frac{1}{2} \sum_{i=1}^{3} \dot{q}^{(i)T} \Gamma_i^T M^{(i)} \Gamma_i \dot{q}^{(i)}$$

$$= \frac{1}{2}\dot{q}^T \left[\sum_{i=1}^{3} H_i^T \Gamma_i^T M^{(i)} \Gamma_i H_i \right] \dot{q}. \tag{4.139}$$

By following the same procedure, we can also determine the potential energy for the complete truss structure by simply replacing $M^{(i)}$ with $K^{(i)}$ and $\dot{q}$ with $\bar{q}$, i.e.,

$$V = \frac{1}{2} \sum_{i=1}^{3} V^{(i)} = \frac{1}{2} \sum_{i=1}^{3} \bar{q}^{(i)T} \Gamma_i^T K^{(i)} \Gamma_i \bar{q}^{(i)}$$

$$= \frac{1}{2}\bar{q}^T \left[\sum_{i=1}^{3} H_i^T \Gamma_i^T K^{(i)} \Gamma_i H_i \right] \bar{q}. \tag{4.140}$$

Recall that Lagrange's equations state that equations of motion of an n-degree-of-freedom structure with coordinates $\bar{q}_i$ can be calculated from the energy of the structure by

$$\frac{\partial}{\partial t} \left(\frac{\partial T}{\partial \dot{\bar{q}}_i} \right) - \frac{\partial T}{\partial \bar{q}_i} + \frac{\partial V}{\partial \bar{q}_i} = F_i, \tag{4.141}$$

where F_i denotes the external forces along the coordinates $\bar{q}_i$. Because there is no term associated with $\bar{q}$ in the kinetic energy T shown in Eq. (4.139), the equality $\partial T / \partial \bar{q}_i = 0$ implies that the second term in Eq. (4.141) vanishes. Calculation of the various derivatives of T relative to $\dot{\bar{q}}_i$ and the derivatives of V with respect to $\bar{q}_i$ required for Lagrange's equation yields the six degrees of freedom for the three-node triangular truss structure as

$$M\ddot{q} + Kq = F, \tag{4.142}$$

where

$$M = \left[\sum_{i=1}^{3} H_i^T \Gamma_i^T M^{(i)} \Gamma_i H_i \right], \tag{4.143}$$

$$K = \left[\sum_{i=1}^{3} H_i^T \Gamma_i^T K^{(i)} \Gamma_i H_i \right]. \tag{4.144}$$

The mass matrix M must be symmetric and positive definite, whereas the stiffness matrix K is symmetric, but it may be positive definite or semidefinite, depending on the boundary conditions.

EXAMPLE 4.10

From Fig. 4.10, let us assume that the three angles α_1, α_2, and α_3 are

$$\alpha_1 = 0, \qquad \alpha_2 = 120°, \qquad \alpha_3 = 240°.$$

The transformation Γ_1 from the first-element local coordinates to the global coordinates is

$$\Gamma_1 = \begin{bmatrix} \cos\alpha_1 & \sin\alpha_1 & 0 & 0 \\ 0 & 0 & \cos\alpha_1 & \sin\alpha_1 \\ -\sin\alpha_1 & \cos\alpha_1 & 0 & 0 \\ 0 & 0 & -\sin\alpha_1 & \cos\alpha_1 \end{bmatrix}$$

$$= \begin{bmatrix} 1 & 0 & 0 & 0 \\ 0 & 0 & 1 & 0 \\ 0 & 1 & 0 & 0 \\ 0 & 0 & 0 & 1 \end{bmatrix}.$$

Similarly, the transformation matrices Γ_2 and Γ_3 are

$$\Gamma_2 = \begin{bmatrix} \cos\alpha_2 & \sin\alpha_2 & 0 & 0 \\ 0 & 0 & \cos\alpha_2 & \sin\alpha_2 \\ -\sin\alpha_2 & \cos\alpha_2 & 0 & 0 \\ 0 & 0 & -\sin\alpha_2 & \cos\alpha_2 \end{bmatrix}$$

$$= \begin{bmatrix} -\dfrac{1}{2} & \dfrac{\sqrt{3}}{2} & 0 & 0 \\ 0 & 0 & -\dfrac{1}{2} & \dfrac{\sqrt{3}}{2} \\ -\dfrac{\sqrt{3}}{2} & -\dfrac{1}{2} & 0 & 0 \\ 0 & 0 & -\dfrac{\sqrt{3}}{2} & -\dfrac{1}{2} \end{bmatrix},$$

$$\Gamma_3 = \begin{bmatrix} \cos\alpha_3 & \sin\alpha_3 & 0 & 0 \\ 0 & 0 & \cos\alpha_3 & \sin\alpha_3 \\ -\sin\alpha_3 & \cos\alpha_3 & 0 & 0 \\ 0 & 0 & -\sin\alpha_3 & \cos\alpha_3 \end{bmatrix}$$

$$= \begin{bmatrix} -\dfrac{1}{2} & -\dfrac{\sqrt{3}}{2} & 0 & 0 \\ 0 & 0 & -\dfrac{1}{2} & -\dfrac{\sqrt{3}}{2} \\ \dfrac{\sqrt{3}}{2} & -\dfrac{1}{2} & 0 & 0 \\ 0 & 0 & \dfrac{\sqrt{3}}{2} & -\dfrac{1}{2} \end{bmatrix}.$$

Assume that all truss elements are identical in material properties, i.e., identical flexural rigidity and mass density. If a linear polynomial is used for the interpolation functions, then the mass and the stiffness matrices derived in Example 4.8 in the local coordinates for the identical truss elements are

$$
M^{(i)} = \frac{\rho \ell}{6} \begin{bmatrix} 1 & 2 & 0 & 0 \\ 2 & 1 & 0 & 0 \\ 0 & 0 & 1 & 2 \\ 0 & 0 & 2 & 1 \end{bmatrix}, \quad i = 1, 2, 3,
$$

$$
K^{(i)} = \frac{EA}{\ell} \begin{bmatrix} 1 & -1 & 0 & 0 \\ -1 & 1 & 0 & 0 \\ 0 & 0 & 0 & 0 \\ 0 & 0 & 0 & 0 \end{bmatrix}, \quad i = 1, 2, 3,
$$

where ρ is the total mass density per length of each truss element, ℓ is the element length, and EA is the flexural rigidity. The mass and the stiffness matrices for the three-node triangular truss structure in the global coordinates are

$$
M = H_1^T \Gamma_1^T M^{(1)} \Gamma_1 H_1 + H_2^T \Gamma_2^T M^{(2)} \Gamma_2 H_2 + H_3^T \Gamma_3^T M^{(3)} \Gamma_3 H_3
$$

$$
= \frac{\rho \ell}{6} \begin{bmatrix} 4 & 0 & 1 & 0 & 1 & 0 \\ 0 & 4 & 0 & 1 & 0 & 1 \\ 1 & 0 & 4 & 0 & 1 & 0 \\ 0 & 1 & 0 & 4 & 0 & 1 \\ 1 & 0 & 1 & 0 & 4 & 0 \\ 0 & 1 & 0 & 1 & 0 & 4 \end{bmatrix},
$$

$$
K = H_1^T \Gamma_1^T K^{(1)} \Gamma_1 H_1 + H_2^T \Gamma_2^T K^{(2)} \Gamma_2 H_2 + H_3^T \Gamma_3^T K^{(3)} \Gamma_3 H_3
$$

$$
= \frac{EA}{\rho} \begin{bmatrix}
\dfrac{5}{4} & \dfrac{\sqrt{3}}{4} & -1 & 0 & \dfrac{-1}{4} & \dfrac{-\sqrt{3}}{4} \\[2mm]
\dfrac{\sqrt{3}}{4} & \dfrac{3}{4} & 0 & 0 & \dfrac{-\sqrt{3}}{4} & \dfrac{-3}{4} \\[2mm]
-1 & 0 & \dfrac{5}{4} & \dfrac{-\sqrt{3}}{4} & \dfrac{-1}{4} & \dfrac{\sqrt{3}}{4} \\[2mm]
0 & 0 & \dfrac{-\sqrt{3}}{4} & \dfrac{3}{4} & \dfrac{\sqrt{3}}{4} & \dfrac{-3}{4} \\[2mm]
\dfrac{-1}{4} & \dfrac{-\sqrt{3}}{4} & \dfrac{-1}{4} & \dfrac{\sqrt{3}}{4} & \dfrac{1}{2} & 0 \\[2mm]
\dfrac{-\sqrt{3}}{4} & \dfrac{-3}{4} & \dfrac{\sqrt{3}}{4} & \dfrac{-3}{4} & 0 & \dfrac{3}{2}
\end{bmatrix}.
$$

The eigenvalues of the mass matrix M are $\rho\ell$, $\rho\ell$, $\rho\ell/2$, $\rho\ell/2$, $\rho\ell/2$, and $\rho\ell/2$, which are all positive, and thus M is positive definite. The eigenvalues of K are $3EA/\ell$, $3EA/2\ell$, $3EA/2\ell$, 0, 0, and 0, and thus K is positive semidefinite. Computation of eigenvalues of matrices M and K gives an initial check to see if the formations of M and K are correct. This check is a necessary condition but not a sufficient condition.

4.6 Concluding Remarks

We have briefly introduced the finite-element method. We first showed how to define the element of a uniform beam and specify the boundary conditions of its interpolation functions. A constraint method was introduced to perform the beam assembly. We then discussed the basic truss elements, including longitudinal motion and rigid-body motion. The energy method was introduced for truss-element assembly. Several examples were given to illustrate the concepts described in this chapter.

The accuracy of the finite-element method depends on the size of the elements and the choice of the interpolation functions. For smaller elements, the interpolation functions can be as simple as a linear first-order polynomial. The smaller the size is, the larger the number of the elements will be to meet the accuracy requirements. With the power of today's computers, finite-element models commonly involve thousands of degrees of freedom.

The finite-element method works with small spatial domains over which the interpolation functions are defined. It is capable of handling a complex structure with nonuniform parameter distributions and irregular geometries. This advantage makes the finite-element method capable of producing solutions where other methods fail. That is why the method has become very popular in the engineering communities for use in system modeling.

4.7 Problems

4.1   Assume that the interpolation function $\phi_1(x)$ for a beam undergoing bending deformation is given by

$$\phi_1(x) = c_1 \cos \beta x + c_2 \sin \beta x + c_3 \cosh \beta x + c_4 \sinh \beta x,$$

where β is an arbitrary constant. The appropriate boundary conditions for the function are given in Eqs. (4.5).

(a) Compute the coefficients c_1, c_2, c_3, and c_4.
(b) Under what conditions are c_1, c_2, c_3, and c_4 not unique?
(c) Use the same interpolation function and solve for $\varphi_1(x)$, $\phi_2(x)$, and $\varphi_2(x)$ by using the boundary conditions shown in Eqs. (4.6)–(4.8).

4.2   Use the third-order polynomial as the interpolation functions to determine the generalized force vector F_d defined in Eq. (4.30) for the triangular force distribution shown in Fig. 4.11, where the maximum force f is a constant value.

4.3   Consider the uniform beam shown in Fig. 4.2 with two forces f_1 and f_2 acting at the ends $x = 0$ and $\text{x} = \ell$, respectively. The kinetic energy for pure bending of a uniform

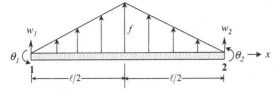

Figure 4.11. A uniform beam with triangular force distribution.

beam is given by

$$T = \frac{1}{2} \int_0^\ell \rho \left[\frac{\partial w(x,t)}{\partial t} \right]^2 dx,$$

and the strain (potential) energy is

$$V = \frac{1}{2} \int_0^\ell EI \left[\frac{\partial^2 w(x,t)}{\partial x^2} \right]^2 dx.$$

Let the bending displacement $w(x,t)$ be approximated by

$$w(x) \approx \chi^T(x) q(t)$$

$$\equiv [\phi_1(x) \quad \varphi_1(x) \quad \phi_2(x) \quad \varphi_2(x)] \begin{bmatrix} w_1(t) \\ \theta_1(t) \\ w_2(t) \\ \theta_2(t) \end{bmatrix},$$

where $\chi(x)$ is a 4×1 column vector containing interpolation functions and $q(t)$ is another 4×1 column vector including displacements and rotation angles at both ends. Use the following Lagrange's equation:

$$\frac{\partial}{\partial t} \left(\frac{\partial T}{\partial \dot{q}_i} \right) - \frac{\partial T}{\partial q_i} + \frac{\partial V}{\partial q_i} = F_i.$$

(a) Derive the equation of motion in the matrix form

$$M\ddot{q} + Kq = F.$$

(b) Prove that the mass matrix M and the stiffness matrix K have the same expressions as shown in Eqs. (4.28) and (4.29).

4.4 Use the energy method to derive Eq. (4.71) for the beam assembly shown in Fig. 4.6. First, write the kinetic energy and potential energy for each beam element. Second, add the total energy of each element to produce the total energy of the assembled beam. Third, apply Lagrange's equations to derive the matrix equation of motion.

4.5 A uniform beam with mass density ρ and rigidity EI is divided into two elements of length $\ell/2$, as shown in Fig. 4.12. Both ends of the beam are simply supported (pinned). The boundary conditions for this case are

$$w(0,t) = 0, \quad \left. \frac{\partial^2 w(x,t)}{\partial x^2} \right|_{x=0} = 0, \quad w(\ell,t) = 0, \quad \left. \frac{\partial^2 w(x,t)}{\partial x^2} \right|_{x=\ell} = 0.$$

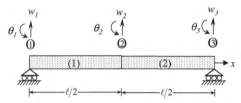

Figure 4.12. A uniform beam with two elements.

The nodes are numbered as shown in Fig. 4.12.

(a) Compute the element mass and stiffness matrices.
(b) Calculate the beam mass and stiffness matrices.

4.6 Repeat Problem 4.5 but change the end at $x = 0$ to be clamped. The boundary conditions for this case are

$$w(0, t) = 0, \quad \left.\frac{\partial w(x, t)}{\partial x}\right|_{x=0} = 0, \quad w(\ell, t) = 0, \quad \left.\frac{\partial^2 w(x, t)}{\partial x^2}\right|_{x=\ell} = 0.$$

(a) Compute the element mass and stiffness matrices.
(b) Calculate the beam mass and stiffness matrices.

4.7 A uniform rod undergoing torsional deformation is shown in Fig. 4.13, where τ_1 and τ_2 are torques applied at $x = 0$ and $x = \ell$, respectively. The kinetic energy and the potential (strain) energy for pure torsion are given by

$$T = \frac{1}{2} \int_0^\ell \rho \left[\frac{\partial \theta(x, t)}{\partial t}\right]^2 dx,$$

$$V = \frac{1}{2} \int_0^\ell GJ \left[\frac{\partial \theta(x, t)}{\partial x}\right]^2 dx,$$

where ρ is the mass density per unit length and GJ is the torsional rigidity. Let the rotation along the element be approximated by

$$w(x) \approx \varphi_1(x)\theta_1(t) + \varphi_2(x)\theta_2(t),$$

where $\varphi_1(x)$ and $\varphi_2(x)$ are interpolation functions and $\theta_1(t)$ and $\theta_2(t)$ are rotation angles at $x = 0$ and $x = \ell$, respectively.

(a) Derive the matrix equation of motion by using Lagrange's equations.
(b) What are the appropriate boundary conditions for the interpolation functions?
(c) Derive the interpolation functions $\varphi_1(x)$ and $\varphi_2(x)$ in terms of a first-order polynomial and compare with the interpolation functions for the longitudinal motion shown in Example 4.6.
(d) Compute mass and stiffness matrices in terms of the first-order polynomial.

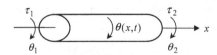

Figure 4.13. A uniform rod undergoing torsional deformation.

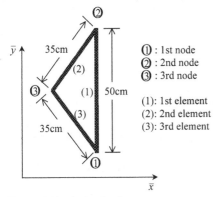

Figure 4.14. A simple plane truss.

4.8 The motion of the uniform rod shown in Fig. 4.13 is governed by the partial differential equation of motion,

$$\rho\frac{\partial^2\theta}{\partial t^2} - GJ\frac{\partial^2\theta}{\partial x^2} = \tau(x,t),$$

where $\tau(x,t)$ is the distributed torque acting on the rod. Use the finite-element method introduced in Subsection 4.4.1 to derive the matrix equation of motion.

4.9 Three bars are connected, as shown in Fig. 4.14, to form a plane truss. The nodes and elements are labeled as indicated on the figure. The global axes $\bar{x}$ and $\bar{y}$ are defined.

(a) What are the transformation matrices defined in Eq. (4.130) for elements (1), (2), and (3)?

(b) What are the matrices H_1, H_2, and H_3 defined in Eqs. (4.134) and (4.136)?

(c) Compute the individual kinetic energy and the potential energy for the elements (1), (2), and (3) in terms of the system global coordinates, assuming that the mass density ρ and the rigidity EA are identical for all three elements.

(d) Compute the mass and the stiffness matrices for the plane truss.

BIBLIOGRAPHY

[1] Bathe, K.-J., *Finite Element Procedures in Engineering Analysis*, Prentice-Hall, Englewood Cliffs, NJ, 1982.

[2] Craig, R. R., *Structural Dynamics*, Wiley, New York, 1981.

[3] Huebuer, K. H. and Thornton, E. A., *The Finite Element Method for Engineering*, Wiley, New York, 1982.

[4] Meirovitch, L., *Principles and Techniques of Vibration*, Prentice-Hall, Upper Saddle River, NJ, 1997.

[5] Przemieniecki, J. S., *Theory of Matrix Structural Analysis*, McGraw-Hill, New York, 1968.

[6] Smith, I. M., *Programming the Finite Element Method with Application to Geomechanics*, Wiley, New York, 1982.

5

Response of Dynamic Systems

5.1 Introduction

In this chapter, the response of linear dynamic systems is studied in more detail. We begin with a study of efficient techniques for solving first-order and second-order single-degree-of-freedom (SDOF) equations of motion. Taking advantage of the simplicity of the first-order equation, we show how to obtain analytical expressions for the system response to harmonic excitation and how to write the steady-state response in terms of the magnitude and phase difference. It leads to the definition of the frequency-response function (FRF) and its analytical expression in terms of a complex variable (Ref. [1]). We then show how two- or more degree-of-freedom equations of motion can be treated similarly. The focus will be on multiple DOF mass–spring–damper systems (lumped parameter systems). The last part of this chapter deals with the bending vibration of flexible beams, which are described by partial differential equations rather than by ordinary differential equations for mass–spring–damper systems (Ref. [2]). The partial differential equations are decoupled by the separation of variable method to yield two sets of ordinary differential equations. From these equations, the common features of flexible structures are introduced, including the natural modes, frequency equations, and modal amplitudes. A discussion of the orthogonality property of the natural modes and the expansion theorem is then presented.

5.2 Single Degree of Freedom

A system of considerable importance in vibrations is the mass–spring–damper system, shown in Fig. 5.1. The differential equation of motion can be derived from Newton's law as

$$m\frac{d^2w}{dt^2} + \zeta\frac{dw}{dt} + kw = f, \tag{5.1}$$

where m, ζ, and k are the mass, damping, and stiffness coefficients, respectively. The variable w represents the displacement of the mass m relative to its equilibrium position, and f is the external force applied to the system. Although simple in form, this equation is one of the most important in vibration analysis and control because even the most complex model describing the linear vibration of a physical system can often be expressed in this form, in which the scalar coefficients are replaced with matrix coefficients. For the moment, we focus our attention on this SDOF system.

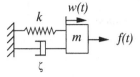

Figure 5.1. A mass–spring–damper system.

5.2.1 First-Order Systems

Consider the special case in which the system is so light that it is practically massless, $m = 0$. The governing equation is

$$\zeta \frac{dw}{dt} + kw = f. \tag{5.2}$$

The dynamic characteristics of the above first-order system can be revealed by a study of its response to harmonic excitation:

$$\zeta \frac{dw}{dt} + kw = kA \cos \omega t, \tag{5.3}$$

where A is a scalar constant. Dividing the equation by ζ, we obtain

$$\frac{dw}{dt} + aw = aA \cos \omega t, \tag{5.4}$$

where

$$a = \frac{k}{\zeta} = \frac{1}{\tau}.$$

The parameter τ is called the time constant of the first-order system. It is easy to solve for the response of such a system. The general solution is the sum of the homogeneous solution and a particular solution. The homogeneous solution is simply

$$w_h(t) = ce^{-at}. \tag{5.5}$$

The particular solution takes the form

$$w_p(t) = c_1 \sin \omega t + c_2 \cos \omega t. \tag{5.6}$$

Substituting the above candidate solution into the differential equation yields

$$ac_1 - \omega c_2 = 0,$$
$$\omega c_1 + ac_2 = aA, \tag{5.7}$$

from which we can find

$$c_1 = \frac{Aa\omega}{a^2 + \omega^2}, \quad c_2 = \frac{Aa^2}{a^2 + \omega^2}. \tag{5.8}$$

Thus the general response is

$$w(t) = ce^{-at} + \frac{Aa}{a^2 + \omega^2}(\omega \sin \omega t + a \cos \omega t). \tag{5.9}$$

Observe that as time progresses, the homogeneous solution decays to zero. For this reason it is called the transient solution. After sufficient time has elapsed, what remains is the particular solution. For this reason the particular solution is known as the steady-state solution. In the steady state, we are now left with

$$w(t) = \frac{Aa}{a^2 + \omega^2}(\omega \sin \omega t + a \cos \omega t). \tag{5.10}$$

The steady-state solution can be conveniently expressed in another form:

$$w(t) = W(\omega)\cos(\omega t - \phi), \tag{5.11}$$

where the amplitude $W(\omega)$ and the phase angle $\phi(\omega)$ can be shown to be

$$W(\omega) = \frac{A}{\sqrt{1 + (\omega/a)^2}},$$

$$\phi = \tan^{-1}\left(\frac{\omega}{a}\right). \tag{5.12}$$

Note that both the amplitude $W(\omega)$ and the phase angle $\phi(\omega)$ are functions of the excitation frequency ω. Thus the steady-state response can be expressed in a form similar to that of the harmonic forcing function in that it has the same frequency but a different amplitude and phase angle. As an example, Fig. 5.2 shows a plot of the input signal $\cos(\omega t)$ and output signal $0.75\cos(\omega t - \phi)$. The two signals have the same frequency ω (hence the same period $2\pi/\omega$), and their amplitude ratio is $1/0.75$. The phase ϕ causes the output signal to lag behind the input signal by an time interval ϕ/ω seconds. Visually it may appear that the output signal is ahead of the input signal, but in fact it is lagging behind because the horizontal axis measures time, not position. Observe that a peak in the input signal will not cause a peak in the output until ϕ/ω seconds later, thus the output is said to lag behind the input.

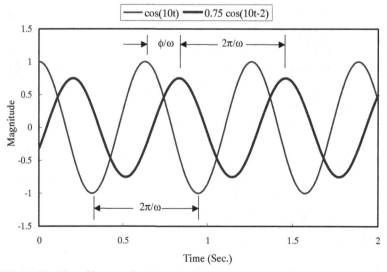

Figure 5.2. Plot of input and output signals.

COMPLEX-VARIABLE APPROACH

The steady-state response to harmonic excitation can be determined with the complex-variable approach. Because $Aa\cos(\omega t)$ is the real part of $Aae^{i\omega t}$, the steady-state response is the real part of the solution to the following complex-variable problem:

$$\frac{dw_c}{dt} + aw_c = Aae^{i\omega t}. \tag{5.13}$$

The steady-state solution (particular solution) to Eq. (5.13) is known to be in the form

$$w_c(t) = W_c(i\omega)e^{i\omega t}. \tag{5.14}$$

Substituting Eq. (5.14) into Eq. (5.13) and solving yields

$$W_c(i\omega) = \frac{Aa}{a + i\omega} = \frac{A}{1 + i(\omega/a)}. \tag{5.15}$$

Let $W_c(i\omega)$ be expressed in the form

$$W_c(i\omega) = \frac{A}{\sqrt{1 + (\omega/a)^2}} e^{-i\phi}, \tag{5.16}$$

where

$$\phi = \tan^{-1}\left(\frac{\omega}{a}\right).$$

The complex solution, Eq. (5.14), can then be rewritten as

$$w_c(t) = W(\omega)e^{i(\omega t - \phi)}, \qquad W(\omega) = \frac{A}{\sqrt{1 + (\omega/a)^2}}. \tag{5.17}$$

Note that the quantities $W(\omega)$ and ϕ are real numbers. The steady-state response to our original physical problem is just the real part of the above complex solution, Eq. (5.17), which is

$$w(t) = W(\omega)\cos(\omega t - \phi),$$

where the amplitude $W(\omega)$ and the phase angle ϕ are exactly the same as shown in Subsection 5.2.1.

FREQUENCY-RESPONSE FUNCTION

The FRF is frequently used in the fields of controls and system identification to describe the dynamic characteristics of the system input and output relationship. Here we briefly show what it means for a first-order system subject to a harmonic excitation. The magnitude of the FRF is the ratio of the steady-state response amplitude and the input amplitude of a sinusoid, i.e.,

$$G(\omega) = \frac{W(\omega)}{aA}$$

$$= \frac{1}{\sqrt{a^2 + \omega^2}}, \tag{5.18}$$

where the last equality is obtained with the insertion of $W(\omega)$ defined in Eq. (5.17). The phase of the FRF is

$$\phi = \tan^{-1}\left(\frac{\omega}{a}\right). \tag{5.19}$$

Together they are referred to as the FRF. The FRF for first-order systems is a function of driving frequency ω and the system parameter a. The FRF give us information about the magnitude and the phase of a steady–state response when the system is driven by harmonic input signals of different frequencies. Such a FRF can also be determined directly from input and output data as shown in Ref. [1].

EXAMPLE 5.1

The amplitude and the phase of the FRF for the first-order system,

$$\frac{dw}{dt} + w = \alpha \cos(\omega t),$$

are

$$G(\omega) = \frac{1}{\sqrt{1+\omega^2}} \quad \text{and} \quad \phi = \tan^{-1}\omega,$$

respectively. Note that the constant α has no influence on the computation of the FRF. Figure 5.3 shows the plots of $G(\omega)$ and ϕ with ω from 0.1 to 10 rad/s. In Fig. 5.3, the excitation frequency ω is shown in radians per second. Commonly,

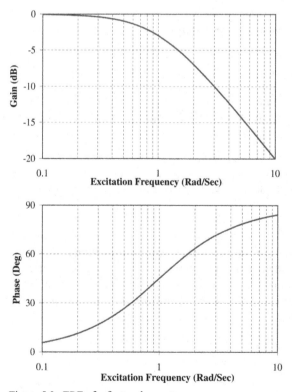

Figure 5.3. FRF of a first-order system.

these are given in hertz (Hz, cycles per second), which is obtained by dividing ω by 2π, i.e., ω (rad/s) $= 2\pi\omega$ (Hz). The magnitude (or gain) is given in decibels (dB). This is a common unit to denote an amplitude ratio, defined as $20\log_{10}$ of the ratio of output-to-input amplitudes, i.e.,

$$G \text{ (dB)} = 20\log_{10}(G).$$

Here, the phase is given in degrees rather than radians, i.e.,

$$\phi \text{ (deg)} = \phi(\text{rad})\frac{180}{\pi}.$$

There is no particular preference as to which unit should be used.

5.2.2 Second-Order Systems

We now extend the results obtained for first-order systems in the previous section to second-order systems. The mass–spring–damper system shown in Fig. 5.1 is a typical second-order system. The system subject to a harmonic excitation can be mathematically expressed by

$$m\frac{d^2 w}{dt^2} + \zeta\frac{dw}{dt} + kw = kA\cos\omega t. \tag{5.20}$$

The reason for writing the forcing input as $kA\cos\omega t$ is that it allows us to simplify later expressions. Dividing both sides of Eq. (5.20) by m yields

$$\frac{d^2 w}{dt^2} + 2\zeta_n\omega_n\frac{dw}{dt} + \omega_n^2 x = \omega_n^2 A\cos\omega t, \tag{5.21}$$

where ω_n is called the natural frequency and ζ_n is the damping factor:

$$\zeta_n = \frac{\zeta}{2m\omega_n}, \qquad \omega_n = \sqrt{\frac{k}{m}}$$

The general solution is the sum of the homogeneous (or transient) solution and the particular solution. When damping is present in the system, the homogeneous solution will decay with time. In the steady state, only the particular solution remains. As before, assume that the particular solution takes the form

$$w_p(t) = c_1\sin\omega t + c_2\cos\omega t.$$

Substituting this candidate solution into Eq. (5.21) yields

$$(\omega_n^2 - \omega^2)(c_1\sin\omega t + c_2\cos\omega t) + 2\zeta_n\omega/\omega_n(c_1\cos\omega t - c_2\sin\omega t)$$
$$= \omega_n^2 A\cos\omega t. \tag{5.22}$$

Equating the coefficients of $\sin\omega t$ and $\cos\omega t$ and solving for c_1 and c_2 yields

$$c_1 = \frac{2A\zeta_n(\omega/\omega_n)}{[1 - (\omega/\omega_n)^2]^2 + (2\zeta_n\omega/\omega_n)^2},$$

$$c_2 = \frac{[1 - (\omega/\omega_n)^2]A}{[1 - (\omega/\omega_n)^2]^2 + (2\zeta_n\omega/\omega_n)^2}. \tag{5.23}$$

The steady-state response is thus

$$w(t) = \frac{A}{[1 - (\omega/\omega_n)^2]^2 + (2\zeta_n\omega/\omega_n)^2}$$
$$\times \left\{ \frac{2\zeta_n\omega}{\omega_n} \sin \omega t + [1 - (\omega/\omega_n)^2] \cos \omega t \right\} \tag{5.24}$$

The steady-state response can be written in compact form as

$$w(t) = W(\omega) \cos(\omega t - \phi), \tag{5.25}$$

where the steady-state amplitude is

$$W(\omega) = \frac{A}{\left\{ [1 - (\omega/\omega_n)^2]^2 + (2\zeta_n\omega/\omega_n)^2 \right\}^{1/2}}, \tag{5.26}$$

and the steady-state phase angle is

$$\phi(\omega) = \tan^{-1} \frac{2\zeta_n(\omega/\omega_n)}{1 - (\omega/\omega_n)^2}. \tag{5.27}$$

From Eq. (5.26), the amplitude of the FRF for a second-order system subject to harmonic excitation is

$$G(\omega) = \frac{W(\omega)}{\omega_n^2 A}$$

$$= \frac{1}{\omega_n^2 \left\{ [1 - (\omega/\omega_n)^2]^2 + (2\zeta_n\omega/\omega_n)^2 \right\}^{1/2}}$$

$$= \frac{1}{\left[(\omega_n^2 - \omega^2)^2 + 4\zeta_n^2\omega^2\omega_n^2 \right]^{1/2}}, \tag{5.28}$$

and its phase angle $\phi(\omega)$ is already given in Eq. (5.27).

EXAMPLE 5.2

The FRF of the second-order system

$$\frac{d^2w}{dt^2} + 0.02\frac{dw}{dt} + w = \cos \omega t$$

includes the amplitude from Eq. (5.28),

$$G(\omega) = \frac{1}{\left[(1 - \omega^2)^2 + (0.02)^2\omega^2 \right]^{1/2}},$$

and the phase angle from Eq. (5.27),

$$\phi(\omega) = \tan^{-1} \frac{0.02\omega}{1 - \omega^2}.$$

Both quantities

$$G \text{ (dB)} = 20 \log[G(\omega)], \qquad \phi \text{ (deg)} = \phi(\omega)\frac{180}{\pi}$$

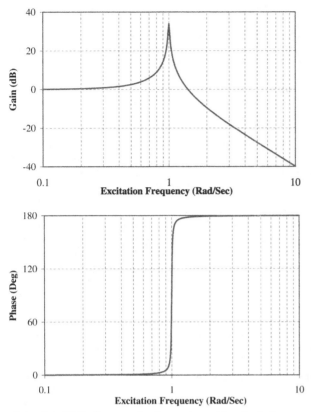

Figure 5.4. FRF of a second-order system with a single frequency.

are plotted in Fig. 5.4, with the excitation frequency ω from 0.1 to 10 rad/s. There is a peak amplitude near the frequency point $\omega = 1$ rad/s. In practice, the FRF can also be computed directly from input and output data. The system natural frequencies can then be approximately identified from the peaks of the amplitude of the computed FRF.

The phase angle changes rapidly from $0°$ to $180°$ near $\omega = 1$ rad/s because of the sign switch of $1 - \omega^2$ in the neighborhood of $\omega = 1$. The sign of $1 - \omega^2$ is positive when $\omega < 1$. It becomes negative for $\omega > 1$.

5.3 Two-Degrees-of-Freedom Systems

We now consider the equation of motion for two-degree-of-freedom systems and see how it differs from that of a SDOF system. Consider a two-degree-of-freedom system shown in Fig. 5.5 where k_1, k_2, and k_3 are the spring stiffness coefficients, ζ_1, ζ_2, and ζ_3 are the damping coefficients, m_1 and m_2 are the masses, $u_1(t)$ and $u_2(t)$ are the forces externally applied to the system, and $w_1(t)$ and $w_2(t)$ are the positions of the two masses measured from their equilibrium positions. Note that we have used $u_1(t)$ and $u_2(t)$ to denote the applied forces rather than $f_1(t)$ and $f_2(t)$. This notation change is made to be consistent with the generally adopted convention in modern control theory of using u to denote the input to the system.

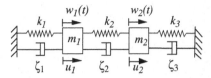

Figure 5.5. Two-degree-of-freedom system.

By writing Newton's law for each of the masses we have

$$m_1\ddot{w}_1(t) = u_1(t) - \zeta_1\dot{w}_1(t) - k_1 w_1(t) + \zeta_2[\dot{w}_2(t) - \dot{w}_1(t)] + k_2[w_2(t) - w_1(t)],$$

$$m_2\ddot{w}_2(t) = u_2(t) - \zeta_3\dot{w}_2(t) - k_3 w_2(t) - \zeta_2[\dot{w}_2(t) - \dot{w}_1(t)] - k_2[w_2(t) - w_1(t)],$$

which can be arranged in the form

$$m_1\ddot{w}_1(t) + (\zeta_1 + \zeta_2)\dot{w}_1(t) - \zeta_2\dot{w}_2(t) + (k_1 + k_2)w_1(t) - k_2 w_2(t) = u_1(t),$$

$$m_2\ddot{w}_2(t) - \zeta_2\dot{w}_1(t) + (\zeta_2 + \zeta_3)\dot{w}_2(t) - k_2 w_1(t) + (k_2 + k_3)w_2(t) = u_2(t). \quad (5.29)$$

We can further simplify the above set of equations by putting them in matrix form:

$$M\ddot{w} + \Xi\dot{w} + Kw = u, \tag{5.30}$$

where

$$w(t) = \begin{bmatrix} w_1(t) \\ w_2(t) \end{bmatrix}, \qquad u(t) = \begin{bmatrix} u_1(t) \\ u_2(t) \end{bmatrix},$$

and the mass, damping, and stiffness matrices are defined as

$$M = \begin{bmatrix} m_1 & 0 \\ 0 & m_2 \end{bmatrix}, \quad \Xi = \begin{bmatrix} \zeta_1 + \zeta_2 & -\zeta_2 \\ -\zeta_2 & \zeta_2 + \zeta_3 \end{bmatrix}, \quad K = \begin{bmatrix} k_1 + k_2 & -k_2 \\ -k_2 & k_2 + k_3 \end{bmatrix},$$

respectively. Thus, for the system considered, the equation of motion of a two-degree-of-freedom system looks very much like that of a SDOF system. Instead of having scalar coefficients, we now have matrix coefficients. Also, observe that the mass, damping, and stiffness matrices are symmetric. In fact, a general two-degree-of-freedom system has the same general structure in that it has a mass matrix, a damping matrix, and a stiffness matrix, all of which are symmetric.

RESPONSE TO HARMONIC EXCITATION

We now investigate the response of a general two-degree-of-freedom system to harmonic excitation. Let us denote the mass, damping, and stiffness matrices by their elements

$$M = \begin{bmatrix} m_{11} & m_{12} \\ m_{12} & m_{22} \end{bmatrix}, \qquad \Xi = \begin{bmatrix} \zeta_{11} & \zeta_{12} \\ \zeta_{12} & \zeta_{22} \end{bmatrix}, \qquad K = \begin{bmatrix} k_{11} & k_{12} \\ k_{12} & k_{22} \end{bmatrix}.$$

The equation of motion in expanded form looks like

$$m_{11}\ddot{w}_1 + m_{12}\ddot{w}_2 + \zeta_{11}\dot{w}_1 + \zeta_{12}\dot{w}_2 + k_{11}w_1 + k_{12}w_2 = u_1(t),$$

$$m_{12}\ddot{w}_1 + m_{22}\ddot{w}_2 + \zeta_{12}\dot{w}_1 + \zeta_{22}\dot{w}_2 + k_{12}w_1 + k_{22}w_2 = u_2(t). \tag{5.31}$$

Let us consider the following harmonic excitation,

$$u_1(t) = U_1 e^{i\omega t}, \qquad u_2(t) = U_2 e^{i\omega t}, \tag{5.32}$$

and correspondingly write the steady-state response as

$$w_1(t) = W_1 e^{i\omega t}, \qquad w_2(t) = W_2 e^{i\omega t}, \tag{5.33}$$

where W_1 and W_2 are generally complex amplitudes. Substituting the expressions for $w_1(t)$ and $w_2(t)$ into Eqs. (5.31) produces

$$(-\omega^2 m_{11} + i\omega\zeta_{11} + k_{11})W_1 + (-\omega^2 m_{12} + i\omega\zeta_{12} + k_{12})W_2 = U_1,$$
$$(-\omega^2 m_{12} + i\omega\zeta_{12} + k_{12})W_1 + (-\omega^2 m_{22} + i\omega\zeta_{22} + k_{22})W_2 = U_2, \tag{5.34}$$

from which W_1 and W_2 can be solved:

$$W_1(\omega) = \frac{Z_{22}(\omega)U_1 - Z_{12}(\omega)U_2}{Z_{11}(\omega)Z_{22}(\omega) - Z_{12}^2(\omega)},$$

$$W_2(\omega) = \frac{-Z_{12}(\omega)U_1 - Z_{11}(\omega)U_2}{Z_{11}(\omega)Z_{22}(\omega) - Z_{12}^2(\omega)}, \tag{5.35}$$

where

$$Z_{11}(\omega) = -\omega^2 m_{11} + i\omega\zeta_{11} + k_{11},$$
$$Z_{12}(\omega) = -\omega^2 m_{12} + i\omega\zeta_{12} + k_{12},$$
$$Z_{22}(\omega) = -\omega^2 m_{22} + i\omega\zeta_{22} + k_{22}. \tag{5.36}$$

Again, $W_1(\omega)$ and $W_2(\omega)$ are generally complex numbers, which have both amplitudes and phase angles. The amplitudes and the phases of $W_1(\omega)$ and $W_2(\omega)$ give us the FRF for the two-degree-of-freedom system given by Eqs. (5.31) subject to harmonic excitation.

EXAMPLE 5.3

Consider a two-degree-of-freedom system with the mass, damping, and stiffness matrices,

$$M = \begin{bmatrix} 1 & 0 \\ 0 & 1 \end{bmatrix}, \qquad \Xi = \begin{bmatrix} 0.02 & -0.01 \\ -0.01 & 0.02 \end{bmatrix}, \qquad K = \begin{bmatrix} 2 & -1 \\ -1 & 2 \end{bmatrix},$$

that correspond to the system shown in Fig. 5.5 with $m_1 = m_2 = 1$, $\zeta_1 = \zeta_2 = \zeta_3 = 0.01$, and $k_1 = k_2 = k_3 = 1$. From Eqs. (5.36), we obtain

$$Z_{11}(\omega) = -\omega^2 + 0.02i\omega + 2,$$
$$Z_{12}(\omega) = -0.01i\omega - 1,$$
$$Z_{22}(\omega) = -\omega^2 + 0.02i\omega + 2.$$

Substituting these values into Eqs. (5.35) produces

$$W_1(\omega) = G_{11}(\omega)U_1 + G_{12}(\omega)U_2,$$
$$W_2(\omega) = G_{21}(\omega)U_1 + G_{22}(\omega)U_2,$$

where

$$G_{11}(\omega) = \frac{-\omega^2 + 0.02i\omega + 2}{(\omega^2 - 0.01i\omega - 1)(\omega^2 - 0.03i\omega - 3)},$$

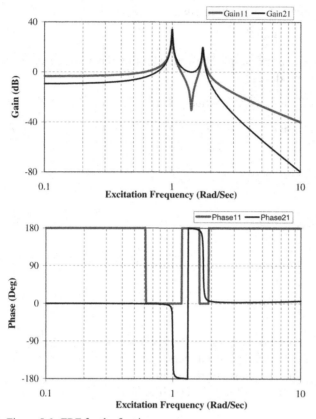

Figure 5.6. FRF for the first input.

$$G_{12}(\omega) = \frac{0.01i\omega + 1}{(\omega^2 - 0.01i\omega - 1)(\omega^2 - 0.03i\omega - 3)},$$

$$G_{21}(\omega) = \frac{0.01i\omega + 1}{(\omega^2 - 0.01i\omega - 1)(\omega^2 - 0.03i\omega - 3)},$$

$$G_{22}(\omega) = \frac{-\omega^2 + 0.02i\omega + 2}{(\omega^2 - 0.01i\omega - 1)(\omega^2 - 0.03i\omega - 3)}.$$

The quantity $G_{jk}(\omega)$ $(j, k = 1, 2)$ is the FRF mapping from the kth input u_k to the jth output w_j. For example, $G_{11}(\omega)$ is the FRF corresponding to the first input u_1 and the first output w_1. Figure 5.6 shows the amplitude (upper plot) and phase angle (lower plot) of $G_{11}(\omega)$ and $G_{21}(\omega)$. The thick curves give the amplitude and the phase angle of $G_{11}(\omega)$. Its amplitude has two positive peaks, with a negative peak between them. The positive peaks occur approximately at $\omega \approx 1$ and $\omega \approx \sqrt{3}$, because the denominator of $G_{11}(\omega)$ is close to zero with ω at these values. The positive peaks are commonly called the poles of the system. The negative peak occurs approximately at $\omega \approx \sqrt{2}$ and is called the zero of the system. The phase angle of $G_{11}(\omega)$ stays either at $0°$ or $180°$ for most of the frequency points.

The thin curves in Fig. 5.6 give the amplitude and the phase angle of $G_{21}(\omega)$. It also has two poles at the same locations as the $G_{11}(\omega)$ because both $G_{11}(\omega)$

and $G_{21}(\omega)$ share the same poles. However, there is no zero for the amplitude of $G_{21}(\omega)$ that can also be observed from its phase plot. The phase angle of $G_{21}(\omega)$ stays either at $0°$ or $\pm180°$ for most of the frequency points.

In this example, $G_{22}(\omega)$ is identical to $G_{11}(\omega)$ and $G_{12}(\omega)$ is the same as $G_{21}(\omega)$. For a multiple-input multiple-output system, the FRFs are considered for each input–output pair at a time. In general, they are all different in amplitude and phase angle but share the same poles.

5.4 Multiple-Degrees-of-Freedom Systems

The generalization of a two-degree-of-freedom system to a general multiple-degree-of-freedom (MDOF) system is quite trivial. The general structure of the mathematical equations is identical for a linear dynamic system that has any number of degrees of freedom. For a general MDOF system, the equations of motion take the following form:

$$M\ddot{w} + \Xi\dot{w} + Kw = B_f u, \tag{5.37}$$

where M, Ξ, and K are called the mass, damping, and stiffness matrices, respectively. In general, M, Ξ, and K are symmetric matrices for flexible structures. The matrix B_f is called the input influence matrix to account for the situation in which the number of inputs is not equal to the number of degrees of freedom. If M is not diagonal, then the equations of motion are coupled through the inertial forces. If Ξ is not diagonal, then the equations are coupled through the damping forces. If K is not diagonal, then the equations are coupled through the elastically restoring forces. The system of equations can be transformed to any coordinates that are convenient. For example, let us consider the linear transformation

$$w = Tq \quad \text{or} \quad q = T^{-1}w \tag{5.38}$$

that transforms the vector w into the vector q. Equation (5.37) becomes

$$MT\ddot{q} + \Xi T\dot{q} + KTq = B_f u. \tag{5.39}$$

Multiplying both sides of Eq. (5.39) by T^T yields

$$\mathcal{M}\ddot{q} + \mathcal{C}\dot{q} + \mathcal{K}q = \mathcal{B}_f u, \tag{5.40}$$

where

$$\mathcal{M} = T^T MT, \quad \mathcal{C} = T^T \Xi T, \quad \mathcal{K} = T^T KT, \quad \mathcal{B}_f = T^T B_f$$

are new mass, damping, stiffness, and input matrices, respectively, in the new coordinates. Note that the form of the equation is still exactly the same as before.

Assume that there is a transformation matrix such that

$$\mathcal{M} = T^T MT = \mathcal{I}, \quad \mathcal{K} = T^T KT = \Omega^2, \tag{5.41}$$

where $\mathcal{I}$ is an identity matrix and Ω^2 is a diagonal matrix with diagonal elements $\omega_1^2, \omega_1^2, \dots, \omega_n^2$. Introduce a column force vector $\bar{u}(t)$ associated with the coordinate vector q and related to the force vector $u(t)$ by

$$\bar{u}(t) = \mathcal{B}_f u(t). \tag{5.42}$$

In view of Eqs. (5.41) and (5.42), Eq. (5.40) can be written as

$$\ddot{q} + C\dot{q} + \Omega^2 q = \bar{u}. \tag{5.43}$$

Consider a special case where the damping matrix Ξ is proportional to the mass matrix M and the stiffness matrix K such that

$$\Xi = \alpha M + \beta K. \tag{5.44}$$

The new damping matrix C becomes

$$C = T^T \Xi T = \alpha \mathcal{I} + \beta \Omega^2, \tag{5.45}$$

where α and β are two arbitrary scalars. Equation (5.43) is rewritten as

$$\ddot{q} + (\alpha \mathcal{I} + \beta \Omega^2)\dot{q} + \Omega^2 q = \bar{u}, \tag{5.46}$$

which represents a set of n uncoupled differential equations of the type

$$\ddot{q}_j + \left(\alpha + \beta \omega_j^2\right)\dot{q}_j + \omega_j^2 q_j = \bar{u}_j, \qquad j = 1, 2, \ldots, n. \tag{5.47}$$

This is precisely the same form as that of the differential equation that describes the motion of a SDOF mass–spring–damper system, which was discussed earlier in this chapter. The uncoupled equation, Eq. (5.47), is considerably easier to solve for response than the original coupled equation, Eq. (5.37). For the case in which $\alpha = \beta = 0$, i.e., no damping matrix, Eq. (5.47) reduces to

$$\ddot{q}_j + \omega_j^2 q_j = \bar{u}_j, \qquad j = 1, 2, \ldots, n. \tag{5.48}$$

This equation can be easily solved by

$$q_j(t) = \frac{1}{\omega_j} \int_0^t \bar{u}_j(t) \sin \omega_j(t - \tau)\, d\tau + q_{j0} \cos \omega_j t + \frac{\dot{q}_{j0}}{\omega_j} \sin \omega_j t, \tag{5.49}$$

where q_{j0} and $\dot{q}_{j0}$ are the initial conditions of $q_j(t)$ and $\dot{q}_j(t)$ at time $t = 0$, respectively. The initial conditions $q(0)$ and $\dot{q}(0)$ are obtained from

$$q(0) = T^{-1} w(0), \qquad \dot{q}(0) = T^{-1} \dot{w}(0). \tag{5.50}$$

The approach of uncoupling the equation of motion, Eq. (5.37), is commonly referred to as modal analysis. The quantities ω_i for $i = 1, 2, \ldots, n$ are called system natural frequencies. The transformation matrix T that satisfies Eqs. (5.41) is the modal matrix and the quantities $q_1, q_2, \ldots, q_n$ are the modal coordinates. The ith column of T is the mode shape associated with the ith natural frequency ω_i. The forces $\bar{u}_j$ for $j = 1, 2, \ldots, n$ are called generalized forces.

EXAMPLE 5.4

Consider the system shown in Fig. 5.7, which consists of discrete masses connected by springs in series. Let w_i denote the displacement of the mass m_i, let k_i represent the ith spring constant, and let x_i give the position of the mass m_i in the x coordinate. From

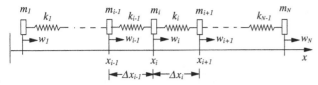

Figure 5.7. A mass–spring–damper system.

Newton's second law, the equations of motion for all masses m_i $(i = 1, 2, \ldots, N)$ are

$$m_1 \frac{d^2 w_1}{dt^2} = -K_1(w_1 - w_2),$$

$$m_i \frac{d^2 w_i}{dt^2} = -K_{i-1}(w_i - w_{i-1}) - K_i(w_i - w_{i+1}),$$

$$m_N \frac{d^2 w_N}{dt^2} = -K_{N-1}(w_N - w_{N-1}).$$

These equations may be written in matrix form as

$$M\ddot{w} + Kw = 0,$$

where

$$w = \begin{bmatrix} w_1 \\ \vdots \\ w_{i-1} \\ w_i \\ w_{i+1} \\ \vdots \\ w_N \end{bmatrix}, \quad M = \begin{bmatrix} m_1 & & & & & \\ & \ddots & & & & \\ & & m_{i-1} & & & \\ & & & m_i & & \\ & & & & m_{i+1} & \\ & & & & & \ddots \\ & & & & & & m_N \end{bmatrix},$$

$$K = \begin{bmatrix} K_1 & & & & \\ & \ddots & & & \\ & K_{i-2} + K_{i-1} & -K_{i-1} & & \\ & -K_{i-1} & K_{i-1} + K_i & -K_i & \\ & & -K_i & K_i + K_{i+1} & \\ & & & & \ddots \\ & & & & & K_N \end{bmatrix}.$$

Let us assume that there is a sufficiently large number $(N \gg 1)$ of springs and that the distance between any two springs is sufficiently small $(\Delta x_i \ll 1)$. Now rewrite the equation of motion for the ith mass such that

$$m_i \frac{d^2 w_i}{dt^2} + (K_{i-1}\Delta x_{i-1})\frac{\Delta w_{i-1}}{\Delta x_{i-1}} - (K_i \Delta x_i)\frac{\Delta w_i}{\Delta x_i} = 0,$$

where

$$\Delta w_{i-1} = w_i - w_{i-1}, \qquad \Delta w_i = w_{i+1} - w_i.$$

Define

$$m_i = \rho(x_i)\Delta x_{i-1}, \qquad EA(x_{i-1}) = K_{i-1}\Delta x_{i-1}, \qquad EA(x_i) = K_i\Delta x_i.$$

The significance of these quantities will be explained later. The equation of motion for the ith mass becomes

$$\rho(x_i)\Delta x_{i-1}\frac{d^2 w_i}{dt^2} + EA(x_{i-1})\frac{\Delta w_{i-1}}{\Delta x_{i-1}} - EA(x_i)\frac{\Delta w_i}{\Delta x_i} = 0$$

or

$$\rho(x_i)\frac{d^2 w_i}{dt^2} + \frac{1}{\Delta x_{i-1}}\left[EA(x_{i-1})\frac{\Delta w_{i-1}}{\Delta x_{i-1}} - EA(x_i)\frac{\Delta w_i}{\Delta x_i}\right] = 0.$$

Taking the limit $\Delta x_{i-1} \to 0$ and $\Delta x_i \to 0$ and noting that the displacement w becomes continuous in the spatial coordinate x in the limit, we find that the above equation becomes

$$\rho(x_i)\frac{\partial^2 w}{\partial t^2}\bigg|_{x=x_i} + \frac{1}{\Delta x_{i-1}}\left[EA(x_{i-1})\frac{\partial w}{\partial x}\bigg|_{x=x_{i-1}} - EA(x_i)\frac{\partial w}{\partial x}\bigg|_{x=x_i}\right] = 0,$$

and furthermore

$$\rho(x_i)\frac{\partial^2 w}{\partial t^2}\bigg|_{x=x_i} - \frac{\partial}{\partial x}EA(x)\frac{\partial w}{\partial x}\bigg|_{x=x_{i-1}} = 0.$$

Note that we obtain the above equation by using following approximation:

$$EA(x_i)\frac{\partial w}{\partial x}\bigg|_{x=x_i} = EA(x_{i-1})\frac{\partial w}{\partial x}\bigg|_{x=x_{i-1}} + \frac{\partial}{\partial x}EA(x)\frac{\partial w}{\partial x}\bigg|_{x=x_{i-1}}\Delta x_{i-1}.$$

Because the variable x_i is arbitrary, the equation of motion in the limit becomes

$$\rho(x)\frac{\partial^2 w}{\partial t^2} = \frac{\partial}{\partial x}\left[EA(x)\frac{\partial w}{\partial x}\right].$$

This is the differential equation of motion for the longitudinal vibration of a thin rod, as shown in Fig. 5.8. The quantity $w(x, t)$ is the axial displacement, $\rho(x)$ is the mass per unit length, and $EA(x)$ represents the axial stiffness of the thin rod. The stiffness $EA(x)$ has the unit of force in which E is the modulus of elasticity and $A(x)$ is the area of the cross section at x.

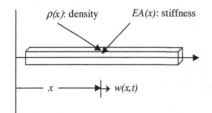

$\rho(x)$: density $EA(x)$: stiffness

$x \longrightarrow$ $w(x,t)$

Figure 5.8. A thin rod.

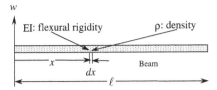

Figure 5.9. A uniform cantilevered beam.

Example 5.4 shows that an elastic thin rod can be considered as an infinite number of springs and mass connected in series. To specify the position of every point in the rod, an infinite number of displacement coordinates is required. Such a system is commonly said to have an infinite number of degrees of freedom. Because its mass and stiffness are distributed, it is also called a distributed parameter system. The dynamic response of a distributed parameter system will be discussed in the following section.

5.5 Bending Vibration of Beams

One of the simplest examples of a continuous (distributed parameter) system is the flexible beam. Let w represent the bending displacement at the location x, ρ the mass density per unit length, and EI the flexural rigidity of the beam. In the particular case in which EI does not vary with x, the equation for free vibration of the beam shown in Fig. 5.9 is (see Ref. [2])

$$EI\left(\frac{\partial^4 w}{\partial x^4}\right) = -\rho\frac{\partial^2 w}{\partial t^2}. \tag{5.51}$$

Assume that the solution of Eq. (5.51) is separable in time and space, i.e.,

$$w(x, t) = \phi(x)q(t). \tag{5.52}$$

Substituting Eq. (5.52) into Eq. (5.51) thus yields

$$\frac{EI}{\rho\phi}\left(\frac{d^4\phi}{dx^4}\right) = -\frac{1}{q}\frac{d^2q}{dt^2}. \tag{5.53}$$

The left-hand side of Eq. (5.53) depends on time x only, whereas the right-hand side depends on t only. Both x and t are independent variables. Equation (5.53) holds only if both sides are equal to a constant, i.e.,

$$\frac{EI}{\rho\phi}\left(\frac{d^4\phi}{dx^4}\right) = -\frac{1}{q}\frac{d^2q}{dt^2} = \omega^2, \tag{5.54}$$

where ω is a real value. A positive value ω^2 is chosen to meet the physical constraint that the bending vibration should be bounded. It will be shown later that the constant ω must satisfy some equation when boundary conditions of a beam are specified. Equation (5.54) produces two differential equations:

$$\frac{d^4\phi}{dx^4} - \beta^4\phi = 0, \qquad \beta^4 = \frac{\omega^2\rho}{EI}; \tag{5.55}$$

$$\frac{d^2q}{dt^2} + \omega^2 q = 0. \tag{5.56}$$

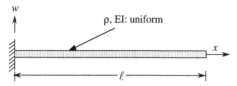

Figure 5.10. Clamped–free beam.

Equation (5.55) is a fourth-order homogeneous ordinary differential equation. The solution of Eq. (5.55) is

$$\phi(x) = c_1 \sin \beta x + c_2 \cos \beta x + c_3 \sinh \beta x + c_4 \cosh \beta x, \tag{5.57}$$

where the constant coefficients c_1, c_2, c_3, and c_4 and the parameter β are to be determined by four (two at each end of the beam) boundary conditions.

FREQUENCY EQUATION

Let us consider the case in which one end of the beam is clamped and the other end is free, as shown in Fig. 5.10. At the clamped end, $x = 0$, the bending displacement and its slope are zero, and thus the boundary conditions are

$$\phi(0) = 0, \tag{5.58}$$

$$\left.\frac{d\phi(x)}{dx}\right|_{x=0} = 0. \tag{5.59}$$

At the free end, $x = \ell$, the external moment and shear force are zero, and thus the boundary conditions are

$$\left.\frac{d^2\phi(x)}{dx^2}\right|_{x=\ell} = 0, \tag{5.60}$$

$$\left.\frac{d^3\phi(x)}{dx^3}\right|_{x=\ell} = 0. \tag{5.61}$$

Equations (5.58)–(5.61) are called natural boundary conditions. Substituting these boundary conditions into Eq. (5.57) produces the following four equations:

$$c_2 + c_4 = 0, \tag{5.62}$$

$$c_1 + c_3 = 0, \tag{5.63}$$

$$-c_1 \sin \beta\ell - c_2 \cos \beta\ell + c_3 \sinh \beta\ell + c_4 \cosh \beta\ell = 0, \tag{5.64}$$

$$-c_1 \cos \beta\ell + c_2 \sin \beta\ell + c_3 \cosh \beta\ell + c_4 \sinh \beta\ell = 0. \tag{5.65}$$

Solving for c_4 and c_3 in terms of c_2 and c_1 from Eqs. (5.62) and (5.63), respectively, and inserting them into Eqs. (5.64) and (5.65) yield

$$c_2 = -c_1 \frac{\sin \beta\ell + \sinh \beta\ell}{\cos \beta\ell + \cosh \beta\ell}, \tag{5.66}$$

$$c_2 = c_1 \frac{\cos \beta\ell + \cosh \beta\ell}{\sin \beta\ell - \sinh \beta\ell}. \tag{5.67}$$

From Eqs. (5.66) and (5.67), the nonzero coefficients c_1 and c_2 exist only when

$$-\frac{\sin \beta\ell + \sinh \beta\ell}{\cos \beta\ell + \cosh \beta\ell} = \frac{\cos \beta\ell + \cosh \beta\ell}{\sin \beta\ell - \sinh \beta\ell},$$

which gives the equality

$$\cos \beta\ell \cosh \beta\ell = -1. \tag{5.68}$$

Equation (5.68) is commonly referred to as the characteristic equation or frequency equation of a cantilevered uniform beam. It has an infinite number of solutions β_k for $k = 1, 2, \ldots, \infty$ that can be obtained only numerically. The first three values of $\beta\ell$ and their corresponding frequency formulas from Eq. (5.55) are

$$\beta_1\ell = 1.875 \implies \omega_1 = 1.875^2 \sqrt{\frac{EI}{\rho\ell^4}},$$

$$\beta_2\ell = 4.694 \implies \omega_2 = 4.694^2 \sqrt{\frac{EI}{\rho\ell^4}},$$

$$\beta_3\ell = 7.855 \implies \omega_3 = 7.855^2 \sqrt{\frac{EI}{\rho\ell^4}}. \tag{5.69}$$

The quantities ω_k $(k = 1, 2, \ldots, \infty)$ are called characteristic values or eigenvalues. They are also known as natural frequencies.

NATURAL MODES (MODE SHAPES)

For each β_k there corresponds a function ϕ_k, as shown in Eq. (5.57), with c_2, c_3, and c_4 expressed in terms of c_1 and β_k, i.e.,

$$\phi_k(x) = c_{1k}[(\sin \beta_k x - \sinh \beta_k x) + \frac{\cos \beta_k\ell + \cosh \beta_k\ell}{\sin \beta_k\ell - \sinh \beta_k\ell}(\cos \beta_k x - \cosh \beta_k x)],$$

$$k = 1, 2, \ldots, \infty \tag{5.70}$$

Here the constant c_{1k} is used to reflect the fact that c_1 has a different value for a different β_k. Equation (5.70) is commonly written as

$$\phi_k(x) = A_k[(\sin \beta_k\ell - \sinh \beta_k\ell)(\sin \beta_k x - \sinh \beta_k x)$$
$$+ (\cos \beta_k\ell + \cosh \beta_k\ell)(\cos \beta_k x - \cosh \beta_k x)],$$

$$k = 1, 2, \ldots, \infty, \tag{5.71}$$

where

$$A_k = \frac{c_{1k}}{\sin \beta_k\ell - \sinh \beta_k\ell}$$

is an arbitrary constant. The functions $\phi_k(x)$ for $k = 1, 2, \ldots, \infty$ are called the natural modes or eigenfunctions. It can be easily proved that the functions $\phi_k(x)$ are orthogonal in the sense that

$$\int_0^\ell \rho\phi_j(x)\phi_k(x) = 0 \quad \text{when} \quad j \neq k, \tag{5.72}$$

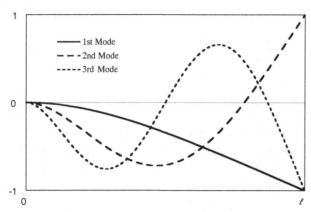

Figure 5.11. Beam mode shapes.

where ρ is, in general, a function of x rather than a constant in our case. This property is very useful when these orthogonal functions are used to discretize a distributed parameter system.

The first three natural modes are plotted in Fig. 5.11, where the coefficients A_k for $k = 1, 2, 3$ are normalized such that the natural modes $\phi_k(x)$ at $x = \ell$ are one, i.e,

$$A_k = \frac{1}{-2 \sin \beta_k \ell \; \sinh \beta_k \ell}.$$

This normalization is performed for the purpose of demonstration.

MODAL COORDINATES

For each β_k there corresponds a frequency ω_k that satisfies the differential equation

$$\frac{d^2 q_k}{dt^2} + \omega_k^2 q_k = 0 \tag{5.73}$$

for $k = 1, 2, \ldots, \infty$. The solution of the differential equation is

$$q_k(t) = q_{k0} \cos \omega_k t + \frac{\dot{q}_{k0}}{\omega_k} \sin \omega_k t, \tag{5.74}$$

where q_{k0} and $\dot{q}_{k0}$ are the initial conditions of $q_k(t)$ and $\dot{q}_k(t)$ at time $t = 0$. The quantity q_k is called the kth modal coordinate associated with the kth natural frequency ω_k and the natural mode ϕ_k. The time history of $q_k(t)$ gives the modal amplitude of the mode ϕ_k with the oscillatory frequency ω_k.

Because there is an infinite number of ω_k, the bending displacement $w(x, t)$ shown in Eq. (5.52) should be written as

$$w(x, t) = \sum_{k=1}^{N} \phi_k(x) q_k(t), \quad \text{with} \quad N \to \infty, \tag{5.75}$$

which includes all natural modes. In practice, the integer N is a finite number that represents the number of modes that sufficiently describe the bending vibration of the beam under excitation. Let us assume that $w(x_i, 0) = w_{i0}$ and $\dot{w}(x_i, 0) = \dot{w}_{i0}$ for $i = 1, 2, \ldots, n$ are given at time $t = 0$. From Eq. (5.75), we obtain the following matrix equation:

$$
\begin{bmatrix}
w_{10} & \dot{w}_{10} \\
w_{20} & \dot{w}_{20} \\
\vdots & \vdots \\
w_{n0} & \dot{w}_{n0}
\end{bmatrix}
=
\begin{bmatrix}
\phi_1(x_1) & \phi_2(x_1) & \cdots & \phi_N(x_1) \\
\phi_1(x_2) & \phi_2(x_2) & \cdots & \phi_N(x_2) \\
\vdots & \vdots & \ddots & \vdots \\
\phi_1(x_n) & \phi_2(x_n) & \cdots & \phi_N(x_n)
\end{bmatrix}
\begin{bmatrix}
q_{10} & \dot{q}_{10} \\
q_{20} & \dot{q}_{20} \\
\vdots & \vdots \\
q_{N0} & \dot{q}_{N0}
\end{bmatrix},
\tag{5.76}
$$

where q_{i0} and $\dot{q}_{i0}$ for $i = 1, 2 \ldots, N$ are the initial conditions of $q_i(t)$ and $\dot{q}_i(t)$ at time $t = 0$. The quantities q_{i0} and $\dot{q}_{i0}$ for $i = 1, 2 \ldots, N$ can then be solved by

$$
\begin{bmatrix}
q_{10} & \dot{q}_{10} \\
q_{20} & \dot{q}_{20} \\
\vdots & \vdots \\
q_{N0} & \dot{q}_{N0}
\end{bmatrix}
=
\begin{bmatrix}
\phi_1(x_1) & \phi_2(x_1) & \cdots & \phi_N(x_1) \\
\phi_1(x_2) & \phi_2(x_2) & \cdots & \phi_N(x_2) \\
\vdots & \vdots & \ddots & \vdots \\
\phi_1(x_n) & \phi_2(x_n) & \cdots & \phi_N(x_n)
\end{bmatrix}^{\dagger}
\begin{bmatrix}
w_{10} & \dot{w}_{10} \\
w_{20} & \dot{w}_{20} \\
\vdots & \vdots \\
w_{n0} & \dot{w}_{n0}
\end{bmatrix},
\tag{5.77}
$$

where $\dagger$ means pseudoinverse. When $n < N$, there is an infinite number of solutions and Eq. (5.77) gives the the minimum-norm solution. When $n > N$, Eq. (5.77) gives the the least-squares solution, which does not satisfy Eq. (5.76). The unique solution occurs only when $n = N$. It becomes clear that the number of modes must be chosen large enough to obtain a solution for initial modal amplitudes. It should be noted that the initial modal amplitudes and velocities should be recomputed when one changes the normalization of natural modes.

To the limit, let us further assume that the initial conditions $w(x, 0)$ and $\dot{w}(x, 0)$ are specified continuously along the spatial coordinate x at time $t = 0$. Then Eq. (5.75) becomes

$$
w(x, 0) = \sum_{k=1}^{N} \phi_k(x) q_k(0),
\tag{5.78}
$$

$$
\dot{w}(x, 0) = \sum_{k=1}^{N} \phi_k(x) \dot{q}_k(0).
\tag{5.79}
$$

Multiplying both side of Eq. (5.78) by $\rho \phi_j(x)$ and integrating from $x = 0$ to $x = \ell$ yields

$$
\int_0^\ell \rho \phi_j(x) w(x, 0) dx = \int_0^\ell \sum_{k=1}^{N} \rho \phi_j(x) \phi_k(x) q_k(0) \, dx
$$

$$
= \sum_{k=1}^{N} \int_0^\ell \rho \phi_j(x) \phi_k(x) \, dx \, q_k(0).
\tag{5.80}
$$

Now, introducing the property of orthogonality, Eq. (5.72), into Eq. (5.80), we obtain

$$
q_k(0) = \frac{\int_0^\ell \rho \phi_k(x) w(x, 0) \, dx}{\int_0^\ell \rho \phi_k(x) \phi_k(x) \, dx}.
\tag{5.81}
$$

If the eigenfunctions are normalized such that

$$\int_0^\ell \rho\phi_k(x)\phi_k(x)\,dx = 1,$$

then Eq. (5.81) becomes

$$q_k(0) = \int_0^\ell \rho\phi_k(x)w(x,0)\,dx. \tag{5.82}$$

Using the similar procedure from Eqs. (5.80)–(5.82) for Eq. (5.79), we obtain an equation similar to Eq. (5.82):

$$\dot{q}_k(0) = \int_0^\ell \rho\phi_k(x)\dot{w}(x,0)\,dx. \tag{5.83}$$

FORCED BENDING RESPONSE OF UNIFORM BEAMS

Now consider a distributing force $f(x,t)$ applied to the beam, as shown in Fig. 5.12. From elementary flexural theory, the partial differential equation of motion is

$$\rho\frac{\partial^2 w}{\partial t^2} + EI\left(\frac{\partial^4 w}{\partial x^4}\right) = f(x,t). \tag{5.84}$$

For the clamped beam, as shown in Fig. 5.12, the boundary conditions are

$$w(0,t) = 0, \qquad \left.\frac{\partial w(x,t)}{\partial x}\right|_{x=0} = 0,$$

$$\left.\frac{\partial^2 w(x,t)}{\partial x^2}\right|_{x=\ell} = 0, \qquad \left.\frac{\partial^3 w(x,t)}{\partial x^3}\right|_{x=\ell} = 0.$$

These boundary conditions must be satisfied when we are solving for $w(x,t)$ from Eq. (5.84).

Free vibrations of the beam were presented in the previous section. The natural modes (eigenfunctions) for the free vibration are given by

$$\phi_k(x) = A_k[(\sin\beta_k\ell - \sinh\beta_k\ell)(\sin\beta_k x - \sinh\beta_k x)$$
$$+ (\cos\beta_k\ell + \cosh\beta_k\ell)(\cos\beta_k x - \cosh\beta_k x)],$$
$$k = 1, 2, \ldots, \infty.$$

Any bending displacement $w(x,t)$ can be obtained by superposition of displacements corresponding to eigenfunctions of bending vibration. Thus bending vibrations caused

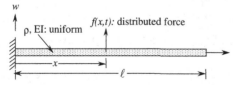

Figure 5.12. Clamped–free beam with a distributing force.

by the distributing force $f(x, t)$ can be represented by the series

$$w(x, t) = \phi_1(x)q_1(t) + \phi_2(x)q_2(t) + \cdots + \phi_N(x)q_N(t)$$

$$= \sum_{k=1}^{N} \phi_k(x)q_k(t), \quad \text{with} \quad N \to \infty, \tag{5.85}$$

where q_k $(k = 1, 2, \ldots, \infty)$ are some unknown functions of time to be determined. In the case of free vibration (i.e., no distributing force), these functions are determined by Eq. (5.74) and initial conditions at time $t = 0$.

To find these functions for the case of forced vibrations, first introduce Eq. (5.85) into Eq. (5.84) to produce

$$\rho \sum_{k=1}^{N} \phi_k(x) \frac{d^2 q_k(t)}{dt^2} + EI \sum_{k=1}^{N} \frac{d^4 \phi_k(x)}{dx^4} q_k(t) = f(x, t). \tag{5.86}$$

Now multiply Eq. (5.86) by $\phi_j(x)$ and integrate over the length of the beam to obtain

$$\sum_{k=1}^{N} \frac{d^2 q_k(t)}{dt^2} \int_0^\ell \rho \phi_j(x)\phi_k(x) \, dx + \sum_{k=1}^{N} q_k(t) \int_0^\ell EI\phi_j(x) \frac{d^4 \phi_k(x)}{dx^4} \, dx$$

$$= \int_0^\ell \phi_j(x) f(x, t) \, dx. \tag{5.87}$$

Observe that the eigenfunctions are solved from Eq. (5.55), i.e.,

$$EI \frac{d^4 \phi_j(x)}{dx^4} - \omega_j^2 \rho \phi_j(x) = 0,$$

and that they are orthogonal and can be normalized such that

$$\int_0^\ell \rho \phi_j(x)\phi_k(x) \, dx = 0 \quad \text{for} \quad j \neq k,$$

$$\int_0^\ell \rho \phi_j(x)\phi_j(x) \, dx = 1 \quad \text{for} \quad j = 1, 2, \ldots, N.$$

Substituting the above three equations into Eq. (5.87), we obtain

$$\frac{d^2 q_j}{dt^2} + \omega_j^2 q_j = f_j(t), \tag{5.88}$$

where

$$f_j(t) = \int_0^\ell \phi_j(x) f(x, t) \, dx, \quad j = 1, 2, \ldots, N,$$

is a generalized force associated with the coordinate $q_j(t)$. For $N \to \infty$, Eq. (5.88) represents an infinite set of uncoupled ordinary equations similar to the SDOF equation. The solution is

$$q_j(t) = \frac{1}{\omega_j} \int_0^t f_j(t) \sin \omega_j(t - \tau) \, d\tau + q_{j0} \cos \omega_j t + \frac{\dot{q}_{j0}}{\omega_j} \sin \omega_j t, \tag{5.89}$$

where q_{j0} and $\dot{q}_{j0}$ are the initial conditions of $q_j(t)$ and $\dot{q}_j(t)$ at time $t = 0$, respectively. The quantities q_{j0} and $\dot{q}_{j0}$ are computed from Eqs. (5.82) and (5.83). The response $w(x, t)$ is obtained when Eq. (5.89) is introduced in conjunction with Eqs. (5.82) and (5.83) into Eq. (5.85). In practice, the number of eigenfunctions is selected to be large enough to compute the response $w(x, t)$. However, the eigenfunctions themselves may be chosen in any order, depending on the frequency content of the excitation force. For example, if there is a single frequency in the excitation force, only those eigenfunctions with frequencies near the excitation frequency need to be included in the computation of $w(x, t)$.

5.6 Concluding Remarks

A simple but extremely important SDOF mass–spring–damper system has been discussed in detail. For this system, the spring, mass, and damper coefficients are scalar. We then focused on the equations of motion of a two-degree-of-freedom system. For this system, these mass, spring, and damper coefficients are matrices (instead of scalars); thus the general structure of the mathematical equations is preserved. In fact, this generalization is true for a linear dynamic system that has any number of degrees of freedom. We also briefly presented a case of a distributed parameter system to show how the treatment is similar to and different from the lumped parameter case.

5.7 Problems

5.1 Consider the two-degree-of-freedom system shown in Fig. 5.13.

(a) Derive the equations of motions and express them in second-order form with mass matrix M, stiffness matrix K, and damping matrix Ξ.

(b) Derive the equations describing the system responses to harmonic excitation,

$$f_1(t) = F_1 \cos \omega t, \qquad f_2(t) = F_2 \cos \omega t,$$

by the complex-variable approach.

5.2 Use the equations derived in Problem 5.1 for the specific case in which

$$m_1 = m_2 = 1, \qquad k_1 = k_2 = 1, \qquad \zeta_1 = \zeta_2 = 0.$$

(a) Plot out the FRFs.

(b) Obtain numerical values of the system natural frequencies and mode shapes.

(c) Let $w = \Phi q$, where Φ is made up of the mode-shape vectors (both w and q are 2×1 column vectors). Obtain the equations of motions for q and verify that they are two uncoupled ordinary differential equations. Solve the ordinary differential equations for q and then transform the answers back to w.

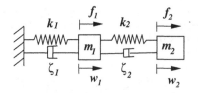

Figure 5.13. Two-degree-of-freedom mass–spring–dashpot system.

5.3 Given an undamped multiple-degree-of-freedom system satisfying the following matrix equation,

$$M\ddot{w} + Kw = 0,$$

with mass matrix M, stiffness matrix K, and displacement vector w, prove that the natural frequencies and the mode shapes can be obtained by solving the following standard eigenvalue–eigenvector problem for $M^{-1}K$,

$$M^{-1}K\phi = \lambda^2\phi$$

where λ^2 is one of the system eigenvalues and ϕ is its corresponding eigenvector.

5.4 Write expressions for the kinetic energy T and the potential energy V for the system shown in Fig. 5.13. Show that from these expressions we can extract the mass and stiffness matrices as well.

5.5 Form the so-called Rayleigh quotient,

$$R(\psi) = \frac{\psi^T K \psi}{\psi^T M \psi},$$

for the system described by

$$M\ddot{w} + Kw = 0,$$

where M is a mass matrix, K is a stiffness matrix, and w is a displacement vector. Verify the following statements:

(a) When ψ is a mode-shape vector, the ratio is the corresponding natural frequency squared.

(b) When ψ is any other vector (pick a few of your choice), the ratio is always larger than the smallest natural frequency (called the fundamental frequency) squared.

5.6 The general solution of a bar hinged at both ends in bending vibration has the form

$$w(x, t) = \sum_{k=1}^{\infty} a_k \phi_k(x) \cos(\omega_k t - \theta_k).$$

Obtain expressions for the amplitudes a_k and the phase angles θ_k by making use of the initial conditions

$$w(x, 0) = g(x), \qquad \frac{\partial w}{\partial t}\bigg|_{t=0} = h(x)$$

and the orthogonality property of the mode shapes,

$$\int_0^\ell \rho(x)\phi_i(x)\phi_j(x)\,dx = \begin{cases} 0 & \text{for } i \neq j \\ q & \text{for } i = j \end{cases},$$

where $\rho(x)$ is a mass density function and q is some nonzero value.

5.7 Consider the free vibration of a flexible string fixed at both ends governed by

$$T\frac{\partial^2 w(x, t)}{\partial x^2} = \rho \frac{\partial^2 w(x, t)}{\partial t^2},$$

where T is the constant tension and ρ is the mass per unit length. Obtain expressions for the system natural frequencies and mode shapes.

BIBLIOGRAPHY

[1] Juang, J.-N., *Applied System Identification*, Prentice-Hall, Englewood Cliffs, NJ, 1994.
[2] Meirovitch, L., *Elements of Vibration Analysis,* 2nd ed., McGraw-Hill, New York, 1986; 6th printing, 1992.

6

Virtual Passive Controllers

6.1 Introduction

When a mass–spring–dashpot is attached to any mechanical system, including flexible space structures, the damping of the system is almost always augmented regardless of the system size. The parameters of the mass–spring–dashpot are arbitrary, model independent, and thus insensitive to the system uncertainties. To satisfy the system performance requirements, we adjust the parameters by using the knowledge of the system model. The more the system is known, the better the parameters of the mass–spring–dashpot may be adjusted to meet the performance requirements. However, no matter what happens, the mass–spring–dashpot will not destabilize the system because it is an energy-dissipative device. The question arises as to whether there are any feedback control designs that use sensors and actuators that behave like the passive mass–spring–dashpot.

We discuss a robust controller design for flexible structures in this chapter by using a set of second-order dynamic equations similar to that describing the passive mass–spring–dashpot. Under certain realistic (practical) conditions, this method provides a stable system in the presence of system uncertainties. For better understanding, two major steps are involved in developing the formulation of the method. First, consider only the direct output feedback for simplicity, implying the absence of dynamics in the feedback controller. Conditions are identified in terms of the number and the type of sensors and their locations to make the system asymptotically stable. Second, assume that the feedback controller contains a set of second-order dynamic equations. It is equivalent to visualizing a virtual passive damping system (Ref. [1–5]), i.e. the feedback controller, which is linked side by side to the real mechanical system. In other words, two sets of second-order dynamic equations are coupled to generate a closed-loop system. Design freedom increases when the dimension of the controller dynamic equations increases. Conditions are derived for the design of a stable closed-loop system having an infinite gain margin. The method takes advantage of the second-order form of equations (instead of transforming to a first-order form), which provides an easy way of discussing and obtaining the stability margin and results in considerable computational efficiency for numerical simulations. Comparisons between the active feedback and the passive mass–spring–dashpot are given through several illustrative examples.

6.2 Direct Feedback

In the analysis and design of dynamics and vibration control of flexible structures, two sets of linear, constant coefficient, ordinary differential equations are frequently used:

$$M\ddot{w} + \Xi\dot{w} + Kw = Bu, \tag{6.1}$$

$$y = C_a\ddot{w} + C_v\dot{w} + C_dw. \tag{6.2}$$

Here w is an $n \times 1$ displacement vector, and M, Ξ, and K are mass, damping, and stiffness matrices, respectively, which generally are symmetric and sparse. The $n \times r$ influence matrix B describes the actuator force distributions for the $r \times 1$ control force vector u. Typically, matrix M is positive definite whereas Ξ and K are positive semidefinite. In the absence of rigid-body motion, K is positive definite. Equation (6.2) is a measurement equation that has y as the $m \times 1$ measurement vector, C_a as the $m \times n$ acceleration influence matrix, C_v as the $m \times n$ velocity, influence matrix, and C_d as the $m \times n$ displacement influence matrix. Note that Eq. (6.1) can be solved for the acceleration in terms of the displacement, velocity, and control force to obtain a new measurement equation in place of Eq. (6.2). However, physical insight is lost in this approach to controller design. As long as the vector w stays in the physical coordinates, the matrices B, C_a, C_v, and C_d are, in general, not functions of the system physical properties including mass, damping, and stiffness.

Measurement equation (6.2) may be used either directly or indirectly for a feedback-controller design. Here we will use direct feedback such as the simple mass–spring system shown in Fig. 6.1. Let the input vector u be

$$u = -\mathcal{F}y = -\mathcal{F}C_a\ddot{w} - \mathcal{F}C_v\dot{w} - \mathcal{F}C_dw \tag{6.3}$$

where $\mathcal{F}$ is a gain matrix to be determined. Substituting Eq. (6.3) into Eq. (6.1) yields

$$(M + B\mathcal{F}C_a)\ddot{w} + (\Xi + B\mathcal{F}C_v)\dot{w} + (K + B\mathcal{F}C_d)w = 0. \tag{6.4}$$

For simplicity, consider the simple case in which $C_a = C_d = 0$ (zero matrix). Assume that the number of sensors m is larger than the number of actuators r. Let the actuators be located such that the row space generated by B^T belongs to the row space generated by C_v. In other words, the actuators are located in such a way that the control influence matrix B can be expressed by

$$B^T = C_bC_v, \tag{6.5}$$

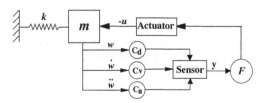

Figure 6.1. A mass–spring system with direct feedback.

where C_b is a $r \times m$ matrix that may be obtained by

$$C_b = B^T C_v^T (C_v C_v^T)^{-1}.$$

Assume that the gain matrix $\mathcal{F}$ is computed by

$$\mathcal{F} = LL^T C_b, \tag{6.6}$$

where L is an $r \times r$ arbitrary nonzero matrix. Substituting Eq. (6.6) into Eq. (6.4) with the assumption $C_a = C_d = 0$ (zero matrix) leads to

$$M\ddot{w} + (\Xi + BLL^T B^T)\dot{w} + Kw = 0. \tag{6.7}$$

The matrix $BLL^T B^T$ is at least positive semidefinite and thus $\Xi + BLL^T B^T$ is at least positive semidefinite for a positive semidefinite matrix Ξ. As a result, the closed-loop system of Eq. (6.7) is stable if $\Xi + BLL^T B^T$ is positive semidefinite or asymptotically stable if $\Xi + BLL^T B^T$ is positive definite. For the case in which Ξ is positive definite, $\Xi + BLL^T B^T$ is positive definite, which yields an asymptotically stable closed-loop system. This leads to a conclusion that, for a structural system with some passive damping, an output-velocity-feedback scheme with noncolocated velocity sensors and actuators may make the closed-loop system asymptotically stable with an infinite gain margin in the sense that the matrix L given in Eq. (6.6) for determination of the gain matrix $\mathcal{F}$ is an arbitrary (nonzero) matrix, as long as the actuators are properly located, satisfying Eq. (6.5). Note that, for colocated sensors and actuators, $B^T = C_v$. Without velocity measurements, the system damping cannot be augmented from direct output feedback alone. However, if there are actuator dynamics involved, the system damping may be augmented by direct displacement or acceleration feedback.

EXAMPLE 6.1
Consider the three-mass-two-spring system illustrated in Fig. 6.2. To simplify the discussion, we neglect any system damping and let $m = k = 1$. The equations of motion are

$$m \begin{bmatrix} 1 & 0 & 0 \\ 0 & 1 & 0 \\ 0 & 0 & 1 \end{bmatrix} \begin{bmatrix} \ddot{w}_1 \\ \ddot{w}_2 \\ \ddot{w}_3 \end{bmatrix} + k \begin{bmatrix} 1 & -1 & 0 \\ -1 & 2 & -1 \\ 0 & -1 & 1 \end{bmatrix} \begin{bmatrix} w_1 \\ w_2 \\ w_3 \end{bmatrix} = \begin{bmatrix} 1 \\ 0 \\ -1 \end{bmatrix} u,$$

$$y = \begin{bmatrix} -1 & 1 & 0 \\ 0 & -1 & 1 \end{bmatrix} \begin{bmatrix} \dot{w}_1 \\ \dot{w}_2 \\ \dot{w}_3 \end{bmatrix},$$

where the first measurement is the relative velocity between $\dot{w}_1$ and $\dot{w}_2$ and the second measurement is the relative velocity between $\dot{w}_2$ and $\dot{w}_3$.

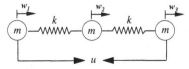

Figure 6.2. Three-mass-two-spring system.

The column of the input influence matrix B are obviously in the space spanned by the rows of the output matrix C_v, i.e.,

$$B^T = \begin{bmatrix} 1 \\ 0 \\ -1 \end{bmatrix}^T = [1 \quad 0 \quad -1] = [-1 \quad -1] \begin{bmatrix} -1 & 1 & 0 \\ 0 & -1 & 1 \end{bmatrix} = C_b C_v,$$

where

$$C_b = [-1 \quad -1], \qquad C_v = \begin{bmatrix} -1 & 1 & 0 \\ 0 & -1 & 1 \end{bmatrix}.$$

Let the control force u be

$$u = -\mathcal{F}y = -\zeta_c C_b y = -\zeta_c [-1 \quad -1] \begin{bmatrix} -1 & 1 & 0 \\ 0 & -1 & 1 \end{bmatrix} \begin{bmatrix} \dot{w}_1 \\ \dot{w}_2 \\ \dot{w}_3 \end{bmatrix}.$$

$$= -\zeta_c [1 \quad 0 \quad -1] \begin{bmatrix} \dot{w}_1 \\ \dot{w}_2 \\ \dot{w}_3 \end{bmatrix},$$

where ζ_c is a positive constant. Here, we have set the control gain to be $\mathcal{F} = \zeta_c C_b$. The force applied to the system becomes

$$Bu = -\zeta_c \begin{bmatrix} 1 \\ 0 \\ -1 \end{bmatrix} [1 \quad 0 \quad -1] \begin{bmatrix} \dot{w}_1 \\ \dot{w}_2 \\ \dot{w}_3 \end{bmatrix}$$

$$= -\zeta_c \begin{bmatrix} 1 & 0 & -1 \\ 0 & 0 & 0 \\ -1 & 0 & 1 \end{bmatrix} \begin{bmatrix} \dot{w}_1 \\ \dot{w}_2 \\ \dot{w}_3 \end{bmatrix}.$$

Substituting the above equation into the state equation yields the closed-loop state equation of motion as

$$m \begin{bmatrix} 1 & 0 & 0 \\ 0 & 1 & 0 \\ 0 & 0 & 1 \end{bmatrix} \begin{bmatrix} \ddot{w}_1 \\ \ddot{w}_2 \\ \ddot{w}_3 \end{bmatrix} + \zeta_c \begin{bmatrix} 1 & 0 & -1 \\ 0 & 0 & 0 \\ -1 & 0 & 1 \end{bmatrix} \begin{bmatrix} \dot{w}_1 \\ \dot{w}_2 \\ \dot{w}_3 \end{bmatrix}$$

$$+ k \begin{bmatrix} 1 & -1 & 0 \\ -1 & 2 & -1 \\ 0 & -1 & 1 \end{bmatrix} \begin{bmatrix} w_1 \\ w_2 \\ w_3 \end{bmatrix} = \begin{bmatrix} 0 \\ 0 \\ 0 \end{bmatrix}.$$

The control force simply adds the damping to the system. In other words, the active control system with the control input $u = -\mathcal{F}y = -\zeta_c C_b y = -\zeta_c \dot{w}_1 + \zeta_c \dot{w}_3$ is equivalent to the passive system with a dashpot of damping coefficient ζ_c. The equivalence between active and passive systems is illustrated in Fig. 6.3.

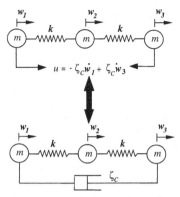

Figure 6.3. Equivalent active and passive systems.

6.3 Controller with Second-Order Dynamics

Assume that the controller to be designed has a set of second-order dynamic equations and measurement equations similar to system equations (6.1) and (6.2):

$$M_c \ddot{w}_c + \Xi_c \dot{w}_c + K_c w_c = B_c u_c, \tag{6.8}$$

$$y_c = C_{ac} \ddot{w}_c + C_{vc} \dot{w}_c + C_{dc} w_c. \tag{6.9}$$

Note that this is a set of equations that do not represent any physical system. In fact, this set of equations basically serves as a filter to shift the phase of measurement signals. Here w_c is the controller state vector of dimension n_c and M_c, Ξ_c, and K_c are thought of as the controller mass, damping, and stiffness matrices, respectively, which generally are symmetric and positive definite to make the controller asymptotically stable. The $n_c \times m$ influence matrix B_c describes the force distributions for the $m \times 1$ input force vector u_c. Equation (6.9) is the controller measurement equation that has y_c as the measurement vector of length r, C_{ac} as the $r \times n_c$ acceleration influence matrix, C_{vc} as the $r \times n_c$ velocity influence matrix, and C_{ad} as the $r \times n_c$ displacement influence matrix. The quantities M_c, Ξ_c, K_c, C_{ac}, C_{dc}, and C_{vc} are the design parameters for the controller.

Let the input vectors u and u_c in Eqs. (6.1) and (6.8) be

$$u = y_c = C_{ac} \ddot{w}_c + C_{vc} \dot{w}_c + C_{dc} w_c, \tag{6.10}$$

$$u_c = y = C_a \ddot{w} + C_v \dot{w} + C_d w. \tag{6.11}$$

Substituting Eq. (6.10) into Eq. (6.1) and Eq. (6.11) into Eq. (6.8) yields

$$M_t \ddot{w}_t + \Xi_t \dot{w}_t + K_t w_t = 0, \tag{6.12}$$

where

$$M_t = \begin{bmatrix} M & -BC_{ac} \\ -B_c C_a & M_c \end{bmatrix}, \quad \Xi_t = \begin{bmatrix} \Xi & -BC_{vc} \\ -B_c C_v & \Xi_c \end{bmatrix},$$

$$K_t = \begin{bmatrix} K & -BC_{dc} \\ -B_c C_d & K_c \end{bmatrix}, \quad w_t = \begin{bmatrix} w \\ w_c \end{bmatrix}.$$

If the design parameters, M_c, Ξ_c, K_c, C_{ac}, C_{dc}, and C_{vc} are chosen such that M_t, Ξ_t, and

K_t are positive definite, the closed-loop system of Eq. (6.12) becomes asymptotically stable, which can be proved as follows.

The total energy of the system described by Eq. (6.12) is

$$\mathcal{E} = \frac{1}{2}\dot{w}_t^T M_t \dot{w}_t + \frac{1}{2}w_t^T K_t w_t, \tag{6.13}$$

where the first term represents the kinetic energy and the second term represents the potential energy. The total energy is positive if the mass matrix M_t is positive definite and the stiffness matrix K_t is at least positive semidefinite. The derivative of the total energy $\mathcal{E}$ becomes

$$\frac{d\mathcal{E}}{dt} = \dot{w}_t^T M_t \ddot{w}_t + \dot{w}_t^T K_t w_t$$

$$= -\dot{w}_t^T \Xi_t \dot{w}_t, \tag{6.14}$$

where the last equality is obtained with closed-loop equation (6.12). When the damping matrix Ξ_t is positive definite, i.e., $\dot{w}_t^T \Xi_t \dot{w}_t > 0$ for any $\dot{w}_t \neq 0$, Eq. (6.14) clearly indicates that the time history of the total energy will decay because the time rate of the energy is negative. If the damping matrix Ξ_t is only positive semidefinite, i.e., $\Xi_t \geq 0$, then the system is stable but not asymptotically stable. In other words, the total energy may not decay at some period of time but also will not increase.

6.3.1   Displacement Feedback

For better understanding of the advantage of the controller's having second-order dynamic equations, consider a special case in which $C_a = C_{ac} = C_v = C_{vc} = 0$. To make K_t symmetric, it is required that

$$BC_{dc} = C_d^T B_c^T \tag{6.15}$$

or

$$\begin{bmatrix} B & -C_d^T \end{bmatrix} \begin{bmatrix} C_{dc} \\ B_c^T \end{bmatrix} = 0. \tag{6.16}$$

For the case in which the sum of the number of actuators r and the number of sensors m is less than the number of states n, the leftmost matrix of Eq. (6.16) is a tall matrix. Unless B is in the space spanned by C_d^T or vice versa, there do not exist any solutions for B_c and C_{dc} in Eq. (6.15). Assume that the number of sensors m is larger than the number of actuators r. Let the actuators be located such that the row space generated by B^T belongs to the row space generated by C_d, i.e., the actuators are located in such a way that the control influence matrix B can be expressed by

$$B^T = Q_b C_d, \tag{6.17}$$

where Q_b is a $r \times m$ matrix that may be obtained by

$$Q_b = B^T C_d^T \left(C_d C_d^T \right)^{-1}.$$

Substituting Eq. (6.17) into Eq. (6.15) yields

$$C_d^T Q_b^T C_{dc} = C_d^T B_c^T. \tag{6.18}$$

Because C_d^T is a tall matrix for $m < n$, the only possible solution is

$$Q_b^T C_{dc} = B_c^T. \tag{6.19}$$

For any given matrix C_{dc}, this equation produces a B_c^T that makes the matrix K_t symmetric, i.e.,

$$K_t = \begin{bmatrix} K & -BC_{dc} \\ -C_{dc}^T B^T & K_c \end{bmatrix} \quad \text{or} \quad K_t = \begin{bmatrix} K & -C_d^T B_c^T \\ -B_c C_d & K_c \end{bmatrix}. \tag{6.20}$$

The next question is how to choose a matrix C_{dc} that makes the closed-loop stiffness matrix K_t positive definite. The matrix K_t is positive definite, generally written as $K_t > 0$, if and only if

$$w_t^T K_t w_t > 0 \tag{6.21}$$

for any real vector w_t, except the null vector. Substituting the definition of K_t from Eq. (6.20) and w_t from Eq. (6.12) into inequality (6.21) yields

$$w_t^T K_t w_t = w^T \left(K - C_d^T B_c^T B_c C_d \right) w$$

$$+ (B_c C_d w - w_c)^T (B_c C_d w - w_c) + w_c^T (K_c - I) w_c. \tag{6.22}$$

This equation is greater than zero if B_c and K_c are chosen such that the following two matrices

$$K - C_d^T B_c^T B_c C_d, \quad K_c - I$$

are positive definite. Note that this is a sufficient condition, but not a necessary condition. To make inequality (6.21) hold, K must be a positive-definite matrix, i.e., $K > 0$, and B_c must be chosen such that

$$K - C_d^T B_c^T B_c C_d > 0.$$

It implies that this controller may not be able to control rigid-body motion because K in this case is only a positive-semidefinite matrix, $K \geq 0$. To release the constraint condition, $K - C_d^T B_c^T B_c C_d > 0$, K must be increased by at least $BC_{dc} C_{dc}^T B^T$. In other words, the system must be stiffened which can be achieved by the addition of displacement feedback shown below.

Instead of using the input force defined in Eq. (6.10), let the input force be changed to

$$u = y_c - \mathcal{F}y = C_{dc} w_c - \mathcal{F} C_d w, \tag{6.23}$$

where $\mathcal{F}$ is a gain matrix to be determined. Note that the velocity feedback is not considered here and neither is the acceleration feedback. When Eq. (6.23) is substituted into system equation (6.1), the closed-loop stiffness matrix, Eq. (6.20), becomes

$$K_t = \begin{bmatrix} K + B\mathcal{F}C_d & -C_d^T B_c^T \\ -B_c C_d & K_c \end{bmatrix}. \tag{6.24}$$

If $\mathcal{F}$ is chosen such that

$$\mathcal{F} = C_{dc}B_c, \tag{6.25}$$

which, from Eq. (6.15), results in

$$B\mathcal{F}C_d = BC_{dc}B_cC_d = C_d^T B_c^T B_c C_d,$$

the closed-loop stiffness matrix, Eq. (6.24), thus becomes

$$K_t = \begin{bmatrix} K + C_d^T B_c^T B_c C_d & -C_d^T B_c^T \\ -B_c C_d & K_c \end{bmatrix}, \tag{6.26}$$

or, from Eq. (6.15),

$$K_t = \begin{bmatrix} K + BC_{dc}C_{dc}^T B^T & -BC_{dc} \\ -C_{dc}^T B^T & K_c \end{bmatrix}$$

which changes Eq. (6.22) to

$$w_t^T K_t w_t = w^T K w + (B_c C_d w - w_c)^T (B_c C_d w - w_c) + w_c^T (K_c - I)w_c$$

$$= w^T K w + (C_{dc}^T B^T w - w_c)^T (C_{dc}^T B^T w - w_c) + w_c^T (K_c - I)w_c. \tag{6.27}$$

Because K_c is a design parameter, the closed-loop system becomes stable as long as K_c is chosen to be larger than I, i.e., $w_c^T(K_c - I)w_c \geq 0$ for any arbitrary vector w_c. An obvious choice is $K_c = I$, where I is an identity matrix of dimension n_c. However, this is not the best choice, which will be discussed later. To this end, it is shown that a stable closed-loop system can be designed with a feedback controller with second-order dynamic equations. The controller has an infinite gain margin in the sense that the matrices M_c, Ξ_c, and K_c, which may be considered as the gain matrices for the controller state vector w_c and its derivatives, can be as large as desired without destabilizing the system as long as they are positive definite and K_c is larger than I.

A little modification of the above design produces a better design, which has physical meaning. Indeed, let

$$B_c = K_c \bar{B}_c \quad \text{or} \quad \bar{B}_c = K_c^{-1} B_c, \tag{6.28}$$

where K_c is assumed to be positive definite so that the solution for $\bar{B}_c$ exists for any given B_c. In addition, let the gain matrix $\mathcal{F}$ in Eq. (6.25) be slightly modified as follows:

$$\mathcal{F} = C_{dc}\bar{B}_c, \tag{6.29}$$

which, with the aid of Eq. (6.15), results in

$$B\mathcal{F}C_d = BC_{dc}\bar{B}_cC_d = C_d^T \bar{B}_c^T K_c \bar{B}_c C_d.$$

The closed-loop stiffness matrix in this case [see Eq. (6.24)] thus becomes

$$K_t = \begin{bmatrix} K + C_d^T \bar{B}_c^T K_c \bar{B}_c C_d & -C_d^T \bar{B}_c^T K_c \\ -K_c \bar{B}_c C_d & K_c \end{bmatrix}, \tag{6.30}$$

which in turn changes Eq. (6.27) to

$$w_t^T K_t w_t = w^T K w + (\bar{B}_c C_d w - w_c)^T K_c (\bar{B}_c C_d w - w_c)$$

$$= w^T K w + (C_{dc}^T \bar{B}^T w - w_c)^T K_c (C_{dc}^T \bar{B}^T w - w_c). \tag{6.31}$$

This equation is obviously positive if K is at least positive semidefinite, i.e., $K \geq 0$. Does this design have any physical meaning? The answer is yes. Consider the special case in which the controller is as large as the system in the sense that the number of system coordinates n is identical to the number of controller coordinates n_c. Furthermore, assume that all the displacements are directly measurable such that $C_d = I$ (identity matrix) and there are n actuators colocated with the displacement sensors, $B^T = C_d = I$. In this case, $Q_b = I$ [Eq. (6.17)], $B_c^T = C_{dc} = K_c$ [Eqs. (6.19) and (6.28) for $\bar{B}_c = I$], and $\mathcal{F} = K_c$ [Eq. (6.29)], which yields, from Eqs. (6.30),

$$K_t = \begin{bmatrix} K + K_c & -K_c \\ -K_c & K_c \end{bmatrix}. \tag{6.32}$$

For a SDOF ($n_c = n = 1$), K_t represents the stiffness matrix for two springs connected in series with spring constants K and K_c.

PHYSICAL INTERPRETATION

For a better understanding of the nature of the dynamic control designs developed here, physical interpretation for some simple systems would help. Three illustrative examples will be shown, starting with a simple spring–mass system.

EXAMPLE 6.2
Consider a single-degree-of-freedom spring–mass system, $n_c = n = 1$, with a displacement measurement of the system mass. The second-order controller for this case reduces to a virtual spring–mass–dashpot system connected in series with the system mass, as shown in Fig. 6.4. Let the position of masses m and m_c be measured from their equilibrium states. The equations of motion for the above system can be derived by application of a force to m and m_c. The force applied to the system mass m in this case is the force u transmitted through the spring k_c. This is precisely the control force applied to the system as given in Eq. (6.23) with $C_{dc} = k_c, C_d = 1$, and $\mathcal{F} = k_c$. Thus the second-order control law is simply

$$u = k_c(w_c - w), \tag{6.33}$$

where w_c is computed from

$$m_c \ddot{w}_c + \zeta_c \dot{w}_c + k_c w_c = k_c w$$

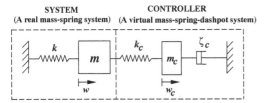

Figure 6.4. A mass–spring system with a SDOF dynamic controller.

The equation of motion that describes the closed-loop behavior of the above system is simply

$$\begin{bmatrix} m & 0 \\ 0 & m_c \end{bmatrix} \begin{bmatrix} \ddot{w} \\ \ddot{w}_c \end{bmatrix} + \begin{bmatrix} 0 & 0 \\ 0 & \zeta_c \end{bmatrix} \begin{bmatrix} \dot{w} \\ \dot{w}_c \end{bmatrix}$$

$$+ \begin{bmatrix} k + k_c & -k_c \\ -k_c & k_c \end{bmatrix} \begin{bmatrix} w \\ w_c \end{bmatrix} = \begin{bmatrix} 0 \\ 0 \end{bmatrix}. \tag{6.34}$$

The above equation verifies Eq. (6.12) with $C_{vc} = C_v = C_{ac} = C_a = 0$ (i.e., no velocity and acceleration measurements) and K_t given in Eq. (6.32). The above set of equations is always asymptotically stable for any positive values of m, k, m_c, k_c, and ζ_c. We now consider various special cases.

CASE 1:

For the controller without damping, $\zeta_c = 0$, the system reduces to two spring–masses connected in series. If k_c is small, the control force given in Eq. (6.33) is small thus the controller exerts little influence on the system. Mathematically, Eq. (6.34) becomes a set of two uncoupled equations of w and w_c, and obviously little change in the response of the controlled system is expected from this controller. If, however, k_c is large, (i.e., the virtual spring is stiff) the relative displacement between the two masses is small. Hence in the limit the two masses move together like a single mass $m + m_c$, and the natural frequency of the system is approaching

$$\omega_n = \sqrt{\frac{k}{m + m_c}}.$$

As a result, for large k_c, changing the design variable m_c will affect the natural frequency of the closed-loop system according to the above equation.

CASE 2:

For $\zeta_c > 0$, the system is always asymptotically stable (unless $k_c = 0$, which, as discussed before, means no control). The energy flows from m to m_c and is dissipated by the damper. Again, for large k_c, the system can be approximated as

$$(m + m_c)\ddot{w} + \zeta_c \dot{w} + kw = 0. \tag{6.35}$$

Introduce the notation

$$\frac{\zeta_c}{m + m_c} = 2\zeta \omega_n, \qquad \frac{k}{m + m_c} = \omega_n^2.$$

Thus

$$\zeta = \frac{1}{2} \frac{\zeta_c}{\sqrt{k(m + m_c)}} \tag{6.36}$$

The design variables in this case are ζ_c and m_c. Various choices of ζ_c and m_c will result in $\zeta > 1$, $\zeta < 1$, or $\zeta = 1$, which corresponds to the cases in which the closed-loop system is overdamped, underdamped, or critically damped, respectively.

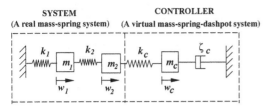

SYSTEM
(A real mass-spring system)

CONTROLLER
(A virtual mass-spring-dashpot system)

Figure 6.5. Two-degree-of-freedom system with a SDOF dynamic controller.

CASE 3:

For general values of k_c, ζ_c, and m_c, the design can be thought of as a virtual vibration absorber. Let the system be excited by some unknown force $Ue^{j\omega_f t}$ and the displacement of the mass m be denoted by $w = We^{j(\omega t + \phi)}$. The typical objective of a vibration-absorber design is to determine the values of k_c, ζ_c, and m_c such that the ratio

$$\gamma = \left| \frac{We^{j(\omega t + \phi)}}{Ue^{j\omega_f t}} \right|$$

is minimized over an interesting range of excitation frequency ω_f.

EXAMPLE 6.3

Consider a two-degree-of-freedom spring–mass system with displacement measurements of the masses m_1 and m_2 from their equilibrium positions, $n = 2$. First consider the case in which the the controller has only one state, $n_c = 1$. The second order controller in this case is simply equivalent to a virtual spring–mass–dashpot system connected in series with the two system masses, as shown in Fig. 6.5.

It can be easily shown that the control force applied to the system is simply

$$u = \begin{bmatrix} u_1 \\ u_2 \end{bmatrix} = \begin{bmatrix} 0 \\ k_c(w_c - w_2) \end{bmatrix}, \tag{6.37}$$

where u_j denotes the force applied to m_j, $j = 1, 2$ and w_c is given by

$$m_c \ddot{w}_c + \zeta_c \dot{w}_c + k_c w_c = k_c w_2.$$

Furthermore, the closed-loop behavior of the above system is governed by

$$\begin{bmatrix} m_1 & 0 & 0 \\ 0 & m_2 & 0 \\ 0 & 0 & m_c \end{bmatrix} \begin{bmatrix} \ddot{w}_1 \\ \ddot{w}_2 \\ \ddot{w}_c \end{bmatrix} + \begin{bmatrix} 0 & 0 & 0 \\ 0 & 0 & 0 \\ 0 & 0 & \zeta_c \end{bmatrix} \begin{bmatrix} \dot{w}_1 \\ \dot{w}_2 \\ \dot{w}_c \end{bmatrix}$$

$$+ \begin{bmatrix} k_1 + k_2 & -k_2 & 0 \\ -k_2 & k_2 + k_c & -k_c \\ 0 & -k_c & k_c \end{bmatrix} \begin{bmatrix} w_1 \\ w_2 \\ w_c \end{bmatrix} = \begin{bmatrix} 0 \\ 0 \\ 0 \end{bmatrix}. \tag{6.38}$$

Note that the above scheme requires only the displacement measurement of the mass m_2. The stiffness matrix is positive definite (see Problem 6.1) for any positive values of k_1, k_2, and k_c.

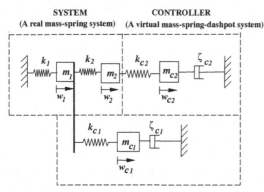

SYSTEM
(A real mass-spring system)

CONTROLLER
(A virtual mass-spring-dashpot system)

Figure 6.6. Two-degree-of-freedom system with a two-degree-of-freedom dynamic controller.

EXAMPLE 6.4

Consider the two-degree-of-freedom system shown in Fig. 6.6 with displacement measurements only, but now displacement measurement of the mass m_1 is also to be used in the controller design. The second-order controller design in this case is simply

$$u = \begin{bmatrix} u_1 \\ u_2 \end{bmatrix} = \begin{bmatrix} k_{c1} & 0 \\ 0 & k_{c2} \end{bmatrix} \begin{bmatrix} w_{c1} - w_1 \\ w_{c2} - w_2 \end{bmatrix}, \tag{6.39}$$

where w_{c1} and w_{c2} are solved by

$$\begin{bmatrix} m_{c1} & 0 \\ 0 & m_{c2} \end{bmatrix} \begin{bmatrix} \ddot{w}_{c1} \\ \ddot{w}_{c2} \end{bmatrix} + \begin{bmatrix} \zeta_{c1} & 0 \\ 0 & \zeta_{c2} \end{bmatrix} \begin{bmatrix} \dot{w}_{c1} \\ \dot{w}_{c2} \end{bmatrix}$$

$$+ \begin{bmatrix} k_{c1} & 0 \\ 0 & k_{c2} \end{bmatrix} \begin{bmatrix} w_{c1} \\ w_{c2} \end{bmatrix} = \begin{bmatrix} k_{c1} & 0 \\ 0 & k_{c2} \end{bmatrix} \begin{bmatrix} w_1 \\ w_2 \end{bmatrix}. \tag{6.40}$$

It is clear that the two controller states w_{c1} and w_{c2} are not coupled in this case. They can be solved independently. The closed-loop system is equivalent to the mass–spring–dashpot system shown in Fig. 6.6, whose behavior is governed by

$$\begin{bmatrix} m_1 & 0 & 0 & 0 \\ 0 & m_2 & 0 & 0 \\ 0 & 0 & m_{c1} & 0 \\ 0 & 0 & 0 & m_{c2} \end{bmatrix} \begin{bmatrix} \ddot{w}_1 \\ \ddot{w}_2 \\ \ddot{w}_{c1} \\ \ddot{w}_{c2} \end{bmatrix} + \begin{bmatrix} 0 & 0 & 0 & 0 \\ 0 & 0 & 0 & 0 \\ 0 & 0 & \zeta_{c1} & 0 \\ 0 & 0 & 0 & \zeta_{c2} \end{bmatrix} \begin{bmatrix} \dot{w}_1 \\ \dot{w}_2 \\ \dot{w}_{c1} \\ \dot{w}_{c2} \end{bmatrix}$$

$$+ \begin{bmatrix} k_1 + k_2 + k_{c1} & -k_2 & -k_{c1} & 0 \\ -k_2 & k_2 + k_{c2} & 0 & -k_{c2} \\ -k_{c1} & 0 & k_{c1} & 0 \\ 0 & -k_{c2} & 0 & k_{c2} \end{bmatrix} \begin{bmatrix} w_1 \\ w_2 \\ w_{c1} \\ w_{c2} \end{bmatrix} = \begin{bmatrix} 0 \\ 0 \\ 0 \\ 0 \end{bmatrix}. \tag{6.41}$$

If velocity measurements are available, say, at the system mass m_1, then a dashpot element may be added in between m_1 and m_{c1}, for example. It should be noted, however, that the controller masses, springs, and dashpots are in fact virtual elements with physical interpretations as such. For ground-based systems, they may represent actual physical elements attached to the ground. But for space-based systems, these are simply controller gains in the control algorithm.

6.3.2 Acceleration Feedback

The displacement controller can be extended to acceleration feedback as well. Consider the system given in Eq. (6.1), but now the measurement vector y in Eq. (6.2) has only acceleration measurements, i.e., $C_v = C_d = C_{vc} = C_{dc} = 0$ in Eq. (6.12). Substituting Eq. (6.10) with $C_{vc} = C_{dc} = 0$ into Eq. (6.1) and Eq. (6.11) with $C_v = C_d = 0$ into Eq. (6.8) yields

$$M_t \ddot{w}_t + D_t \dot{w}_t + K_t w_t = 0,$$

where

$$M_t = \begin{bmatrix} M & -BC_{ac} \\ -B_c C_a & M_c \end{bmatrix}, \qquad \Xi_t = \begin{bmatrix} \Xi & 0 \\ 0 & \Xi_c \end{bmatrix},$$

$$K_t = \begin{bmatrix} K & 0 \\ 0 & K_c \end{bmatrix}, \qquad w_t = \begin{bmatrix} w \\ w_c \end{bmatrix}.$$

To make M_t symmetric, it is required that $BC_{ac} = C_a^T B_c^T$, as discussed in Eqs. (6.15)–(6.19). All the discussions regarding the positive definiteness of K_t from Eqs. (6.20)–(6.22) for displacement feedback also apply to M_t.

Additional coupling in the closed-loop mass matrix M_t can be achieved if the input u in Eq. (6.10) is allowed to include direct acceleration feedback and output feedback, i.e.,

$$u = y_c - \mathcal{F}_a y = C_{ac} \ddot{w}_c - \mathcal{F}_a y, \tag{6.42}$$

which makes M_t become

$$M_t = \begin{bmatrix} M + B\mathcal{F}_a C_a & -BC_{ac} \\ -B_c C_a & M_c \end{bmatrix}.$$

As before, M_t can be made symmetric and positive definite by proper choices of C_{ac}, B_c, and $\mathcal{F}_a$. Let

$$B_c = M_c \bar{B}_c \quad \text{or} \quad \bar{B}_c = M_c^{-1} B_c, \tag{6.43}$$

where M_c is positive definite so that the solution to $\bar{B}_c$ exists for any given B_c. Note that Eq. (6.43) is identical to Eq. (6.28), with K_c replaced with M_c. Let $\mathcal{F}_a$ be chosen such that

$$\mathcal{F}_a = C_{ac} \bar{B}_c = C_{ac} M_c^{-1} B_c, \tag{6.44}$$

which, with the aid of the equality $BC_{ac} = C_a^T B_c^T$, results in

$$B\mathcal{F}_a C_a = BC_{ac} \bar{B}_c C_a = C_a^T B_c^T \bar{B}_c C_a = C_a^T \bar{B}_c^T M_c \bar{B}_c C_a. \tag{6.45}$$

The closed-loop mass matrix in this case becomes

$$M_t = \begin{bmatrix} M + C_a^T \bar{B}_c^T M_c \bar{B}_c C_a & -C_a^T \bar{B}_c^T M_c \\ -M_c \bar{B}_c C_a & M_c \end{bmatrix}. \tag{6.46}$$

This is a positive-definite matrix, as discussed in Eq. (6.30) for K_t, regardless of the value of M as long as M is positive definite. The closed-loop in this case becomes

$$\begin{bmatrix} M + C_a^T \bar{B}_c^T M_c \bar{B}_c C_a & -C_a^T \bar{B}_c^T M_c \\ -M_c \bar{B}_c C_a & M_c \end{bmatrix} \begin{bmatrix} \ddot{w} \\ \ddot{w}_c \end{bmatrix}$$

$$+ \begin{bmatrix} \Xi & 0 \\ 0 & \Xi_c \end{bmatrix} \begin{bmatrix} \dot{w} \\ \dot{w}_c \end{bmatrix} + \begin{bmatrix} K & 0 \\ 0 & K_c \end{bmatrix} \begin{bmatrix} w \\ w_c \end{bmatrix} = 0. \tag{6.47}$$

The procedure for deriving the second-order controller with acceleration feedback is identical to that for displacement feedback. Mathematically, both controllers are identical in the sense that the closed-loop mass matrix M_t for acceleration feedback can be obtained if K is replaced with M in the closed-loop stiffness matrix K_t for displacement feedback and the subscript d with a. In other words, both displacement and acceleration feedback are dual concepts. However, significant differences between both controllers appear when they are implemented either actively or passively, which will be shown in the following example.

EXAMPLE 6.5

Consider a SDOF spring–mass system with acceleration measurement of the system mass. The second-order controller is a virtual spring–mass–dashpot connected in series with the system mass, as shown in Fig. 6.7. Note that the vector w_c here means the position of m_c relative to the position of m. In this case, C_a and $\bar{B}_c$ in Eq. (6.47) are chosen to be $\bar{B}_c = -C_a = 1$. The control force applied at the mass m based on Newton's law is

$$u = -m_c(\ddot{w}_c + \ddot{w}),$$

where w_c is computed from

$$m_c \ddot{w}_c + \zeta_c \dot{w}_c + k_c w_c = -m_c \ddot{w}.$$

The closed-loop system can then be rewritten as

$$\begin{bmatrix} m + m_c & m_c \\ m_c & m_c \end{bmatrix} \begin{bmatrix} \ddot{w} \\ \ddot{w}_c \end{bmatrix} + \begin{bmatrix} 0 & 0 \\ 0 & \zeta_c \end{bmatrix} \begin{bmatrix} \dot{w} \\ \dot{w}_c \end{bmatrix} + \begin{bmatrix} k & 0 \\ 0 & k_c \end{bmatrix} \begin{bmatrix} w \\ w_c \end{bmatrix} = \begin{bmatrix} 0 \\ 0 \end{bmatrix}.$$

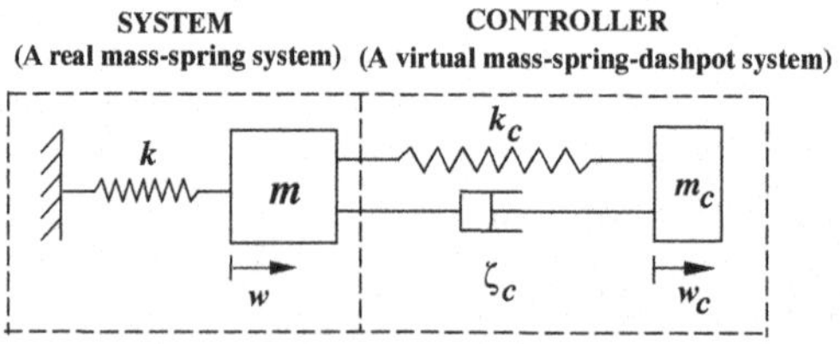

SYSTEM   CONTROLLER
(A real mass-spring system)   (A virtual mass-spring-dashpot system)

Figure 6.7. SDOF system with acceleration feedback.

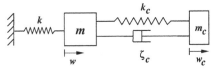

Figure 6.8. A SDOF system with a SDOF controller.

A comparison of Figs. 6.4 and 6.7 reveals the difference between the acceleration and displacement feedback controllers. The controller for acceleration feedback does not have a virtual ground attached to the control mass and thus cannot control the position of the system.

6.3.3 Frequency-Matched Virtual Passive Controller

The virtual passive controller is robust in stability. It by no means implies that it is also robust in performance. To meet performance requirements, the controller design parameters must be carefully tuned. This subsection introduces a method to tune the controller parameters to suppress the unwanted structural vibration. The controller frequency is matched to the driving frequency of the actuator for a desired system damping ratio, ζ_{ds}; hence the unwanted vibration energy in the system is absorbed. The strategy is to match the coefficient terms of the actual and desired closed-loop characteristic equations.

For simplicity without losing generality, let us consider only the SDOF system with a SDOF controller shown in Fig. 6.8. The equations of motion for the system are

$$m\ddot{w} + \zeta\dot{w} + kw - \zeta_c\dot{w}_c - k_c w_c = 0, \tag{6.48}$$

$$m_c\ddot{w}_c + \zeta_c\dot{w}_c + k_c w_c + m_c\ddot{w} = 0, \tag{6.49}$$

where w_c here means the position of m_c relative to the position of m. The closed-loop characteristic equation of this system becomes

$$s^4 + s^3\left(\frac{\zeta_c}{m} + \frac{\zeta_c}{m_c} + \frac{\zeta}{m}\right) + s^2\left(\frac{\zeta\zeta_c}{m_c m} + \frac{k_c}{m_c} + \frac{k_c}{m} + \frac{k}{m}\right)$$
$$+ s\left(\frac{k_c\zeta}{m_c m} + \frac{k\zeta_c}{m_c m}\right) + \frac{kk_c}{m_c m} = 0. \tag{6.50}$$

The frequency-matched desired system and controller characteristic equation is written as

$$(s^2 + 2\zeta_{ds}\omega s + \omega^2)(s^2 + 2\zeta_{dc}\omega s + \omega^2) = 0, \tag{6.51}$$

and its expanded form is

$$s^4 + s^3(2\zeta_{dc}\omega + 2\zeta_{ds}\omega) + s^2(2\omega^2 + 4\zeta_{ds}\zeta_{dc}\omega^2)$$
$$+ s(2\zeta_{ds}\omega^3 + 2\zeta_{dc}\omega^3) + \omega^4 = 0. \tag{6.52}$$

The quantity ζ_{ds} is the desired system damping ratio whereas ζ_{dc} represents the desired controller damping ratio. Now the coefficient terms are matched to define the controller parameters. The s^0 term is

$$\frac{k_c}{m_c} = \frac{\omega^4}{\omega_s^2} \quad \text{with} \quad \omega_s^2 = \frac{k}{m}, \tag{6.53}$$

where ω_s is the natural frequency of the system. The s^1 term is

$$\frac{\zeta_c}{m_c} = \frac{2}{\omega_s^2}\left(\zeta_{ds}\omega^3 + \zeta_{dc}\omega^3 - \zeta_s\frac{\omega^4}{\omega_s}\right) \quad \text{with} \quad \frac{\zeta}{m} = 2\zeta_s\omega_s, \tag{6.54}$$

where ζ_s is the damping ratio of the system. The s^2 term is

$$\zeta_{dc} = \frac{\left(1 + \mu_c - 4\zeta_s^2\right)f^4 + 4\zeta_s\zeta_{ds}f^3 - 2f^2 + 1}{4\zeta_{ds}f^2 - 4\zeta_s f^3}, \tag{6.55}$$

where

$$\mu_c = \frac{m_c}{m}, \qquad f = \frac{\omega}{\omega_s}$$

are the mass ratio and the frequency ratio, respectively, between the system and the controller. The s^3 term is

$$f^6[-(1 + \mu_c)^2] + f^5[4\zeta_{ds}\zeta_s(1 + \mu_c)] + f^4[(1 + \mu_c)(3 - 4\zeta_{ds}^2) - 4\zeta_s^2]$$
$$+ f^2(4\zeta_{ds}^2 + 4\zeta_s^2 - 3 - \mu_c) + f(-4\zeta_{ds}\zeta_s) + 1 = 0, \tag{6.56}$$

where ζ_s is the actual system damping ratio. Equation (6.56) is a sixth-order polynomial relative to the frequency ratio f. Given the system damping ratio ζ_s, an assumed ratio μ_c, and the desired damping ratio ζ_{ds}, Eq. (6.56) produces frequency ratio f. The frequency ratio f is then used to calculate the desired controller damping ratio ζ_{dc} from Eq. (6.55).

The optimal ζ_{dc} is defined as occurring when the difference between ζ_{ds} and ζ_{dc} is less than 5%. The optimal ζ_{dc} is achieved by varying μ_c. From Eqs. (6.53) and (6.54), the actual optimal controller parameters can now be defined through the optimal desired closed-loop parameters as

$$\omega_c = \omega_s f^2, \tag{6.57}$$

$$\zeta_c = (\zeta_{ds} + \zeta_{dc})f - \zeta_s f^2. \tag{6.58}$$

Here, ω_c and ζ_c are the optimal controller natural frequency and damping ratio, respectively. The desired system damping ratio ζ_{ds} is selected to avoid actuator saturation as well as to optimize the controller damping.

6.4 Concluding Remarks

This chapter formulates a robust second-order dynamic stabilization controller design for second-order dynamic systems. The design is passive in the sense that it contains mechanisms that serve only to transfer and dissipate energy of the system. The controller interacts with the physical system only through spring, mass, and dashpot elements, and therefore it can be implemented actively or passively. In other words, stabilization can be accomplished either by a controller with gains interpreted as virtual mass, spring, and dashpot elements, or by actual physical masses, springs, and dashpots connected to the system.

The passive design means that the controller does not destabilize the system. As far as stability is concerned, the controller is model independent, and this is a robust design. Specifically, overall closed-loop stability is guaranteed independently of the system structural uncertainty and variations in the structural parameters. It should be emphasized that this is a robustness result with respect to structural uncertainty in the absence of measurement uncertainty and other contributing factors.

However, control performance, unlike stability robustness, is dependent on the system characteristics. Knowledge of the system model can always help improve a controller design. In this method, the controller order and/or controller gains can be adjusted to meet the desired performance. Physical interpretation of the controller gains as virtual masses, springs, and dashpots provides convenient rules of thumb as to how these should be adjusted to meet a certain desired performance objective.

6.5 Problems

6.1 Consider the matrix

$$K = \begin{bmatrix} k_1 + k_2 & -k_2 \\ -k_2 & k_2 + k_c \end{bmatrix},$$

where k_1, k_2, and k_c are positive scalars. Prove that K is positive definite. Consider another matrix,

$$K = \begin{bmatrix} k_1 + k_2 & -k_2 & 0 \\ -k_2 & k_2 + k_c & -k_c \\ 0 & -k_c & k_c \end{bmatrix}.$$

Prove that K is also a positive-definite matrix.

6.2 Consider the matrix

$$K = \begin{bmatrix} K_1 + C^T K_2 C & -C^T K_2 \\ -K_2 C & K_2 \end{bmatrix},$$

where K_1 of dimension $n \times n$ and K_2 of dimension $m \times m$ are positive definite and C is a $m \times n$ arbitrarily rectangular matrix. Prove that K is positive definite.

6.3 Prove that if every eigenvalue of the symmetric matrix K is greater than unity, then $K - I > 0$, i.e.,

$$x^T K x > x^T x$$

for all nonzero vectors x.

6.4 Consider the matrix

$$M = \begin{bmatrix} M_1 + M_2 & M_2 \\ M_2 & M_2 \end{bmatrix},$$

where M_1 and M_2 of dimension $n \times n$ are both symmetric and positive definite. Is this matrix always positive definite?

6.5 Given the following matrix equation,

$$\begin{bmatrix} m & 0 \\ 0 & m_c \end{bmatrix} \begin{bmatrix} \ddot{w} \\ \ddot{w}_c \end{bmatrix} + \begin{bmatrix} 0 & 0 \\ 0 & \zeta_c \end{bmatrix} \begin{bmatrix} \dot{w} \\ \dot{w}_c \end{bmatrix} + \begin{bmatrix} k + k_c & -k_c \\ -k_c & k_c \end{bmatrix} \begin{bmatrix} w \\ w_c \end{bmatrix} = \begin{bmatrix} 0 \\ 0 \end{bmatrix},$$

SYSTEM   CONTROLLER
(A real mass-spring system)   (A virtual mass-spring-dashpot system)

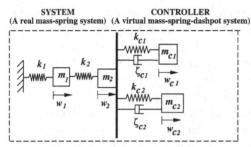

Figure 6.9. A mass–spring system with a two-degree-of-freedom dynamic controller in parallel.

prove that the system is asymptotically stable for any positive values of m, m_c, ζ_c, k, and k_c, i.e., its system eigenvalues have negative real parts.

6.6   Consider the two-degree-of-freedom mass–spring system shown in Fig. 6.9. The second-order controller is a virtual mass–spring–dashpot system connected in parallel to the second mass m_2. Let w_1 be the position of the first mass m_1 from its equilibrium state and w_2 be the position of the second mass m_2 from its equilibrium state relative to the first mass. The quantities w_{c1} and w_{c2} are positions relative to the second mass m_2. Assume that there exists only an accelerometer attached to the second mass to measure its absolute acceleration $\ddot{w}_1 + \ddot{w}_2$.

The equations of motion for the system can be derived by application of a force to m_2 and its reaction force to m_{c1} and m_{c2}. The force applied to the system mass m_2 is the control force u transmitted through the springs k_{c1} and k_{c2} and the dashpots ζ_{c1} and ζ_{c2}.

(a) Treat the whole system as a passive system and derive its equation of motion.
(b) Separate the equation of motion into two parts, i.e., the system and the controller.
(c) What are the values of C_a and $\bar{B}_c$ in Eq. (6.47)? With the choices of C_a and $\bar{B}_c$, use Eqs. (6.42) and (6.44) to express the control force u.

6.7   Similar to Problem 6.6, Fig. 6.10 shows a two-degree-of-freedom mass-spring system connected in series with a virtual mass–spring–dashpot to the second mass m_2. The quantity w_{c2} is the position of mass m_{c2} relative to the mass m_{c1}.

The equations of motion for the system can be derived by application of a force to m_2 and its reaction force to m_{c1}. The force applied to the system mass m_2 is the control force u transmitted through the spring k_{c1} and the dashpot ζ_{c1}.

(a) Treat the whole system as a passive system and derive its equation of motion.
(b) Separate the equation of motion into two parts, i.e., the system and the controller.

SYSTEM   CONTROLLER
(A real mass-spring system)   (A virtual mass-spring-dashpot system)

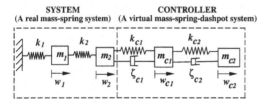

Figure 6.10. A mass–spring system with a two-degree-of-freedom dynamic controller in series.

(c) What are the values of C_a and $\bar{B}_c$ in Eq. (6.47)? With the choices of C_a and $\bar{B}_c$, use Eqs. (6.42) and (6.44) to express the control force u.

(d) Discuss this controller design in comparison with that derived from Problem 6.6.

BIBLIOGRAPHY

[1] Bupp, R. T., Bernstein, D. S., Chellaboina, V.-S., and Haddad, W. M., "Resetting Virtual Absorbers for Vibration Control," *J. Vib. Control*, Vol. 6, pp. 61–83, 2000.

[2] Juang, J. N. and Phan, M., "Robust Controller Designs for Second-Order Dynamic Systems: A Virtual Passive Approach," *J. Guid. Control Dyn.*, Vol. 15, pp. 1192–1198. 1992.

[3] Juang, J. N., Wu, S. C., Phan, M., and Longman, R. W., "Passive Dynamic Controllers for Non-Linear Mechanical Systems," *J. Guid. Control Dyn.*, Vol. 16, pp. 845–851, 1993.

[4] Lum, K.-Y., Bhat, S. P., Coppola, V. T., and Bernstein, D. S., "Adaptive Virtual Autobalancing for a Magnetic Rotor with Unknown Mass Imbalance: Theory and Experiment," *Trans. ASME J. Vib. Acoust.*, Vol. 120, pp. 557–570, 1998.

[5] Morris, K. A. and Juang, J. N., "Dissipative Controller Designs for Second-Order Dynamic Systems," *IEEE Trans. Autom. Control*, Vol. 39, pp. 1056–1063, 1994.

SUGGESTED READING

Hill, D. J. and Moylan, P. J., "Stability of Non-Linear Dissipative Systems," *IEEE Trans. Autom. Control*, Vol. AC-21, pp. 708–711, 1976.

Hill, D. J. and Moylan, P. J., "Dissipative Dynamical Systems: Basic Input-Output and State Properties," *J. Franklin Inst.*, Vol. 309, pp. 327–357, 1980.

Willems, J. C., "Dissipative Dynamical Systems," *Arch. Ration. Mech. Anal.*, Vol. 45, pp. 321–393, 1972.

7

State–Space Models

7.1 Introduction

This chapter shows the reader how to rewrite the second-order equations of motion for a general multiple-degree-of-freedom system into the form of a first-order matrix differential equation. What we have learned from Chap. 1 about the mathematics of a first-order matrix differential equation then applies. The first-order matrix differential equation is known as the state–space model, which is a fundamental equation on which modern control theory is based.

In this chapter, we will also study a sampled-data (or discrete-time) representation of the continuous-time state–space model. Assuming a constant input at each sampling interval, it is possible to represent the continuous-time state–space model that is in the form of a first-order matrix differential equation by a discrete-time first-order matrix difference equation. At the sampling points, the corresponding discrete-time state–space model describes exactly the continuous-system with a constant input at each sampling interval without any kind of approximation (Ref. [1–3]). The corresponding discrete-time model is also of the first order. As a result, control methods that are originally developed in the continuous-time domain can be converted almost trivially to the discrete-time domain for digital control implementation (Ref. [2]). Digital control is flexible in that changing a control strategy amounts to writing a different program (software) rather than constructing a different analog control circuitry (hardware). For the same reason, different signal processing techniques such as filtering, identification (Ref. [1]), etc., can be incorporated much more conveniently into the discrete-time digital format. In this chapter, elementary results in both continuous-time and discrete-time formats will be presented in parallel. This presentation will make very clear the strong parallelism between the two state–space representations.

We will also study how to obtain a system dynamic response by using different models when the system is driven by some general input function. In the continuous-time case, the solution can be obtained simply by integration of the first-order matrix differential equation. We have already studied various techniques to obtain the solution for such a differential equation in the first two chapters. It is now just a matter of adapting these results to the specific problem at hand. In the discrete-time case, it is even easier to compute the dynamic response to a general input. Only elementary matrix operations such as addition, subtraction, multiplication, etc., are required. The elementary linear algebra reviewed in Chap. 2 directly applies here. How a coordinate transformation affects the state–space presentation will also be shown. Finally, aliasing that is known to be an intrinsic problem in a sampled-data system will be briefly discussed.

7.2 Continuous-Time State–Space Model

So far, we have studied the equations of motion in second-order form, i.e., the highest derivative in the differential equations is of the second order. Let us now make the next step of expressing the same set of equations in first-order form so that theory for first-order matrix differential equations can be applied. Consider the equations of motion for a finite-dimensional linear dynamic system of the form

$$M\ddot{w} + \Xi\dot{w} + Kw = B_f u. \tag{7.1}$$

Assuming that the matrix M is invertible, we can solve for $\ddot{w}$ as

$$\ddot{w} = -M^{-1}\Xi\dot{w} - M^{-1}Kw + M^{-1}B_f u. \tag{7.2}$$

The original second-order equations can now be expressed in first-order form as

$$\frac{d}{dt}\begin{bmatrix} w(t) \\ \dot{w}(t) \end{bmatrix} = \begin{bmatrix} 0 & I \\ -M^{-1}K & -M^{-1}\Xi \end{bmatrix}\begin{bmatrix} w(t) \\ \dot{w}(t) \end{bmatrix} + \begin{bmatrix} 0 \\ M^{-1}B_f \end{bmatrix}u(t). \tag{7.3}$$

To simplify the equations further, define the state vector $x(t)$ as

$$x(t) = \begin{bmatrix} w(t) \\ \dot{w}(t) \end{bmatrix}. \tag{7.4}$$

Equation (7.2) can then be written in the form

$$\frac{dx}{dt} = A_c x + B_c u, \tag{7.5}$$

where

$$A_c = \begin{bmatrix} 0 & I \\ -M^{-1}K & -M^{-1}\Xi \end{bmatrix}, \qquad B_c = \begin{bmatrix} 0 \\ M^{-1}B_f \end{bmatrix}. \tag{7.6}$$

We have studied this first-order matrix differential equation in Chap. 1. Also, note that the nonzero forcing terms to the system are grouped together as an input vector u.

In a control problem, besides the equations of motions describing the dynamics of the system, there is another set of equations describing the output or measured quantities in terms of the variables that describe the system dynamics. For example, consider the spring–mass–damper system shown in Fig. 7.1. The output variable of interest may be the position of the first mass or the velocity of the second mass, etc. In general, the output variable(s) is denoted by the vector y, which is a function of the state variable x:

$$y = Cx, \tag{7.7}$$

where the matrix C is commonly called the output (or measurement) influence matrix. State equation (7.5) and output equation (7.7) are called the continuous-time state–space model of the system.

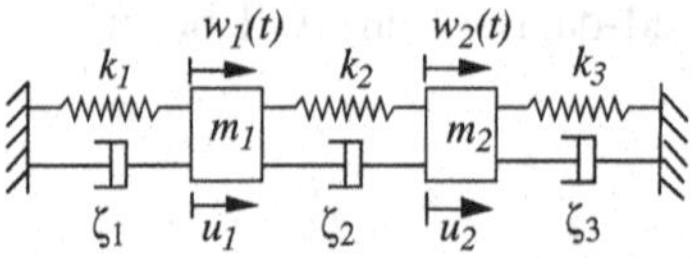

Figure 7.1. Two-degree-of-freedom system.

EXAMPLE 7.1

In the following, we illustrate a number of state–space models of the spring–mass–damper system shown in Fig. 7.1 for a number of combinations of input and output variables. In these example the vector $w(t)$ is made up of the positions of two masses:

$$w(t) = \begin{bmatrix} w_1(t) \\ w_2(t) \end{bmatrix}.$$

The second-order matrix equation is

$$M\ddot{w} + \Xi\dot{w} + Kw = u,$$

where the displacement vector w and the force vector u are

$$w(t) = \begin{bmatrix} w_1(t) \\ w_2(t) \end{bmatrix}, \qquad u(t) = \begin{bmatrix} u_1(t) \\ u_2(t) \end{bmatrix},$$

and the mass, damping, and stiffness matrices are defined as

$$M = \begin{bmatrix} m_1 & 0 \\ 0 & m_2 \end{bmatrix}, \quad \Xi = \begin{bmatrix} \zeta_1 + \zeta_2 & -\zeta_2 \\ -\zeta_2 & \zeta_2 + \zeta_3 \end{bmatrix}, \quad K = \begin{bmatrix} k_1 + k_2 & -k_2 \\ -k_2 & k_2 + k_3 \end{bmatrix}.$$

Let the state vector be defined as

$$x(t) = \begin{bmatrix} w_1(t) \\ w_2(t) \\ \dot{w}_1(t) \\ \dot{w}_2(t) \end{bmatrix}.$$

The state equation is

$$\frac{dx}{dt} = A_c x + B_c u,$$

where

$$A_c = \begin{bmatrix} 0 & 0 & 1 & 0 \\ 0 & 0 & 0 & 1 \\ -(k_1 + k_2)/m_1 & k_2/m_1 & -(\zeta_1 + \zeta_2)/m_1 & \zeta_2/m_1 \\ k_2/m_2 & -(k_2 + k_3)/m_2 & \zeta_2/m_2 & -(\zeta_2 + \zeta_3)/m_2 \end{bmatrix}.$$

The state matrix A_c is a function of the system parameters including m_j, ζ_j, and k_j for $j = 1, 2$. Before determining the input matrix B_c, we need to know the number of inputs u and their locations. In this example, the maximum number of independent inputs are two, i.e., one at the location of m_1 and another at m_2. Three cases are shown in the following.

(a) *Force input to the first mass:*
This is a single-input system. There is only one input force u_1 applied at the mass m_1. Therefore the quantity $B_c u$ in state equation becomes $B_c u_1$, where

$$B_c = \begin{bmatrix} 0 \\ 0 \\ \dfrac{1}{m_1} \\ 0 \end{bmatrix}.$$

The force unit must be a physical unit such as a newton. Otherwise there is a conversion factor from the command unit to the physical unit such as electrical volts, amps, etc. In that case, the nonzero element of B_c should be multiplied by the conversion factor.

(b) *Force input to the second mass:*
This is again a single-input system. The quantity $B_c u$ in the state equation becomes $B_c u_2$, where

$$B_c = \begin{bmatrix} 0 \\ 0 \\ 0 \\ \dfrac{1}{m_2} \end{bmatrix}.$$

(c) *Force input to both masses:*
This is a two-input system. There are two inputs, and the input matrix B_c is

$$B_c = \begin{bmatrix} 0 & 0 \\ 0 & 0 \\ \dfrac{1}{m_1} & 0 \\ 0 & \dfrac{1}{m_2} \end{bmatrix}.$$

Now let us look at the output equation

$$y = Cx$$

The output matrix C depends on the number and location of the sensors used to measure the system outputs.

(1) *Position output of the first mass:*
This is a single-output system. The output matrix C is

$$C = [1 \quad 0 \quad 0 \quad 0].$$

Here it is assumed that the measurement unit is a physical unit. Otherwise, there is a conversion factor from the physical unit to the measurement unit such as volts, amps, etc. In such a case, the first element of C will be the conversion factor rather than unity.

(2) *Velocity output of the first mass:*
This is a single-output system. The output matrix C is

$$C = [0 \quad 0 \quad 1 \quad 0].$$

(3) *Position output of the second mass:*
This is a single-output system. The output matrix C is

$$C = [0 \quad 1 \quad 0 \quad 0].$$

(4) *Velocity output of the second mass:*
This is a single-output system. The output matrix C is

$$C = [0 \quad 0 \quad 0 \quad 1].$$

(5) *Position and velocity outputs of the first mass:*
This is a two-output system. The output matrix C has two rows, i.e.,

$$C = \begin{bmatrix} 1 & 0 & 0 & 0 \\ 0 & 0 & 1 & 0 \end{bmatrix}.$$

(6) *Position and velocity outputs of the second mass:*
This is a two-output system. The output matrix C has two rows, i.e.,

$$C = \begin{bmatrix} 0 & 1 & 0 & 0 \\ 0 & 0 & 0 & 1 \end{bmatrix}.$$

(7) *Position and velocity outputs of both masses:*
This is a four-output system. The output matrix C has four rows, i.e.,

$$C = \begin{bmatrix} 1 & 0 & 0 & 0 \\ 0 & 1 & 0 & 0 \\ 0 & 0 & 1 & 0 \\ 0 & 0 & 0 & 1 \end{bmatrix}.$$

Any combination of the state equation and the output equation represent a state–space model. For example, the combination of (a) and (5) represents a single-input and two-output state–space model. Depending on whether there is one or two inputs to the system and which outputs are considered to be the variables of interest, we can easily construct a corresponding state–space model representation. In all cases, we see that the state–space model is represented by three matrices, A_c, B_c, C, which are called the state matrix, the input influence matrix, and the output influence matrix, respectively.

7.2.1 Direct-Transmission Term

Note that the output vector y in Eq. (7.7) is limited to information concerning the displacement $w(t)$ or the velocity $\dot{w}(t)$, or at most a linear combination of both. Often, in practice, acceleration measurements are also available. Therefore, we are interested in incorporating acceleration information in the output equation as well. From Eq. (7.2), it is seen that the acceleration $\ddot{w}(t)$ is a function of not only the displacement vector but also of the input vector. Thus, to include acceleration measurements in the output equation, the acceleration measurement equation may be written as

$$y = C_a \ddot{w}. \tag{7.8}$$

The matrix C_a describes the relationship between the vector $\ddot{w}$ and the measurement vector y and thus contains the conversion factors between physical units such as meters and electrical units such as volts. Substituting Eq. (7.2) for $\ddot{w}$ into Eq. (7.8) yields

$$y = C_a M^{-1}[B_f u - \Xi \dot{w} - K w].$$

Thus, a more general form takes the form

$$y = Cx + Du, \tag{7.9}$$

where

$$C = [-C_a M^{-1} K \quad -C_a M^{-1} \Xi], \qquad D = C_a M^{-1} B_f.$$

Here, C is the output influence matrix for the state vector x. The D term in Eq. (7.9) is called the direct-transmission term. It directly transmits the input u to the output y without having to go through the intermediate state equation.

EXAMPLE 7.2

Consider the spring–mass–damper system shown in Example 7.1; the second-order matrix equation can be written as

$$
\begin{bmatrix} \ddot{w}_1 \\ \ddot{w}_2 \end{bmatrix} =
\begin{bmatrix} -\dfrac{k_1 + k_2}{m_1} & \dfrac{k_2}{m_1} \\[2ex] \dfrac{k_2}{m_2} & -\dfrac{k_2 + k_3}{m_2} \end{bmatrix}
\begin{bmatrix} w_1 \\ w_2 \end{bmatrix}
$$

$$
+ \begin{bmatrix} -\dfrac{\zeta_1 + \zeta_2}{m_1} & \dfrac{\zeta_2}{m_1} \\[2ex] \dfrac{\zeta_2}{m_2} & -\dfrac{\zeta_2 + \zeta_3}{m_2} \end{bmatrix}
\begin{bmatrix} \dot{w}_1 \\ \dot{w}_2 \end{bmatrix}
+ \begin{bmatrix} \dfrac{u_1}{m_1} \\[2ex] \dfrac{u_2}{m_2} \end{bmatrix}.
$$

Assume that the output measurements consist of position of the first mass and the velocity and the acceleration of the second mass, i.e.,

$$
y = \begin{bmatrix} y_1 \\ y_2 \\ y_3 \end{bmatrix} = \begin{bmatrix} w_1 \\ \dot{w}_2 \\ \ddot{w}_2 \end{bmatrix}.
$$

The component y_3 corresponds to the acceleration measurement at the second mass. The influence matrix C_a shown in Eq. (7.8) for the acceleration measurement has the value $C_a = [0 \ 1]$ because

$$y_3 = \ddot{w}_2 = [0 \ \ 1] \begin{bmatrix} \ddot{w}_1 \\ \ddot{w}_2 \end{bmatrix}.$$

Here the conversion factor from the physical unit to the electrical measurement unit is assumed to be unity. Application of Eq. (7.9) thus yields

$$y_3 = \begin{bmatrix} \dfrac{k_2}{m_2} & -\dfrac{k_2 + k_3}{m_2} \end{bmatrix} \begin{bmatrix} w_1 \\ w_2 \end{bmatrix} + \begin{bmatrix} \dfrac{\varsigma_2}{m_2} & -\dfrac{\varsigma_2 + \varsigma_3}{m_2} \end{bmatrix} \begin{bmatrix} \dot{w}_1 \\ \dot{w}_2 \end{bmatrix} + \dfrac{1}{m_2} u_2.$$

The corresponding measurement equation for all three outputs becomes

$$y = \begin{bmatrix} 1 & 0 & 0 & 0 \\ 0 & 0 & 0 & 1 \\ \dfrac{k_2}{m_2} & -\dfrac{k_2 + k_3}{m_2} & \dfrac{\varsigma_2}{m_2} & -\dfrac{\varsigma_2 + \varsigma_3}{m_2} \end{bmatrix} \begin{bmatrix} w_1 \\ w_2 \\ \dot{w}_1 \\ \dot{w}_2 \end{bmatrix} + \begin{bmatrix} 0 & 0 \\ 0 & 0 \\ 0 & \dfrac{1}{m_2} \end{bmatrix} \begin{bmatrix} u_1 \\ u_2 \end{bmatrix}.$$

Obviously, the direct-transmission matrix is

$$D = \begin{bmatrix} 0 & 0 \\ 0 & 0 \\ 0 & \dfrac{1}{m_2} \end{bmatrix}.$$

7.2.2 Coordinate Transformation

Basically, the state–space model describes the relationship between the inputs and the outputs of a system by means of an intermediate variable called the state vector. The state–space model is coordinate dependent. For example, let the state vector be transformed by a new set of coordinates,

$$z = Tx \quad \text{or} \quad x = T^{-1}z. \tag{7.10}$$

Multiply both sides of the original state–space model, Eq. (7.5), by T,

$$T\frac{dx}{dt} = TA_cx + TB_cu \tag{7.11}$$

and recognize that

$$T\frac{dx}{dt} = \frac{d}{dt}Tx, \qquad T^{-1}T = I,$$

where I is an identity matrix. We can express the state–space model of the same system with z as a new state vector as

$$\frac{dz}{dt} = (T A_c T^{-1})z + T B_c u,$$
$$y = (C T^{-1})z + Du. \tag{7.12}$$

With a new state matrix and new input and output influence matrices

$$\bar{A}_c = T A_c T^{-1}, \quad \bar{B}_c = T B_c, \quad \bar{C} = C T^{-1},$$

the state–space model becomes

$$\frac{dz}{dt} = \bar{A}_c z + \bar{B}_c u,$$
$$y = \bar{C}z + Du. \tag{7.13}$$

In the new coordinates, the state–space model still preserves its general form. The new state–space model is related to the original state–space model by a similarity transformation in the sense that it preserves the eigenvalues of the state matrix. The direct-transmission term D is coordinate independent.

7.2.3 Dynamic Response to General Input

One major advantage of putting the equations in the state–space form is that the system of dynamic equations is now in the form of a first-order matrix differential equation. To obtain the state of the system at any time t, we simply integrate the state equation

$$\frac{dx}{dt} = A_c x + B_c u$$

to solve for $x(t)$ by using the matrix exponential approach shown in Chap. 2:

$$x(t) = e^{A_c(t-t_0)}x(t_0) + \int_{t_0}^{t} e^{A_c(t-\tau)} B_c u(\tau)d\tau, \tag{7.14}$$

where $x(t_0)$ is the initial state of the system at time $t = t_0$. Once the state of the system is determined, the output of the system is directly given by the output equation

$$y(t) = Cx(t) + Du(t).$$

If we are not interested in the state, then we can obtain the output directly from

$$y(t) = Ce^{A_c(t-t_0)}x(t_0) + \int_{t_0}^{t} Ce^{A_c(t-\tau)} B_c u(\tau)d\tau + Du(t). \tag{7.15}$$

The system can be represented by the following block diagram:

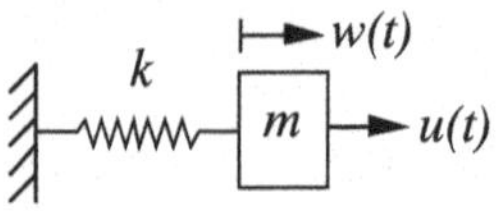

Figure 7.2. Single-degree-of-freedom mass–spring system.

EXAMPLE 7.3

Consider the single-degree-of-freedom mass–spring system shown in Fig. 7.2, where $m = k = 1$. The second-order equation of motion is

$$\ddot{w}(t) + w(t) = u(t),$$

where $w(t)$ denotes the position of the mass. Suppose that an accelerometer is located at the mass to measure its acceleration, i.e., the output equation is

$$y(t) = \ddot{w}(t).$$

Define a state vector $x(t)$ as

$$x(t) = \begin{bmatrix} w(t) \\ \dot{w}(t) \end{bmatrix}.$$

A continuous-time state–space model of the system is

$$\frac{d}{dt}\begin{bmatrix} w(t) \\ \dot{w}(t) \end{bmatrix} = \begin{bmatrix} 0 & 1 \\ -1 & 0 \end{bmatrix}\begin{bmatrix} w(t) \\ \dot{w}(t) \end{bmatrix} + \begin{bmatrix} 0 \\ 1 \end{bmatrix}u(t),$$

$$y(t) = \begin{bmatrix} -1 & 0 \end{bmatrix}\begin{bmatrix} w(t) \\ \dot{w}(t) \end{bmatrix} + u(t).$$

Thus, the state–space system matrices are

$$A_c = \begin{bmatrix} 0 & 1 \\ -1 & 0 \end{bmatrix}, \quad B_c = \begin{bmatrix} 0 \\ 1 \end{bmatrix}, \quad C = \begin{bmatrix} -1 & 0 \end{bmatrix}, \quad D = 1.$$

For any given forcing input $u(t)$ and initial conditions $w(0)$ and $\dot{w}(0)$, integrating the state equation yields the state vector:

$$
\begin{aligned}
x(t) &= e^{A_c t}x(0) + \int_0^t e^{A_c(t-\tau)}B_c u(\tau)d\tau \\
&= \begin{bmatrix} \cos(t) & \sin(t) \\ -\sin(t) & \cos(t) \end{bmatrix}\begin{bmatrix} w(0) \\ \dot{w}(0) \end{bmatrix} \\
&\quad + \int_0^t \begin{bmatrix} \cos(t-\tau) & \sin(t-\tau) \\ -\sin(t-\tau) & \cos(t-\tau) \end{bmatrix}\begin{bmatrix} 0 \\ 1 \end{bmatrix}u(\tau)d\tau \\
&= \begin{bmatrix} w(0)\cos(t) + \dot{w}(0)\sin(t) + \int_0^t \sin(t-\tau)u(\tau)d\tau \\ -w(0)\sin(t) + \dot{w}(0)\cos(t) + \int_0^t \cos(t-\tau)u(\tau)d\tau \end{bmatrix}.
\end{aligned}
$$

The acceleration output is therefore

$$y(t) = -w(0)\cos(t) - \dot{w}(0)\sin(t) - \int_0^t \sin(t - \tau)u(\tau)d\tau + u(t).$$

For the case in which $w(0) = \dot{w}(0) = 0$ and $u(t) = 1$ for $t \geq 0$, the acceleration output becomes

$$y(t) = -\int_0^t \sin(t - \tau)d\tau + 1 = -\cos(t - \tau)\big|_{\tau=0}^{\tau=t} + 1 = \cos(t).$$

This is the acceleration response to a unit step forcing input at the end mass.

7.3 Discrete-Time State–Space Model

Physical quantities such as displacement, velocity, and acceleration are continuous in nature and thus change continuously with time. Most sensors measuring such physical quantities generate an analog signal that changes continuously with time. The analog signal must be sampled for a digital computer to be able to handle it. In this section, we study how a continuous-time state–space model can be converted into a discrete-time (sampled-data) representation for digital control. The specific mechanism considered here is a sample and hold device (zero-order hold).

7.3.1 A Zero-Order Hold (or Sample and Hold)

A zero-order hold takes a continuous signal and turns it into a stepwise signal in which the signal is sampled and held for a certain interval of time, called the sampling interval. The sampling frequency is the inverse of the sampling interval. With a zero-order hold on the input, the continuous-time state–space equation can be converted to a discrete-time one. A typical discretized input is shown in Fig. 7.3, where Δt is the sampling interval and k is an integer denoting the time index.

7.3.2 Continuous-Time to Discrete-Time Conversion

With a zero-order-hold input, as shown in Fig. 7.3, a continuous-time system can be represented by a discrete-time one that is exact at the sampling instants. Consider the discrete sampling intervals $0, \Delta t, 2\Delta t, \ldots, (k + 1)\Delta t$. To see how $x(t)$ changes from one time step to the next, substitute $t_0 = k\Delta t$ and $t = (k + 1)\Delta t$ into Eq. (7.15). Furthermore, with a zero-order hold, i.e., the input $u(k\Delta t)$ is held constant over the

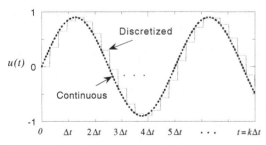

Figure 7.3. Continuous vs. discrete signals.

time interval from $k\Delta t$ to $(k+1)\Delta t$, we obtain from Eq. (7.15) that

$$x[(k+1)\Delta t] = e^{A_c(k\Delta t)}x(k\Delta t) + \int_{k\Delta t}^{(k+1)\Delta t} e^{A_c[(k+1)\Delta t - \tau]}B_c u(\tau)d\tau$$

$$= e^{A_c(k\Delta t)}x(k\Delta t) + \int_0^{\Delta t} e^{A_c\tau'}d\tau' B_c u(k\Delta t), \tag{7.16}$$

where $\tau' = (k+1)\Delta t - \tau$ has been used to obtain the last equality. The output equation at each sampled instant is

$$y(k\Delta t) = Cx(k\Delta t) + Du(k\Delta t). \tag{7.17}$$

Using a simplified notation k for the time argument $k\Delta t$, [for example, $x(k)$ means $x(k\Delta t)$], we now have a discrete-time (sampled-data) representation of the original continuous-time state–space model:

$$x(k+1) = Ax(k) + Bu(k),$$
$$y(k) = Cx(k) + Du(k). \tag{7.18}$$

The discrete-time system matrix A and the input influence matrix B are

$$A = e^{A_c(k\Delta t)}, \qquad B = \int_0^{\Delta t} e^{A_c\tau'}d\tau' B_c \tag{7.19}$$

and matrices C and D are the same as in the continuous-time representation.

The system can be represented by the following block diagram shown in Fig. 7.4: The discrete-time model gives the exact value of the response of the system at the sampling instants, when the input signal is turned into a staircase signal by a sample-and-hold device before entering the system. It is commonly misunderstood that the discrete-time model is only an approximation of the original continuous-time model in the sense that the time derivative is approximated by a finite difference that is never exact, no matter how small the approximation error is. Clearly this is not the case here because we obtain the discrete-time model by actually integrating the system equation over each successive time interval. The only requirement being made here is that the input signal be held constant during the sampling interval by a sample-and-hold device. It is still true that our discrete-time model only approximates the actual response, but to the extent that it does not give us information in between sampling instants. What it tells us at the sampling instants, however, is correct (exact). If we had tried to represent the continuous model by some finite-difference scheme, then we in general will not get the correct information at the sampling instants no matter how good the finite-difference scheme is. Furthermore, a better finite-difference scheme will likely result in a discrete-time model of a much higher order (note that here the discrete-time model is first order in the time step). In practice, when the control system is implemented by a computer, the inclusion of a sample-and-hold device is routine.

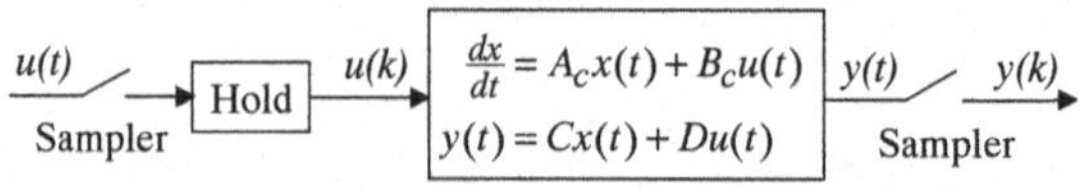

Figure 7.4. Continuous system with a zero-order-hold.

To calculate the discrete-time system matrix A and the input matrix B, we can use Eqs. (7.19) directly for a simple system. For a more complex system, however, we may compute the discrete-time matrices numerically by using the matrix exponential approach. The calculation for A is straightforward:

$$A = e^{A_c \Delta t} = I + A_c \Delta t + \frac{1}{2!}(A_c \Delta t)^2 + \frac{1}{3!}(A_c \Delta t)^3 + \cdots. \qquad (7.20)$$

The discrete-time input influence matrix B can be computed from

$$B = \int_0^{\Delta t} e^{A_c \tau} d\tau \, B_c$$

$$= \int_0^{\Delta t} \left(I + A_c \tau + \frac{1}{2!}(A_c \tau)^2 + \frac{1}{3!}(A_c \tau)^3 + \cdots \right) d\tau \, B_c$$

$$= \left[I \Delta t + \frac{1}{2!} A_c (\Delta t)^2 + \frac{1}{3!} A_c^2 (\Delta t)^3 + \cdots \right] B_c. \qquad (7.21)$$

If none of the eigenvalues of A_c are zero, then A^{-1} exists and the expression for B can be simplified further as

$$B = \int_0^{\Delta t} e^{A_c \tau} d\tau \, B_c = A_c^{-1} e^{A_c \tau} \big|_0^{\Delta t} = A_c^{-1}(A - I)B_c. \qquad (7.22)$$

EXAMPLE 7.4

Consider the same mass–spring system shown in Fig. 7.2 for Example 7.3. The second-order differential equation for this system is

$$\ddot{w} + w = u(t).$$

The continuous-time first-order model becomes

$$\frac{d}{dt} \begin{bmatrix} w \\ \dot{w} \end{bmatrix} = \begin{bmatrix} 0 & 1 \\ -1 & 0 \end{bmatrix} \begin{bmatrix} w \\ \dot{w} \end{bmatrix} + \begin{bmatrix} 0 \\ 1 \end{bmatrix} u,$$

$$y = [-1 \quad 0] \begin{bmatrix} w \\ \dot{w} \end{bmatrix} + u,$$

The state-space matrices are

$$A_c = \begin{bmatrix} 0 & 1 \\ -1 & 0 \end{bmatrix}, \quad B_c = \begin{bmatrix} 0 \\ 1 \end{bmatrix}, \quad C = [-1 \quad 0], \quad D = 1.$$

To derive the discrete-time model, first note that

$$A_c^2 = A_c A_c = \begin{bmatrix} 0 & 1 \\ -1 & 0 \end{bmatrix} \begin{bmatrix} 0 & 1 \\ -1 & 0 \end{bmatrix} = \begin{bmatrix} -1 & 0 \\ 0 & -1 \end{bmatrix},$$

$$A_c^3 = A_c A_c^2 = \begin{bmatrix} 0 & 1 \\ -1 & 0 \end{bmatrix} \begin{bmatrix} -1 & 0 \\ 0 & -1 \end{bmatrix} = \begin{bmatrix} 0 & -1 \\ 1 & 0 \end{bmatrix},$$

$$A_c^4 = A_c A_c^3 = \begin{bmatrix} 0 & 1 \\ -1 & 0 \end{bmatrix} \begin{bmatrix} 0 & -1 \\ 1 & 0 \end{bmatrix} = \begin{bmatrix} 1 & 0 \\ 0 & 1 \end{bmatrix}.$$

The discrete-time matrices A and B can be obtained as

$$A = I + A_c \Delta t + \frac{1}{2!}(A_c \Delta t)^2 + \frac{1}{3!}(A_c \Delta t)^3 + \cdots$$

$$= \begin{bmatrix} 1 - \frac{1}{2!}\Delta t^2 + \frac{1}{4!}\Delta t^4 + \cdots & \Delta t - \frac{1}{3!}\Delta t^3 + \frac{1}{5!}\Delta t^5 + \cdots \\ -\left(\Delta t - \frac{1}{3!}\Delta t^3 + \frac{1}{5!}\Delta t^5 + \cdots\right) & 1 - \frac{1}{2!}\Delta t^2 + \frac{1}{4!}\Delta t^4 + \cdots \end{bmatrix}$$

$$= \begin{bmatrix} \cos \Delta t & \sin \Delta t \\ -\sin \Delta t & \cos \Delta t \end{bmatrix},$$

$$B = \left[I\Delta t + \frac{1}{2!}A_c(\Delta t)^2 + \frac{1}{3!}A_c^2(\Delta t)^3 + \cdots \right] B_c$$

$$= \begin{bmatrix} \Delta t - \frac{1}{3!}\Delta t^3 + \frac{1}{5!}\Delta t^5 + \cdots & 1 - \left(1 - \frac{1}{2!}\Delta t^2 + \frac{1}{4!}\Delta t^4 - \cdots\right) \\ -1 + \left(1 - \frac{1}{2!}\Delta t^2 + \frac{1}{4!}\Delta t^4 + \cdots\right) & \Delta t - \frac{1}{3!}\Delta t^3 + \frac{1}{5!}\Delta t^5 + \cdots \end{bmatrix} \begin{bmatrix} 0 \\ 1 \end{bmatrix}$$

$$= \begin{bmatrix} \sin \Delta t & 1 - \cos \Delta t \\ -1 + \cos \Delta t & \sin \Delta t \end{bmatrix} \begin{bmatrix} 0 \\ 1 \end{bmatrix}$$

$$= \begin{bmatrix} 1 - \cos \Delta t \\ \sin \Delta t \end{bmatrix}.$$

Therefore the discrete-time model becomes

$$\begin{bmatrix} x_1(k+1) \\ x_2(k+1) \end{bmatrix} = \begin{bmatrix} \cos \Delta t & \sin \Delta t \\ -\sin \Delta t & \cos \Delta t \end{bmatrix} \begin{bmatrix} x_1(k) \\ x_2(k) \end{bmatrix} + \begin{bmatrix} 1 - \cos \Delta t \\ \sin \Delta t \end{bmatrix} u(k),$$

$$y(k) = [-1 \quad 0] \begin{bmatrix} x_1(k) \\ x_2(k) \end{bmatrix} + u(k).$$

Observe that the discrete-time model varies as a function of sampling interval Δt. For a fixed sampling interval, the discrete-time model is time invariant if the original continuous-time model is also time invariant.

7.3.3 Coordinate Transformation

Similar to the continuous-time state–space model, the discrete-time state–space model is also coordinate dependent. Let the state vector in Eq. (7.18) be transformed by a new set of coordinates,

$$x = Tz \quad \text{or} \quad z = T^{-1}x, \tag{7.23}$$

where T is a transformation matrix. Premultiplying the original state–space model, Eqs. (7.18), by T^{-1} and using the equality $TT^{-1} = I$, we obtain

$$T^{-1}x(k+1) = T^{-1}ATT^{-1}x(k) + T^{-1}Bu(k),$$

$$y(k) = CTT^{-1}x(k) + Du(k). \tag{7.24}$$

The state–space model of the same system can then be expressed with z as a new state vector:

$$z(k+1) = \bar{A}z(k) + \bar{B}u(k),$$

$$y(k) = \bar{C}z(k) + Du(k), \qquad (7.25)$$

where

$$\bar{A} = T^{-1}AT, \quad \bar{B} = T^{-1}B, \quad \bar{C} = CT. \qquad (7.26)$$

Thus, in the new coordinates, the discrete-time state–space model still preserves its general form. The new state matrix, new input influence matrix, and new output influence matrix are related to the original ones by a similarity transformation. Note that the direct-transmission term D is unaffected by this transformation.

7.3.4 Dynamic Response of a Discrete-Time Model

For the discrete-time model, however, the computation of the dynamic response to a general input $u(t)$ is much simpler because the integration action is already built into the model. Starting with an initial state $x(0)$, the state of the system at each successive time step can be computed directly as

$$x(1) = Ax(0) + Bu(0),$$

$$x(2) = Ax(1) + Bu(1)$$
$$= A^2x(0) + ABu(0) + Bu(1),$$

$$x(3) = Ax(2) + Bu(2)$$
$$= A^3x(0) + A^2Bu(0) + ABu(1) + Bu(2),$$

$$\vdots \qquad (7.27)$$

where $u(k)$ is the value of $u(t)$ at time $t = k\Delta t$. In general, the state of the system at any time $t = k\Delta t$ is simply

$$x(k) = A^k x(0) + \sum_{i=1}^{k} A^{i-1} Bu(k-i). \qquad (7.28)$$

Correspondingly, the output response at any time $t = k\Delta t$ is

$$y(k) = CA^k x(0) + \sum_{i=1}^{k} CA^{i-1} Bu(k-i) + Du(k). \qquad (7.29)$$

It is important to realize that for the *same* input function $u(t)$, the solution from integrating the differential equation (continuous-time model) and that from the corresponding difference equation (discrete-time model) are *not* the same. In the continuous-time case, the solution is the response of the system to the original input function $u(t)$. In the discrete-time case, it is the response of the system to the discretized version of $u(t)$ generated by a sampling-and-hold device with a sampling interval Δt.

7.4 Sampling and Aliasing

In theory, it is desirable to sample a signal sufficiently fast (i.e., Δt is sufficiently small) so that the sampled signal captures the information present in the original signal. If the sampling interval is too coarse, then high-frequency information in the original signal may be lost. More precisely, a high-frequency signal will appear as a low-frequency signal if the sampling interval is too large. This phenomenon is called aliasing, which is clearly not desirable. The frequency at which aliasing starts occurring is called the Nyquist frequency. The Nyquist frequency can be shown to be one half of the sampling frequency. Recall that the sampling frequency is the reciprocal of the sampling interval, $f_s = 1/\Delta t$.

For example, if we decide to sample a signal with a sampling interval of $\Delta t = 0.1$ s, then the sampling frequency is $f_s = 1/\Delta t = 10$ Hz. The Nyquist frequency in this case is $f_N = f_s/2 = 5$ Hz. If a signal to be sampled contains frequencies higher than 5 Hz, then these frequencies will appear (or fold back) as low frequencies when they are sampled. To avoid this problem, either we must make sure that the signal to be sampled does not contain frequencies higher than the Nyquist frequency or we must increase the sampling frequency. Alternatively, we can filter the signal by a low-pass filter to remove any unwanted high frequencies before sampling. However, a low-pass filter will also distort the original signal, which should be carefully handled.

Figure 7.5 shows a 5-Hz signal $y = \cos(\omega t)$, where $\omega = 2\pi f$ with $f = 5$ Hz. This signal is sampled at a sampling frequency $f_s = 10$ Hz (the sampling interval is $\Delta t = 0.1$ s). Because $f = 5$ Hz is exactly the Nyquist frequency, it is expected that the sampled signal is barely able to resolve the frequency of the original signal (at best), as shown in Fig. 7.6. The dots are connected only as a visual aid. This sampling frequency will not be able to resolve a signal higher than 5 Hz. To illustrate this case, Fig. 7.7 shows a 12.5-Hz signal. When sampled at $f_s = 10$ Hz, this signal appears as if it is a 2.5-Hz signal. The aliasing effect is shown in Fig. 7.8.

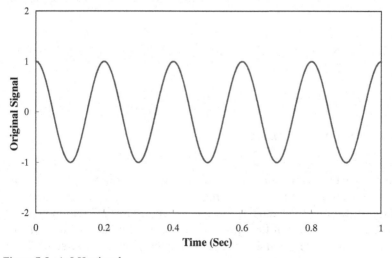

Figure 7.5. A 5-Hz signal.

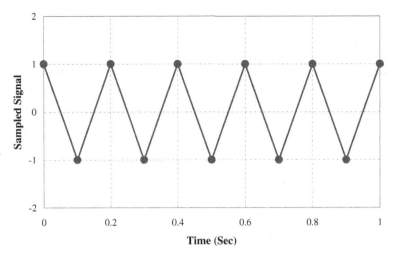

Figure 7.6. A 5-Hz signal sampled at 10 Hz.

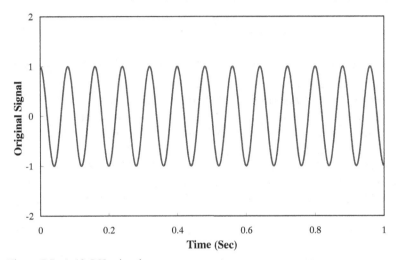

Figure 7.7. A 12.5-Hz signal.

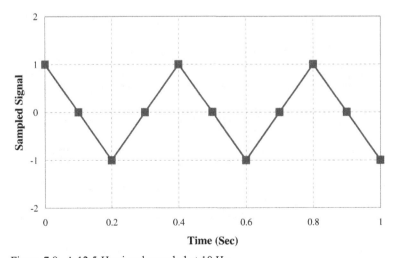

Figure 7.8. A 12.5-Hz signal sampled at 10 Hz.

179

7.5 System Eigenvalues

Let λ denote an eigenvalue of the continuous-time system matrix A_c and z denote the corresponding eigenvalue of the corresponding discrete-time system A with a sampling interval Δt. Because $A = e^{A_c \Delta t}$, it follows that

$$z = e^{\lambda \Delta t}. \tag{7.30}$$

Conversely,

$$\lambda = \frac{\ln z}{\Delta t}. \tag{7.31}$$

It is important to note that the transformation from the discrete-time model to the continuous-time model is not unique. The imaginary part of the natural logarithm of a complex number can be adjusted by the addition of any multiple of 2π, which allows λ to take on different values, i.e.,

$$z = e^{\lambda \Delta t} = e^{\lambda \Delta t + 2ik\pi} \implies \lambda + i\frac{2k\pi}{\Delta t} = \frac{\ln z}{\Delta t}, \tag{7.32}$$

where k is an arbitrary integer. This corresponds to the fact that any two frequencies that differ by a multiple of $2\pi/\Delta t$ are indistinguishable when observed at the sampling instants. Therefore, in practice, if we wish to interpret the natural frequencies of the physical system, either the sampling interval Δt must be sufficiently short or a filter must be added to prevent the frequencies beyond the Nyquist frequency from being misinterpreted as real frequencies.

The mapping $z = e^{\lambda \Delta t}$ maps the left half (or right half) of the complex plane of λ into the area inside (or outside) unit circle of the complex plane of z, as shown in Fig. 7.9.

EXAMPLE 7.5

Consider the system in which

$$A_c = \begin{bmatrix} 0 & 1 \\ -1 & -1 \end{bmatrix},$$

which has the complex eigenvalues $\lambda_1 = -0.5 + 0.866i$ and $\lambda_2 = -0.5 - 0.866i$.

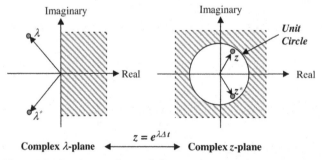

Figure 7.9. Continuous-time and discrete-time mapping.

When sampled at $\Delta t = 0.01$ s, its discrete-time version is

$$A = \begin{bmatrix} 1 & 0.01 \\ -0.01 & 0.99 \end{bmatrix},$$

which has the complex eigenvalues $z_1 = -0.995 + 0.0086i$ and $z_2 = -0.995 - 0.0086i$. It can be verified that

$$z_1 = e^{\lambda_1 \Delta t}, \qquad z_2 = e^{\lambda_2 \Delta t},$$

$$\lambda_1 = \frac{\ln z_1}{\Delta t}, \qquad \lambda_2 = \frac{\ln z_2}{\Delta t}.$$

7.6 Concluding Remarks

This chapter describes the first-order state-space model used in modern control in both continuous-time and discrete-time settings. Although it is natural to represent a physical system in continuous time, the discrete-time version is often more useful in the digital world for system characterization and feedback control. With regard to the control problem, there are two approaches for controller designs. The first approach is to use the continuous-time model to design a continuous-times controller and discretize it for actual implementation with a digital computer. The continuous-time model may be derived analytically from physical principles. The second approach is to design a discrete-time controller for the discrete-time model directly without any discretization. The discrete-model may be obtained by discretizing the continuous-time model or by system identification directly from input and output data that will be shown later in Chap. 10.

7.7 Problems

7.1 Consider the dynamical equation of motion:

$$m\ddot{w} + \zeta \dot{w} + kw = u.$$

Assume that

$$m = 1, \qquad k = 0$$

and ζ is an arbitrary constant. Let the sampling time interval be

$$\Delta t = 1.$$

(a) What is its first-order continuous model? Is the model unique? If not, why?
(b) What is its discrete model? Is the model unique? If not, why?
(c) Is the discrete model controllable? Explain.
(d) Let the measurement equation be $y = \ddot{w}$. Is this observable? Explain.
(e) Let the measurement equation be $y = \dot{w}$. Is this observable? Explain.
(f) Let the measurement equation be $y = w$. Is this observable? Explain.
(g) Choose any one which is controllable and observable, and complete the following items.

   (1) Given an initial pulse such that $u(0) = 1$, compute pulse-response functions.

(2) Compute a deadbeat-observer gain G such that $(A + GC)^2 = 0$, where A is the discrete-time state matrix and C is the output matrix.

7.2 Prove that the discrete-time state matrix A and the continuous-time state matrix A_c share the same eigenvectors. The relationship between A and A_c is shown in Eq. (7.20).

7.3 Rewrite the following mass–spring–damper system,

$$\ddot{w} + 0.1\dot{w} + w = u,$$

in first-order state-space format A, B, C, D using the following output(s):

(a) velocity of the mass, $\dot{w}$,
(b) acceleration of the mass, $\ddot{w}$,
(c) position and velocity of the mass, w and $\dot{w}$,
(d) position and acceleration of the mass, w and $\ddot{w}$,
(e) position, velocity, and acceleration of mass, w, $\dot{w}$, and $\ddot{w}$.

7.4 Consider the system given by

$$\ddot{w}(t) + w(t) = u(t)$$

with $u(t) = 1$ corresponding to a step input and initial conditions $w(0) = 0$ and $\dot{w}(0) = 1$.

(a) Solve for $w(t)$ directly from the second-order differential equation and then take the time derivative to solve for $\dot{w}(t)$.
(b) First put the system in state–space format and then solve for the system state consisting of $w(t)$ and $\dot{w}(t)$ by using the matrix exponential approach.

7.5 Rewrite the following fourth-order scalar differential equation

$$\frac{d^4 w(t)}{dt^4} + 2\frac{d^3 w(t)}{dt^3} + 3\frac{d^2 w(t)}{dt^2} + 4\frac{dw(t)}{dt} + 5w(t) = u(t)$$

as a first-order matrix differential equation, where the state variable is made up of

$$w(t), \quad \frac{dw(t)}{dt}, \quad \frac{d^2 w(t)}{dt^2}, \quad \frac{d^3 w(t)}{dt^3}.$$

7.6 For the same system considered in Problem 7.4, convert the continuous-time model into a discrete-time one with a sampling interval $\Delta t = 0.1$ s by using the power-series expressions subject to a zero-order hold on the input. Do this by hand calculation including up to, say, five terms or more.

7.7 Using the discrete-time model derived in Problem 7.6, compute the system response $y(t)$ (position and velocity) to the input $u(t) = \sin(2t)$ sampled at time interval $\Delta t = 0.1$ s. The system has zero initial conditions. Do this by hand calculation for the first 10 sampling instants. Now make the following coordinate transformation to change the state to a new coordinate z defined by

$$z = Tx, \quad T = \begin{bmatrix} 1 & 1 \\ -1 & 1 \end{bmatrix}.$$

Derive the new state–space model in the new coordinates for z. Use this new state–space model to compute the response when driven by the input $u(t) = \sin(2t)$ sampled at time interval $\Delta t = 0.1$ s. Do we expect to have the same answer compared with the result before transformation?

7.8 Observe that in the system

$$\frac{d}{dt}\begin{bmatrix} x_1(t) \\ x_2(t) \end{bmatrix} = \begin{bmatrix} -3 & 1 \\ 1 & -3 \end{bmatrix}\begin{bmatrix} x_1(t) \\ x_2(t) \end{bmatrix} + \begin{bmatrix} 1 \\ 1 \end{bmatrix}u(t),$$

$$y(t) = \begin{bmatrix} 1 & 0 \end{bmatrix}\begin{bmatrix} x_1(t) \\ x_2(t) \end{bmatrix},$$

the state variables $x_1(t)$ and $x_2(t)$ are coupled through the system matrix A:

$$A = \begin{bmatrix} -3 & 1 \\ 1 & -3 \end{bmatrix}.$$

Transform this state–space model to a new coordinate system in which the new state variables $z_1(t)$ and $z_2(t)$ become uncoupled in the new (transformed) A matrix. Also compute the transformed B and C matrices.

7.9 Suppose that a continuous-time signal is made up of three distinct frequencies, 1, 5, and 10 Hz. We wish to sample this signal at discrete-time instants. Discuss the consequences of selecting each of the following sampling frequencies: 1, 2, 10, 20, 50, and 100 Hz. Be as precise and clear as possible in your answers. (The sampling frequency is defined to be $1/\Delta t$, where Δt is the sampling interval).

BIBLIOGRAPHY

[1] Juang, J.-N., *Applied System Identification*, Prentice-Hall, Englewood Cliffs, NJ, 1994.

[2] Ogata, K., *Discrete-Time Control Systems*, Prentice-Hall, Upper Saddle River, NJ, 1995.

[3] Petkov, P. Hr., Christov, N. D., and Konstantinov, M. M., *Computational Methods for Linear Control Systems*, Prentice-Hall, Upper Saddle River, NJ, 1991.

8

State-Feedback Control

8.1 Introduction

The objective of a control system is to influence the dynamic system to make it behave in a desirable manner (Ref. [1–9]). Typical objectives of a control system are regulation and tracking. In a regulation problem, the system is controlled so that its output is maintained at a certain set point. In a tracking problem, the system is controlled so that its output follows a particular desired trajectory. A special case of the regulation problem is the stabilization problem, in which a control system is designed to bring the system to rest from any nonzero initial conditions (i.e., the desirable set point is zero). For a flexible structure that may be subjected to unwanted vibration, this is usually the most important goal of a controlled system. Stabilization is the focus of this chapter. In particular, we consider a special but very important class of control systems, namely state-feedback control, in which the control input is some function of the system states. For the moment we assume that there are enough sensors to measure the state of the system at any point in time to be used in computing the control input. If the state of the system cannot be measured directly, then a state observer is needed to estimate the system state from the measurements. The estimated state is then used in a state-feedback-control law. This subject of state estimation will be dealt with in the next chapter.

8.2 Controllability and Observability

In the following we will address the concepts of controllability and observability for a discrete-time model. It is simpler to explain these concepts in the discrete-time domain. The treatment for a continuous-time model is similar. Consider the discrete-time state–space representation of a dynamic system:

$$x(k + 1) = Ax(k) + Bu(k), \tag{8.1a}$$

$$y(k) = Cx(k) + Du(k). \tag{8.1b}$$

The order of the state–space model is n, i.e., A is an $n \times n$ matrix. The system has r inputs and m outputs. Thus the dimensions of B are $n \times r$, the dimensions of C are $m \times n$, and the dimensions of D are $m \times r$.

8.2.1 Controllability in the Discrete-Time Domain

A state $x(k)$ of a system is said to be controllable if this state can be reached from any initial state $x(0)$ of the system in a finite time interval by some control action. If all states are controllable, the system is said to be completely controllable or simply controllable. Starting with some initial state $x(0)$, the state of the system at any time step $k = p$ is given by

$$x(1) = Ax(0) + Bu(0), \tag{8.2a}$$

$$x(2) = Ax(1) + Bu(1)$$
$$= A^2 x(0) + ABu(0) + Bu(1), \tag{8.2b}$$

$$x(3) = Ax(2) + Bu(2)$$
$$= A^3 x(0) + A^2 Bu(0) + ABu(1) + Bu(2), \tag{8.2c}$$

$$\vdots$$

$$x(p) = A^p x(0) + A^{p-1} Bu(0) + \cdots + ABu(p-2) + Bu(p-1)$$

$$= A^p x(0) + [A^{p-1} B \quad \cdots \quad AB \quad B] \begin{bmatrix} u(0) \\ \vdots \\ u(p-2) \\ u(p-1) \end{bmatrix}. \tag{8.2d}$$

Starting with an initial state $x(0)$ and specifying any desired state $x(p)$ at time step $k = p$, we find that the control input time history that will bring $x(0)$ to $x(p)$ must be solved from the following equation:

$$[A^{p-1} B \quad \cdots \quad AB \quad B] \begin{bmatrix} u(0) \\ \vdots \\ u(p-2) \\ u(p-1) \end{bmatrix} = x(p) - A^p x(0). \tag{8.3}$$

The question of whether or not there exists a control history that will bring the system from an initial state $x(0)$ to some desired state $x(p)$ depends on the rank of the controllability matrix:

$$C_p = [A^{p-1} B \quad \cdots \quad AB \quad B]. \tag{8.4}$$

Let us define the vector u_p by

$$u_p = \begin{bmatrix} u(0) \\ \vdots \\ u(p-2) \\ u(p-1) \end{bmatrix}. \tag{8.5}$$

Equation (8.3) becomes

$$C_p u_p = x(p) - A^p x(0). \tag{8.6}$$

Assuming that there are r inputs; the column vector u_p thus contains rp elements and the controllability matrix C_p has $n \times rp$ elements. As a result, Eq. (8.6) has n equations with rp unknowns in u_p. Equation (8.6) has a solution for any desired final state $x(p)$ only when rp is larger than or equal to n and C_p has rank n. For the case in which C_p has a rank less than n, i.e., the number of equations is larger than the number of unknowns, there exists no solution that satisfies Eq. (8.6) for any desired final state $x(p)$. When C_p has rank n and $rp > n$, Eq. (8.6) has more unknowns than number of equations. For this case, the input vector u_p solved from Eq. (8.6) is not unique,

$$u_p = C_p^\dagger [x(p) - A^p x(0)] + [I_{rp} - C_p^\dagger C_p] \alpha, \tag{8.7}$$

where † means pseudoinverse:

$$C_p^\dagger = C_p^T (C_p C_p^T)^{-1}. \tag{8.8}$$

I_{rp} is an identity matrix of $rp \times rp$, and α is an $rp \times 1$ arbitrary vector. Normally, we would choose the minimum-norm solution in which $\alpha = 0$, i.e.,

$$u_p = C_p^\dagger [x(p) - A^p x(0)], \tag{8.9}$$

because it minimizes the norm of the solution vector u_p.

As a summary, a linear, finite-dimensional, discrete-time, time-invariant dynamical system, Eq. (8.1), of the order of n is controllable if and only if the controllability matrix C_p has rank n.

One may ask what value of p should be chosen to construct the controllability matrix C_p . It turns out that it is sufficient to include the terms up to $A^{n-1}B$ for a test of controllability, i.e.,

$$C_n = [A^{n-1}B \quad \cdots \quad AB \quad B]. \tag{8.10}$$

This can be explained by means of the Caley–Hamilton theorem, which states that an $n \times n$ matrix A satisfies its own characteristic equation,

$$A^n + \alpha_{n-1} A^{n-1} + \cdots + \alpha_1 A + \alpha_0 I = 0, \tag{8.11}$$

where

$$\lambda^n + \alpha_{n-1} \lambda^{n-1} + \cdots + \alpha_1 \lambda + \alpha_0 = 0 \tag{8.12}$$

is the characteristic equation of the matrix A. A consequence of this result is that the product $A^n B$ can be expressed as a linear combination of $A^{n-1}B, \ldots, AB, B$:

$$A^n B = -\alpha_{n-1} A^{n-1} B - \cdots - \alpha_1 AB - \alpha_0 B. \tag{8.13}$$

By making repeated use of the above equation, we can also express the product $A^k B$ for any $k \geq n$ as a linear combination of $A^{n-1}B, \ldots, AB, B$. Adding additional terms beyond $A^{n-1}B$ in the controllability matrix will not improve its rank. For this reason, the controllability matrix is usually defined up to the term $A^{n-1}B$.

Perhaps it is easier to understand the concept of controllability by looking at a special case in which the system has only one input and the matrix A has a full set of distinct eigenvalues. We can put the original state–space equation in a special set of coordinates that diagonalizes A by using the following coordinate transformation:

$$x = \Psi z, \tag{8.14}$$

where Ψ is a (nonsingular) matrix made up of the eigenvectors of A. The state–space equation for $z(k)$ is

$$z(k+1) = \Psi^{-1}A\Psi z(k) + \Psi^{-1}Bu(k)$$
$$= \Lambda z(k) + \bar{B}u(k), \tag{8.15}$$

where we have made use of the property $\Psi^{-1}A\Psi = \Lambda$ and Λ is a diagonal matrix containing the eigenvalues of A:

$$\Lambda = \begin{bmatrix} \lambda_1 & & & \\ & \lambda_2 & & \\ & & \ddots & \\ & & & \lambda_n \end{bmatrix}. \tag{8.16}$$

Furthermore, let the elements of $\bar{B} = \Psi^{-1}B$ be denoted by

$$\bar{B} = \begin{bmatrix} b_1 \\ b_2 \\ \vdots \\ b_n \end{bmatrix}. \tag{8.17}$$

The diagonal matrix Λ implies that all the elements $z_1(k), z_2(k), \ldots, z_n(k)$ of the state $z(k)$ are decoupled:

$$z_1(k+1) = \lambda_1 z_1(k) + b_1 u(k), \tag{8.18a}$$

$$z_2(k+1) = \lambda_2 z_2(k) + b_2 u(k), \tag{8.18b}$$

$$\vdots$$

$$z_n(k+1) = \lambda_n z_n(k) + b_n u(k). \tag{8.18c}$$

The controllability matrix of the system in the new coordinates becomes

$$\bar{C}_n = \begin{bmatrix} \lambda_1^{n-1}b_1 & \cdots & \lambda_1 b_1 & b_1 \\ \lambda_2^{n-1}b_2 & \cdots & \lambda_2 b_2 & b_2 \\ \vdots & \ddots & \vdots & \vdots \\ \lambda_n^{n-1}b_n & \cdots & \lambda_n b_n & b_n \end{bmatrix}. \tag{8.19}$$

If any of the elements of $\bar{B}$ is zero, say, the first element $b_1 = 0$, then $\bar{C}_n$ becomes a rank-deficient matrix (its rank is now $n-1$):

$$\bar{C}_n = \begin{bmatrix} 0 & \cdots & 0 & 0 \\ \lambda_2^{n-1} b_2 & \cdots & \lambda_2 b_2 & b_2 \\ \vdots & \ddots & \vdots & \vdots \\ \lambda_n^{n-1} b_n & \cdots & \lambda_n b_n & b_n \end{bmatrix}. \tag{8.20}$$

This implies that it is impossible to influence the first element of the state vector z by any input because of the first row of zeros in $\bar{C}_n$:

$$\begin{bmatrix} 0 & \cdots & 0 & 0 \\ \lambda_2^{n-1} b_2 & \cdots & \lambda_2 b_2 & b_2 \\ \vdots & \ddots & \vdots & \vdots \\ \lambda_n^{n-1} b_n & \cdots & \lambda_n b_n & b_n \end{bmatrix} \begin{bmatrix} u(0) \\ \vdots \\ u(n-2) \\ u(n-1) \end{bmatrix} = z(p) - \Lambda^n z(0). \tag{8.21}$$

This element of the state is said to be uncontrollable.

8.2.2 Observability in the Discrete-Time Domain

A state $x(p)$ of the system at time step $k = p$ is said to be observable if knowledge of the input $u(k)$ and output $y(k)$ over a finite time interval $0 \le k \le p-1$ can be used to determine the state $x(p)$. If all states are observable, the system is said to be completely observable or simply observable.

To determine observability, however, it is sufficient to find if the initial state $x(0)$ can be determined from the output $y(k)$ alone, $0 \le k \le p-1$, when the input $u(k)$ is zero. Once the initial state $x(0)$ is known, the state of the system at any time step $k > 0$ can (in theory) be obtained from the state equation $x(k+1) = Ax(k) + Bu(k)$.

From the discrete-time model we have for $u(k) = 0$:

$$y(0) = Cx(0), \tag{8.22a}$$

$$y(1) = Cx(1) = CAx(0), \tag{8.22b}$$

$$\vdots$$

$$y(p-1) = Cx(p-1) = CA^{p-1}x(0). \tag{8.22c}$$

In matrix form, these equations can be written as

$$y_p = \mathcal{O}_p x(0), \tag{8.23}$$

where

$$y_p = \begin{bmatrix} y(0) \\ y(1) \\ \vdots \\ y(p-1) \end{bmatrix}, \qquad \mathcal{O}_p = \begin{bmatrix} C \\ CA \\ \vdots \\ CA^{p-1} \end{bmatrix}. \tag{8.24}$$

We can solve for the initial state $x(0)$ uniquely if and only if the observability matrix

$\mathcal{O}_p$ of $mp \times n$ has rank n. The unique solution should be

$$x(0) = \mathcal{O}_p^\dagger y_p. \tag{8.25}$$

If $\mathcal{O}_p$ is not full rank, then the initial state cannot be solved uniquely because there are more unknowns than number of equations. Again, by making use of the Caley–Hamilton theorem, it is necessary to include terms only up to CA^{n-1} in the observability matrix to determine its rank.

Again we can easily understand the concept of observability by looking at a special case in which the system has only one output. Assuming that A has a full set of distinct eigenvalues, we can follow the same procedure described earlier to put the original state–space equation in a new set of coordinates in which all the elements of the state $z(k)$ are decoupled, as shown in Eq. (8.18). The corresponding output influence matrix in the new coordinates is

$$\bar{C} = C\Psi = [c_1 \quad c_2 \quad \cdots \quad c_n], \tag{8.26}$$

so that the observability matrix becomes

$$\bar{\mathcal{O}}_p = \begin{bmatrix} \bar{C} \\ \bar{C}\Lambda \\ \vdots \\ \bar{C}\Lambda^{n-1} \end{bmatrix} = \begin{bmatrix} c_1 & c_2 & \cdots & c_n \\ \lambda_1 c_1 & \lambda_2 c_2 & \cdots & \lambda_n c_n \\ \vdots & \vdots & \ddots & \vdots \\ \lambda_1^{n-1} c_1 & \lambda_2^{n-1} c_2 & \cdots & \lambda_n^{n-1} c_n \end{bmatrix}. \tag{8.27}$$

The output of the system is

$$y(k) = c_1 z_1(k) + c_2 z_2(k) + \cdots + c_n z_n(k). \tag{8.28}$$

If any element of $\bar{C}$ is zero, say $c_1 = 0$, then the observability matrix of the system in the new coordinates is rank deficient (rank $n - 1$). The output of the system does not contain any information about $z_1(k)$ at all. This element of the state is said to be unobservable.

EXAMPLE 8.1

Consider a rigid body of mass m with an applied force f acting along the direction of motion. The equation of motion is

$$m\ddot{w}(t) = f(t),$$

where w is the displacement of the mass. This equation can be written in the first-order state–space form,

$$\begin{bmatrix} \dot{w} \\ \ddot{w} \end{bmatrix} = \begin{bmatrix} 0 & 1 \\ 0 & 0 \end{bmatrix} \begin{bmatrix} w \\ \dot{w} \end{bmatrix} + \begin{bmatrix} 0 \\ 1 \end{bmatrix} \frac{f}{m},$$

or

$$\dot{x} = A_c x + B_c u,$$

where

$$x = \begin{bmatrix} w \\ \dot{w} \end{bmatrix}, \qquad A_c = \begin{bmatrix} 0 & 1 \\ 0 & 0 \end{bmatrix}, \qquad B_c = \begin{bmatrix} 0 \\ 1 \end{bmatrix}, \qquad u = \frac{f}{m}.$$

Assume that the displacement w is sampled on every time interval Δt and the applied force does not vary over the interval Δt. The continuous-time model can be converted into the following discrete-time model:

$$x(k+1) = Ax(k) + Bu(k),$$

where

$$A = e^{A_c \Delta t} = \left[I + A_c \Delta t + \frac{1}{2}(A_c \Delta t)^2 + \cdots \right] = \begin{bmatrix} 1 & \Delta t \\ 0 & 1 \end{bmatrix},$$

$$B = \int_0^{\Delta t} e^{A_c \Delta \tau} d\tau \, B_c = \Delta t \left[I + \frac{1}{2}A_c \Delta t + \frac{1}{3}(A_c \Delta t)^2 + \cdots \right] B_c$$

$$= \begin{bmatrix} \Delta t & \frac{1}{2}\Delta t^2 \\ 0 & \Delta t \end{bmatrix} \begin{bmatrix} 0 \\ 1 \end{bmatrix} = \begin{bmatrix} \frac{1}{2}\Delta t^2 \\ \Delta t \end{bmatrix}.$$

Note that $A_c^2 = A_c^3 = \cdots = A_c^\infty = 0$.

The discrete-time model is controllable because the controllability matrix,

$$C_2 = [B \quad AB] = \begin{bmatrix} \frac{1}{2}\Delta t^2 & \frac{3}{2}\Delta t^2 \\ \Delta t & \Delta t \end{bmatrix} = \Delta t \begin{bmatrix} \frac{1}{2}\Delta t & \frac{3}{2}\Delta t \\ 1 & 1 \end{bmatrix},$$

always has rank 2 for $\Delta t \neq 0$.

Let y be the direct measurement of the displacement w. The measurement equation becomes

$$y(t) = w(t) = [1 \quad 0] \begin{bmatrix} w \\ \dot{w} \end{bmatrix}$$

Assume that the displacement w is sampled on every time interval Δt. The discrete measurement equation becomes

$$y(k) = Cx(k) \quad \text{with} \quad C = [1 \quad 0].$$

The discrete-time model is observable because the observability matrix,

$$\mathcal{O}_2 = \begin{bmatrix} C \\ CA \end{bmatrix} = \begin{bmatrix} 1 & 0 \\ 1 & \Delta t \end{bmatrix},$$

has rank 2 for $\Delta t \neq 0$. Physically, this implies that knowledge of displacement alone determines the state x, which includes the displacement and the velocity. The velocity can be obtained by differentiating the displacement.

With direct velocity measurement, the measurement equation becomes

$$y(t) = \dot{w}(t) = [0 \quad 1] \begin{bmatrix} w \\ \dot{w} \end{bmatrix}$$

Assume that the velocity $\dot{w}$ is sampled on every time interval Δt. The discrete measurement equation becomes

$$y(k) = Cx(k) \quad \text{with} \quad C = [0 \quad 1].$$

The discrete-time model is not observable because the observability matrix,

$$\mathcal{O}_2 = \begin{bmatrix} C \\ CA \end{bmatrix} = \begin{bmatrix} 0 & 1 \\ 0 & 1 \end{bmatrix},$$

has rank 1. Physically this means that knowledge of the velocity alone cannot determine the state x, which includes the displacement w and the velocity $\dot{w}$. Indeed, it is not possible to integrate velocity to produce displacement unless the initial displacement is known.

With direct acceleration measurement, the measurement equation becomes

$$y(t) = \ddot{w}(t) = \frac{f(t)}{m}.$$

Assume that the acceleration $\ddot{w}$ is sampled on every time interval Δt. The discrete measurement equation becomes

$$y(k) = Cx(k) + Du(k) \quad \text{with} \quad C = [0 \quad 0], \quad D = 1, \quad u(k) = \frac{f(k)}{m}.$$

The discrete-time model is not observable because the observability matrix,

$$\mathcal{O}_2 = \begin{bmatrix} C \\ CA \end{bmatrix} = \begin{bmatrix} 0 & 0 \\ 0 & 0 \end{bmatrix},$$

has rank 0. This indicates that knowledge of the acceleration alone cannot determine the state x, which includes the displacement w and the velocity $\dot{w}$. Indeed, it is not possible to integrate acceleration to produce displacement and velocity unless the initial displacement and velocity are known.

8.3 Continuous-Time State Feedback

In linear systems, the simplest state–feedback control is that the control input is a linear function of the state vector. If the system dynamics is described by a state–space model in continuous-time format,

$$\frac{dx}{dt} = A_c x(t) + B_c u(t), \tag{8.29}$$

then a linear state-feedback controller has the form

$$u(t) = \mathcal{F}_c x(t), \tag{8.30}$$

where $\mathcal{F}_c$ is called the controller gain. For a single-input system, $\mathcal{F}_c$ is a row vector. For a multiple-input system, $\mathcal{F}_c$ is a matrix.

What should the controller gain matrix $\mathcal{F}_c$ be such that the system is stabilizable with the above control law? To answer this question, let us examine the closed-loop system equation we obtain by substituting Eq. (8.30) into Eq. (8.29):

$$\frac{dx}{dt} = A_c x(t) + B\mathcal{F}_c x(t)$$

$$= (A_c + B\mathcal{F}_c) x(t). \tag{8.31}$$

This equation describes the dynamic response of the system when its input is governed by the control law, Eq. (8.30). Let us denote

$$\bar{A}_c = (A_c + B\mathcal{F}_c) \tag{8.32}$$

and call $\bar{A}_c$ the (continuous-time) closed-loop system matrix. Thus the closed-loop dynamics is governed by the equation

$$\frac{dx}{dt} = \bar{A}_c x(t), \tag{8.33}$$

which is a first-order matrix differential equation. The solution of this matrix equation subject to a nonzero initial condition describes the response of the closed-loop system. At the moment, we are concerned with only stability, i.e., whether or not the controlled response will grow unbounded or remain stable or converge to zero. It is necessary to examine only the eigenvalues of the closed-loop system matrix $\bar{A}_c$. If all eigenvalues of $\bar{A}_c$ have negative real parts, then the closed-loop response will converge to zero in the steady state. This means that the dynamic system will be brought to rest by the controller when it is disturbed from the equilibrium position. For asymptotic stability of a time-invariant dynamic system by use of the control law $u(t) = \mathcal{F}_c x(t)$, the controller gain $\mathcal{F}_c$ must be designed such that all the eigenvalues of the closed-loop system matrix $(A_c + B_c \mathcal{F}_c)$ have negative real parts. The eigenvalues of $\bar{A}_c$ are commonly called (continuous-time) closed-loop eigenvalues or closed-loop poles.

If the state of the system is brought to zero, what will happen to its output? Mathematically, we can answer this question rather easily by examining the measurement equation:

$$y(t) = Cx(t) + Du(t) = (C + D\mathcal{F}_c)x(t). \tag{8.34}$$

The output $y(t)$ will reduce to zero as soon as the state $x(t)$ diminishes.

8.4 Discrete-Time State Feedback

The state feedback in the continuous-time domain has its counterpart in the discrete-time domain. Consider the discrete-time model,

$$x(k+1) = Ax(k) + Bu(k) \tag{8.35}$$

Similar to Eq. (8.30), the linear state-feedback law in the discrete-time format is

$$u(k) = \mathcal{F}x(k). \tag{8.36}$$

Substituting Eq. (8.36) into Eq. (8.35) yields the resultant closed-loop system

$$x(k+1) = Ax(k) + B\mathcal{F}x(k)$$
$$= (A + B\mathcal{F})x(k). \tag{8.37}$$

This equation governs the closed-loop response of the controlled system. Similarly, let us denote

$$\bar{A} = A + B\mathcal{F} \tag{8.38}$$

and call it the (discrete-time) closed-loop system matrix. Thus the closed-loop dynamics is governed by the equation

$$x(k + 1) = \bar{A}x(k), \tag{8.39}$$

which is the first-order matrix difference equation that has been studied previously. The solution of this matrix equation subject to a nonzero initial condition describes the response of the closed-loop system in the discrete-time domain. As far as stability is concerned, it is necessary to examine only the eigenvalues of the closed-loop system matrix $\bar{A}$. If magnitudes of all eigenvalues of $\bar{A}$ (which can be real or complex) are less than one, then the closed-loop response will converge to zero in the steady state. Thus the controller gain $\mathcal{F}$ must be designed such that the magnitudes of all eigenvalues of $A + B\mathcal{F}$ are less than one. The eigenvalues of $\bar{A}$ are commonly called (discrete-time) closed-loop eigenvalues or closed-loop poles. Similar to the continuous-time case, the state $x(k)$ and the output $y(k)$ reduces to zero simultaneously because

$$y(k) = Cx(k) + Du(k) = (C + D\mathcal{F})x(k). \tag{8.40}$$

Convergence of the state to zero implies convergence of the control input and system output to zero.

8.5   Placement of Closed-Loop Eigenvalues

In both continuous-time and discrete-time domains, the controller gain that we can design to stabilize a dynamic system is not unique. The closed-loop behavior is largely determined by the locations of the closed-loop eigenvalues commonly called closed-loop poles. For a controllable system, it can be shown that any real or complex-conjugate pairs of eigenvalues in the complex plane may be achieved. This result can be mathematically stated as follows:

   Consider the system

$$\frac{dx}{dt} = A_c x(t) + B_c u(t)$$

with any preassigned real or complex-conjugate pairs of the closed-loop eigenvalues (poles). There exists a state-feedback-control law of the form

$$u(t) = \mathcal{F}_c x(t)$$

such that the poles of the closed-loop system

$$\frac{dx}{dt} = (A_c + B_c \mathcal{F}_c) x(t)$$

lie in the specified locations if and only if the system is controllable.

   The proof of this statement will be given later in this section. In the discrete-time case, the same result applies without alteration. Recall that in the discrete-time case, the closed-loop poles are less than one in magnitude for stability as opposed to the continuous-time case in which they must all have negative real parts. A particularly interesting closed-loop system results (in the discrete-time case) if the state-feedback gain is chosen such that all closed-loop poles are placed at the origin (corresponding to zero magnitudes). In the absence of external input, this means that the state of the closed-loop system will be driven to zero identically in at most n steps, where n is the order of the state–space model. This is known as a deadbeat-control system. A deadbeat-control system is the minimum-time control system, i.e., the controller will bring the state of the system to zero identically in the minimum amount of time. This is a remarkable result that occurs only for a discrete-time model but not for a continuous-time one. Because minimum-time control tends to require excessive control input, a deadbeat controller is not necessarily useful in practice. Yet, it is theoretically beautiful that the state of the closed-loop system can actually be brought to zero identically in a finite number of time steps (rather than asymptotically as time tends to infinity in a continuous-time model). Furthermore, this result is physically significant in light of the fact that a linear (physical) system in continuous time can be represented by a discrete-time model exactly without approximation with the aid of an zero-order hold on the input.

EXAMPLE 8.2

Let us consider how state-feedback control can be used to augment damping to a mass–spring system. With the absence of damping, the mass–spring system when perturbed will oscillate indefinitely. The objective of the feedback-control system is to bring the system back to rest, i.e., the equilibrium position. The mass–spring system is shown in Fig. 8.1.

The equation of motion for this system with $k = 1$, $m = 1$ is simply

$$\ddot{w}(t) + \dot{w}(t) = u(t),$$

which can be written in a state–space format as

$$\frac{d}{dt} \begin{bmatrix} w(t) \\ \dot{w}(t) \end{bmatrix} = \begin{bmatrix} 0 & 1 \\ -1 & 0 \end{bmatrix} \begin{bmatrix} w(t) \\ \dot{w}(t) \end{bmatrix} + \begin{bmatrix} 0 \\ 1 \end{bmatrix} u(t)$$

$$= A_c x(t) + B_c u(t),$$

where

$$x(t) = \begin{bmatrix} w(t) \\ \dot{w}(t) \end{bmatrix}, \qquad A_c = \begin{bmatrix} 0 & 1 \\ -1 & 0 \end{bmatrix}, \qquad B_c = \begin{bmatrix} 0 \\ 1 \end{bmatrix}.$$

The open-loop system eigenvalues are $\pm i$. We assume that the state of the system, which includes both position and velocity information, can be directly measurable. A state-feedback controller takes the form

$$u(t) = \mathcal{F}_c x(t)$$

$$= [f_1 \quad f_2] \begin{bmatrix} w(t) \\ \dot{w}(t) \end{bmatrix}$$

$$= f_1 w(t) + f_2 \dot{w}(t).$$

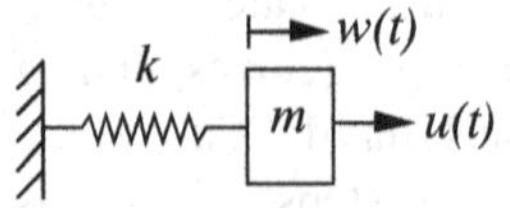

Figure 8.1. A mass–spring system.

Formally, the controller gain matrix $\mathcal{F}_c$ should be designed such that the eigenvalues of the resultant closed-loop system matrix $A_c + B_c\mathcal{F}_c$ have negative real parts. This step will be carried out later, but for now it is perhaps more beneficial to examine how this controller influences the system to be controlled. Note that the feedback-control input is a linear combination of position and velocity measurements; hence the closed-loop (or controlled) system is

$$\ddot{w}(t) + \dot{w}(t) = f_1 w(t) + f_2 \dot{w}(t),$$

which can be rearranged into

$$\ddot{w}(t) - f_2 \dot{w}(t) + (1 - f_1)\, w(t) = 0.$$

This is an equation of motion of a spring–mass–damper system with the same mass but an augmented spring stiffness $1 - f_1$ and a new damping coefficient determined by $-f_2$. Thus this controller in effect turns a marginally stable (i.e., zero damping) open-loop system into a closed-loop system that emulates a damped oscillator. By proper choices of f_1 and f_2 any desired characteristics of the closed-loop system can be obtained. For example, we can choose $f_1 = 0$, $f_2 = -1$. This controller has the effect of adding a damper to the system. The characteristic equation of the closed-loop system is

$$\lambda^2 + \lambda + 1 = 0,$$

which yields two roots, $\lambda_{1,2} = -0.50 \pm 0.866i$, having negative real parts as expected.

With a desired set of closed-loop eigenvalues, it is very easy to determine the necessary gain values f_1 and f_2. As far as stability of the closed-loop system is concerned, any set of closed-loop eigenvalues with negative real parts will suffice. To illustrate the pole-placement technique, let us take the above design and work backward. Assume that the desired eigenvalues are $\lambda_{1,2} = -0.50 \pm 0.866i$. In the complex domain, the desired eigenvalues must be a pair including its complex conjugate. The corresponding characteristic equation must then be

$$(\lambda - \lambda_1)(\lambda - \lambda_2) = \lambda^2 + \lambda + 1 = 0.$$

The characteristic equation of the closed-loop system matrix,

$$\bar{A}_c = A_c + B_c\mathcal{F} = \begin{bmatrix} 0 & 1 \\ -1 & 0 \end{bmatrix} + \begin{bmatrix} 0 \\ 1 \end{bmatrix}[f_1 \quad f_2]$$

$$= \begin{bmatrix} 0 & 1 \\ -1 + f_1 & f_2 \end{bmatrix},$$

is obtained from

$$\begin{aligned}
|\bar{A} - \lambda I| &= \begin{vmatrix} -\lambda & 1 \\ -1 + f_1 & f_2 - \lambda \end{vmatrix} \\
&= -\lambda(f_2 - \lambda) - (-1 + f_1) \\
&= \lambda^2 - f_2\lambda + (1 - f_1) = 0.
\end{aligned}$$

By equating like powers of λ, $-f_2 = 1$, and $1 - f_1 = 1$, we obtain $f_1 = 0$ and $f_2 = -1$, as expected.

We now illustrate the pole-placement theorem in the discrete-time domain. Let the sampling interval be 0.01 s, $\Delta t = 0.01$ s. The discrete-time model for this system (with a zero-order hold on the input) is

$$A = \begin{bmatrix} \cos \Delta t & \sin \Delta t \\ -\sin \Delta t & \cos \Delta t \end{bmatrix}, \qquad B = \begin{bmatrix} 1 - \cos \Delta t \\ \sin \Delta t \end{bmatrix}, \qquad \Delta t = 0.01.$$

The equivalent discrete-time closed-loop eigenvalues corresponding to the continuous-time eigenvalues,

$$\lambda_{1,2} = -0.50 \pm 0.866i,$$

are

$$z_{1,2} = e^{\lambda_{1,2}\Delta t} = 0.995 \pm 0.0086i.$$

The desired characteristic equation for the discrete-time closed-loop system is therefore

$$(z - z_1)(z - z_2) = z^2 - 1.990\lambda + 0.990 = 0.$$

The discrete-time state-feedback-control law has the form

$$u(k) = \mathcal{F}x(k)$$

$$= [f_1 \quad f_2]\begin{bmatrix} w(k) \\ \dot{w}(k) \end{bmatrix}$$

$$= f_1 w(k) + f_2 \dot{w}(k).$$

The discrete-time closed-loop system matrix becomes

$$\bar{A} = A + B\mathcal{F}$$

$$= \begin{bmatrix} \cos \Delta t & \sin \Delta t \\ -\sin \Delta t & \cos \Delta t \end{bmatrix} + \begin{bmatrix} 1 - \cos \Delta t \\ \sin \Delta t \end{bmatrix}[f_1 \quad f_2]$$

$$= \begin{bmatrix} \cos \Delta t + (1 - \cos \Delta t)f_1 & \sin \Delta t + (1 - \cos \Delta t)f_2 \\ -\sin \Delta t + f_1 \sin \Delta t & \cos \Delta t + f_2 \sin \Delta t \end{bmatrix}.$$

The characteristic equation of the discrete-time closed-loop system matrix is obtained from

$$|\bar{A} - zI| = \begin{vmatrix} \cos \Delta t + (1 - \cos \Delta t)f_1 - z & \sin \Delta t + (1 - \cos \Delta t)f_2 \\ -\sin \Delta t + f_1 \sin \Delta t & \cos \Delta t + f_2 \sin \Delta t - z \end{vmatrix} = 0.$$

With $\Delta t = 0.01$, the characteristic equation for $\bar{A}$ (which involves f_1 and f_2) is computed. By equating like powers of z to make this characteristic equation the same as the desired characteristic equation, we will again obtain $f_1 = 0$ and $f_2 = -1$, as expected.

Figure 8.2 shows the open-loop [without control, $u(k) = 0$] versus closed-loop [with control, $u(k) = -\dot{w}(k)$] positions and velocities. Initially, the system is disturbed by an initial displacement, $w(0) = 1$ and $\dot{w}(0) = 0$. Also shown is the closed-loop control input.

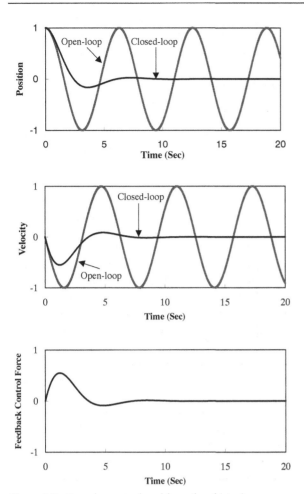

Figure 8.2. Open-loop vs. closed-loop time histories.

EXAMPLE 8.3

A classic example of stabilizing an inverted pendulum on a moving cart will be used (Ref. [4]). The system is shown in Fig. 8.3. The linearized equation of motion of the above system (for small angle motion of the inverted pendulum) is

$$
\frac{d}{dt}\begin{bmatrix} z \\ \dot{z} \\ \theta \\ \dot{\theta} \end{bmatrix} = \begin{bmatrix} 0 & 1 & 0 & 0 \\ 0 & 0 & -\dfrac{mg}{M} & 0 \\ 0 & 0 & 0 & 1 \\ 0 & 0 & \dfrac{(M+m)g}{M\ell} & 0 \end{bmatrix} \begin{bmatrix} z \\ \dot{z} \\ \theta \\ \dot{\theta} \end{bmatrix} + \begin{bmatrix} 0 \\ \dfrac{1}{M} \\ 0 \\ -\dfrac{1}{M\ell} \end{bmatrix} u(t).
$$

Suppose that the system parameters are

$$M = 1\,\text{Kg}, \quad m = 0.1\,\text{Kg}, \quad \ell = 1\,\text{m},$$

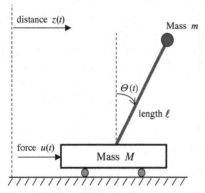

Figure 8.3. Inverted pendulum.

and, with $g = 10$ m/s^2, the system state–space model becomes

$$\frac{d}{dt}\begin{bmatrix} z(t) \\ \dot{z}(t) \\ \theta(t) \\ \dot{\theta}(t) \end{bmatrix} = \begin{bmatrix} 0 & 1 & 0 & 0 \\ 0 & 0 & -1 & 0 \\ 0 & 0 & 0 & 1 \\ 0 & 0 & 11 & 0 \end{bmatrix}\begin{bmatrix} z(t) \\ \dot{z}(t) \\ \theta(t) \\ \dot{\theta}(t) \end{bmatrix} + \begin{bmatrix} 0 \\ 1 \\ 0 \\ -1 \end{bmatrix} u(t).$$

Thus the system matrices are

$$A_c = \begin{bmatrix} 0 & 1 & 0 & 0 \\ 0 & 0 & -1 & 0 \\ 0 & 0 & 0 & 1 \\ 0 & 0 & 11 & 0 \end{bmatrix}, \qquad B_c = \begin{bmatrix} 0 \\ 1 \\ 0 \\ -1 \end{bmatrix}.$$

The poles of the open-loop system are the eigenvalues of A_c that we obtain by solving

$$|A_c - \lambda I| = \lambda^4 - 11\lambda^2 = 0.$$

The four eigenvalues are

$$\lambda_1 = \lambda_2 = 0, \qquad \lambda_3 = -\sqrt{11}, \qquad \lambda_4 = \sqrt{11}.$$

The system has two repeated roots at the origin, one stable and one unstable root. The homogeneous solution is thus quite unstable because it grows unbounded as a function of time. This is consistent with physical reasoning. We wish to stabilize this open-loop system with a state-feedback controller of the form

$$u(t) = \mathcal{F}_c x(t)$$
$$= f_1 z(t) + f_2 \dot{z}(t) + f_3 \theta(t) + f_4 \dot{\theta}(t),$$

where

$$\mathcal{F}_c = [f_1 \quad f_2 \quad f_3 \quad f_4], \qquad x(t) = \begin{bmatrix} z(t) \\ \dot{z}(t) \\ \theta(t) \\ \dot{\theta}(t) \end{bmatrix}.$$

The controller gain $\mathcal{F}_c$ must be such that all closed-loop poles have negative real parts for stability. Suppose that closed-loop poles are placed at

$$\lambda = -1, \ -2, \ -1 \pm i;$$

then the desired characteristic equation must be

$$p(\lambda) = (\lambda + 1)(\lambda + 2)(\lambda + 1 + i)(\lambda + 1 - i)$$
$$= \lambda^4 + 5\lambda^3 + 10\lambda^2 + 10\lambda + 4 = 0.$$

Substituting the control law $u(t) = \mathcal{F}_c x(t)$ into the open-loop state–space equation produces the closed-loop system

$$\frac{dx}{dt} = (A + B\mathcal{F}_c)x(t).$$

The closed-loop system matrix is then

$$\bar{A} = A + B\mathcal{F}_c$$

$$= \begin{bmatrix} 0 & 1 & 0 & 0 \\ 0 & 0 & -1 & 0 \\ 0 & 0 & 0 & 1 \\ 0 & 0 & 11 & 0 \end{bmatrix} + \begin{bmatrix} 0 \\ 1 \\ 0 \\ -1 \end{bmatrix} \begin{bmatrix} f_1 & f_2 & f_3 & f_4 \end{bmatrix}$$

$$= \begin{bmatrix} 0 & 1 & 0 & 0 \\ f_1 & f_2 & f_3 - 1 & f_4 \\ 0 & 0 & 0 & 1 \\ -f_1 & -f_2 & 11 - f_3 & -f_4 \end{bmatrix}.$$

The closed-loop characteristic equation is

$$|\bar{A} - \lambda I| = \lambda^4 + (f_4 - f_2)\lambda^3 + (f_3 - f_1 - 11)\lambda^2 + 10f_2\lambda + 10f_1 = 0.$$

Setting this closed-loop characteristic equation to be the desired one,

$$p(\lambda) = \lambda^4 + 5\lambda^3 + 10\lambda^2 + 10\lambda + 4 = 0,$$

yields the feedback-controller gain vector by equating like powers of λ :

$$f_1 = 0.4,$$
$$f_2 = 1,$$
$$f_3 = 10 + 11 + f_1 = 21.4,$$
$$f_4 = 5 + f_2 = 6.$$

Thus, the feedback gain $\mathcal{F} = \begin{bmatrix} 0.4 & 1 & 21.4 & 6 \end{bmatrix}$ will stabilize the inverted pendulum system by bringing the cart to its origin as well as keeping the pendulum from falling. To implement this controller, however, we need the linear position and the linear velocity of the cart as well as the angular position and angular velocity of the pendulum.

8.5.1 Null-Space Technique

Let A_c be the $n \times n$ state matrix and B_c be the $n \times r$ influence matrix of the r control inputs for a linear, time-invariant, constant-gain feedback system. Assume that full state feedback is used to design the controller, with gain matrix $\mathcal{F}_c$ of dimension $r \times n$. The eigensolution of the closed-loop system $A_c + B_c \mathcal{F}_c$ can then be written as

$$(A_c + B_c \mathcal{F}_c)\psi_k = \psi_k \lambda_k, \quad k = 1, \ldots, n, \tag{8.41}$$

where ψ_k is the $n \times 1$ eigenvector corresponding to the desired closed-loop eigenvalue λ_k. Both λ_k and ψ_k may be in the complex domain. Rearranging Eq. (8.41) in a compact matrix form produces

$$[A_c - \lambda_k I \quad B_c] \begin{bmatrix} \psi_k \\ \mathcal{F}_c \psi_k \end{bmatrix} = 0_{(n+r) \times 1}, \tag{8.42}$$

where I is an $n \times n$ identity matrix and $0_{(n+r) \times 1}$ is an $(n + r) \times 1$ zero vector. Equation (8.42) also holds for the case of repeated complex-conjugate pairs of eigenvalues. The only assumption required in Eq. (8.42) is that the system be nondefective, i.e., that a full set of eigenvectors corresponding to the eigenvalues to be assigned must exist.

To obtain the nontrivial solution space of the homogeneous equation, Eq. (8.42), the singular-value decomposition (SVD) is applied to the complex matrix Γ_k of $n \times (n + r)$ defined by

$$\Gamma_k = [A_c - \lambda_k I \quad B_c], \tag{8.43}$$

yielding [see Eq. (2.26)]

$$\Gamma_k = U_k \Sigma_k V_k^* = U_k \begin{bmatrix} \sigma_k & 0_{q \times (n+r-q)} \\ 0_{(n-q) \times q} & 0_{(n-q) \times (n+r-q)} \end{bmatrix} \begin{bmatrix} V_{\sigma k}^* \\ V_{ok}^* \end{bmatrix}, \tag{8.44}$$

where q is the number of nonzero singular values in the $q \times q$ diagonal matrix σ_k, r is number of inputs, and $*$ means complex-conjugate transpose. Note that Γ_k is a complex matrix for a complex eigenvalue λ_k. In general, $q = n$ for the case in which the closed-loop eigenvalue λ_k is not equal to any of the open-loop eigenvalues of A_c, because $A_c - \lambda_k I$ should not be rank deficient in this case. On the other hand, if λ_k is an eigenvalue of A_c, then $A_c - \lambda_k I$ will be rank deficient and thus $q < n$. The degree of rank deficiency depends on the multiplicity of the eigenvalue λ_k.

Equation (8.44) implies that

$$\Gamma_k [V_{\sigma k} \quad V_{ok}] = U_k \begin{bmatrix} \sigma_k & 0_{q \times (n+r-q)} \\ 0_{(n-q) \times q} & 0_{(n-q) \times (n+r-q)} \end{bmatrix}, \tag{8.45}$$

which gives

$$\Gamma_k V_{ok} = 0_{n \times (n+r-q)}. \tag{8.46}$$

The $(n + r) \times (n + r - q)$ matrix V_{ok} represents a set of orthogonal basis vectors spanning the null space of the matrix Γ_k, so that

$$\Gamma_k V_{ok} c_k = 0_{n \times 1}, \tag{8.47}$$

where c_k is an arbitrary nonzero vector of dimension $(n + r - q) \times 1$. The size of Γ_k is $n \times (n + r)$, i.e., the number of columns is larger than the number of rows. For $r \geq 1$, there always exists at least one column of orthogonal basis vectors in V_{ok}.

There are a total of n sets of V_{ok} for $k = 1, 2, \ldots, n$ for an $n \times n$ state matrix A_c. Each V_{ok} may have a different number of columns, depending on the dimension of the null space of Γ_k. Let us now choose a particular set of vector ϕ_k ($k = 1, \ldots, n$) satisfying Eq. (8.47), corresponding to some choice c_k, i.e.,

$$\phi_k = V_{ok} c_k \implies \Gamma_k \phi_k = 0_{n \times 1}. \tag{8.48}$$

Partition the vector ϕ_k into two components such that

$$\Gamma_k \phi_k = \Gamma \begin{bmatrix} \bar{\phi}_k \\ \hat{\phi}_k \end{bmatrix}$$

$$= [A_c - \lambda_k I \quad B_c] \begin{bmatrix} \bar{\phi}_k \\ \hat{\phi}_k \end{bmatrix}$$

$$= 0_{n \times 1}, \quad k = 1, \ldots, n, \tag{8.49}$$

where $\bar{\phi}_k$ is an $n \times 1$ vector and $\hat{\phi}_k$ is an $r \times 1$. Observation of Eqs. (8.42) and (8.49) indicates that

$$\mathcal{F}_c \bar{\phi}_k = \hat{\phi}_k, \quad k = 1, \ldots, n. \tag{8.50}$$

Form the matrix equation

$$\mathcal{F}_c [\bar{\phi}_1 \quad \bar{\phi}_2 \quad \cdots \quad \bar{\phi}_n] = [\hat{\phi}_1 \quad \hat{\phi}_2 \quad \cdots \quad \hat{\phi}_n], \tag{8.51}$$

which produces

$$\mathcal{F}_c = [\hat{\phi}_1 \quad \hat{\phi}_2 \quad \cdots \quad \hat{\phi}_n][\bar{\phi}_1 \quad \bar{\phi}_2 \quad \cdots \quad \bar{\phi}_n]^{-1}. \tag{8.52}$$

A matrix inversion is required in the computation of the gain matrix $\mathcal{F}_c$. To ensure that Eq. (8.52) is well conditioned for inversion, the condition number of the matrix to be inverted must be the smallest possible. Interestingly, the numerical requirement for well conditioning of the matrix inversion problem corresponds exactly to the eigenvalue conditioning problem because $\bar{\phi}_1, \ldots, \bar{\phi}_n$ are eigenvectors corresponding to the desired closed-loop eigenvalues $\lambda_1, \ldots, \lambda_n$. For those open-loop eigenvalues that are not to be moved, the closed-loop eigenvectors can be chosen as the open-loop ones (Ref. [10]).

Because the control force $u = \mathcal{F}_c x$ and the state vector x are in the real domain, the control gain matrix $\mathcal{F}_c$ must be a real value. When complex closed-loop eigenvalues are specified in pairs, their corresponding closed-loop eigenvectors must also be chosen in pairs. For example, if the eigenvalue λ_2 is the complex conjugate of λ_1, i.e., $\lambda_2 = \lambda_1^*$, then $\bar{\phi}_2 = \bar{\phi}_1^*$ and $\hat{\phi}_2 = \hat{\phi}_1^*$ must also be satisfied. To avoid the computation of $\mathcal{F}_c$ in the complex domain, we may choose only the real part and the imaginary part of $\bar{\phi}_1$ and $\hat{\phi}_1$ to replace $\bar{\phi}_1$ and $\bar{\phi}_2$ and $\hat{\phi}_1$ and $\hat{\phi}_2$ in Eq. (8.52), respectively. The same operation applies for any other complex eigenvalues. The gain matrix $\mathcal{F}_c$ thus computed will then be surely real.

For a single-input system, such a controller gain matrix is unique because the basis vector spanning the null space of Γ_k corresponding to each λ_k is one dimensional and unique. For a multiple-input system, however, such a controller gain is not unique because the null space of Γ_k is multidimensional and thus the closed-loop eigenvectors are not unique. We may make use of this extra freedom to find a gain that possesses further useful properties, such as minimum norm or robustness with respect to parameter uncertainty (Ref. [10]).

EXAMPLE 8.4

Consider the mass–spring system given in Example 8.2. The equation of motion for this system is

$$\dot{x} = A_c x + B_c u,$$

where

$$A_c = \begin{bmatrix} 0 & 1 \\ -1 & 0 \end{bmatrix}, \qquad B_c = \begin{bmatrix} 0 \\ 1 \end{bmatrix}.$$

The open-loop system eigenvalues are $\pm i$.

Let the desired closed-loop eigenvalues be $\lambda_{1,2} = -0.50 \pm 0.866i$, having negative real parts. With the desired set of closed-loop eigenvalues, construct the matrix Γ_1 shown in Eq. (8.43):

$$\Gamma_1 = [A_c - \lambda_1 I \quad B_c] = \begin{bmatrix} 0.5 - 0.866j & 1 & 0 \\ -1 & 0.5 - 0.866j & 1 \end{bmatrix}.$$

Taking the SVD of Γ_1 yields

$$\Gamma_1 = U_1 \Sigma_1 V_1^* = U_1 [\Sigma_1 \quad 0] \begin{bmatrix} V_{\sigma1}^* \\ V_{o1}^* \end{bmatrix},$$

where

$$U_1 = \begin{bmatrix} 0.13347 - 0.58610j & -0.77922 + 0.17746j \\ -0.77922 - 0.17746j & -0.13347 - 0.58610j \end{bmatrix},$$

$$\Sigma_1 = \begin{bmatrix} 2.0743 & 0 & 0 \\ 0 & 0.83500 & 0 \end{bmatrix},$$

$$V_1 = \begin{bmatrix} 0.65252 & -0.49808 & 0.57735 \\ -0.49392 - 0.65065j & -0.40525 - 0.27687j & -0.28868 + 0.5j \\ -0.37565 - 0.85549j & -0.15985 - 0.70191j & 0.28868 - 0.5j \end{bmatrix}.$$

The vector

$$V_{o1} = \begin{bmatrix} 0.57735 \\ -0.28868 + 0.5j \\ 0.28868 - 0.5j \end{bmatrix}$$

represents a basis vector spanning the null space of the matrix Γ_1, i.e.,

$$\Gamma_1 V_{o1} c_1 = 0,$$

where c_1 can be any complex number.

For the eigenvalue $\lambda_2 = -0.50 - 0.866i$, a similar process can be used to compute V_{o2}. Because $\lambda_2 = -0.50 - 0.866i$ is the complex conjugate of

$\lambda_1 = -0.50 + 0.866i$, V_{o2} is the complex conjugate of V_{o1}, i.e.,

$$V_{o2} = \begin{bmatrix} 0.57735 \\ -0.28868 - 0.5j \\ 0.28868 + 0.5j \end{bmatrix}.$$

Let us now choose a particular set of vectors ϕ_k ($k = 1, 2$) satisfying Eq. (8.47), corresponding to the choice $c_1 = c_2 = 1$, i.e.,

$$\phi_k = V_{ok} \implies \Gamma_k \phi_k = 0, \quad k = 1, 2.$$

Partition the vector ϕ_1 into two components such that

$$\Gamma_1 \phi_1 = [A_c - \lambda_1 I \quad B_c] \begin{bmatrix} \bar{\phi}_1 \\ \hat{\phi}_1 \end{bmatrix}$$

$$= \begin{bmatrix} 0.5 - 0.866j & 1 & 0 \\ -1 & 0.5 - 0.866j & 1 \end{bmatrix} \begin{bmatrix} 0.57735 \\ -0.28868 + 0.5j \\ 0.28868 - 0.5j \end{bmatrix}$$

$$= \begin{bmatrix} 0 \\ 0 \\ 0 \end{bmatrix}.$$

A similar partition can also be done for the vector ϕ_2 such that

$$\Gamma_2 \phi_2 = [A_c - \lambda_2 I \quad B_c] \begin{bmatrix} \bar{\phi}_2 \\ \hat{\phi}_2 \end{bmatrix}$$

$$= \begin{bmatrix} 0.5 + 0.866j & 1 & 0 \\ -1.0 & 0.5 + 0.866j & 1 \end{bmatrix} \begin{bmatrix} 0.57735 \\ -0.28868 - 0.5j \\ 0.28868 + 0.5j \end{bmatrix}$$

$$= \begin{bmatrix} 0 \\ 0 \\ 0 \end{bmatrix}.$$

The gain matrix $\mathcal{F}_c$ can then be solved with Eq. (8.51),

$$\mathcal{F}_c [\bar{\phi}_1 \quad \bar{\phi}_2] = [\hat{\phi}_1 \quad \hat{\phi}_2]$$

or

$$\mathcal{F}_c \begin{bmatrix} 0.57735 & 0.57735 \\ -0.28868 + 0.5j & -0.28868 - 0.5j \end{bmatrix}$$

$$= [0.28868 - 0.5j \quad 0.28868 + 0.5j],$$

which produces

$$\mathcal{F}_c = [0 \quad -1].$$

To avoid the complex arithmetic, we may use only the real part and the imaginary part of $\bar{\phi}_1$ and $\hat{\phi}_1$ to form the matrix equation

$$\mathcal{F}_c \begin{bmatrix} 0.57735 & 0 \\ -0.28868 & 0.5 \end{bmatrix} = [0.28868 \quad -0.5],$$

which also produces

$$\mathcal{F}_c = [0 \quad -1].$$

The result is identical to that shown in Example 8.2.

8.5.2 Gain Minimization

For each specific closed-loop eigenvalue λ_i, we may also specify a desired eigenvector $\bar{\phi}_{dk}$ to form the vector ϕ_{dk}, where the subscript d signifies the desired one. As shown in Eq. (8.49) that the first n rows of ϕ_{dk} constitute the closed-loop eigenvector corresponding to the eigenvalue λ_k. The last r rows, i.e., $\hat{\phi}_{dk}$, of ϕ_{dk} are used in Eq. (8.52) to compute the feedback gain $\mathcal{F}_c$. To minimize the gain matrix $\mathcal{F}_c$, we should choose the minimum value of $\hat{\phi}_{dk}$, i.e., $\hat{\phi}_{dk} = 0$. Thus, the vector ϕ_{dk} should be chosen as

$$\phi_{dk} = \begin{bmatrix} \bar{\phi}_{dk} \\ 0 \end{bmatrix}, \qquad k = 1, 2, \ldots, n, \tag{8.53}$$

where $\bar{\phi}_{dk}$ is the desired eigenvector corresponding to the desired eigenvalue λ_k. One obvious choice for the closed-loop eigenvector $\bar{\phi}_{dk}$ is the open-loop eigenvector associated with the open-loop eigenvalue as close as possible to the desired closed-loop eigenvalue λ_k. The intuitive reason is that it requires a larger-value gain matrix for moving both eigenvalue and eigenvector simultaneously from their open-loop positions. For example, the control gain matrix would be zero if none of the open-loop eigenvalues and eigenvectors are moved. Mathematically, it can also be explained as follows. From Eq. (8.52), the matrix that needs to be inverted is formed from the closed-loop eigenvectors. To minimize the control gain matrix $\mathcal{F}_c$, the condition number of the closed-loop eigenvector matrix must be as close as possible to one. This implies that all closed-loop eigenvectors have to be as orthogonal as possible. Because open-loop eigenvectors for structures are orthogonal for zero or proportional damping, they are a good choice for the desired closed-loop eigenvectors.

The vector ϕ_{dk} may not be in the null space spanned by the column vectors of V_{ok}. For the closed-loop eigenvalue λ_k, then we want

$$\phi_{dk} \approx V_{ok} c_k, \qquad k = 1, 2, \ldots, n. \tag{8.54}$$

Equality occurs only if ϕ_{dk} is in the space spanned by V_{ok} (which is usually not the case). The least-squares solution that best approximates the desired vector ϕ_{dk} yields

$$c_k = V_{ok}^{\dagger} \phi_{dk} = [V_{ok}^* V_{ok}]^{-1} V_{ok}^* \phi_{dk} = V_{ok}^* \phi_{dk}, \qquad k = 1, 2, \ldots, n, \tag{8.55}$$

where $\dagger$ indicates the pseudoinverse. The last equality is true only when $V_{ok}^* V_{ok} = I_{n+r-q}$. The vector c_k provides the admissible vector

$$\phi_k = V_{ok} c_k, \qquad k = 1, 2, \ldots, n \tag{8.56}$$

for λ_k, which is as close as possible to the desired one, ϕ_{dk}, in the sense of the minimum error norm between ϕ_{dk} and ϕ_k. Thus, inserting all admissible vectors $\phi_1, \phi_2, \ldots, \phi_n$

into Eq. (8.52) produces the control gain $\mathcal{F}_c$ that gives the closed-loop eigenvalues as desired.

As a summary, the typical procedure for the pole-placement technique includes the following steps (Ref. [10]).

- Derive an analytical model or use a model identified from experimental data. The model provides the state matrix A_c and the input matrix B_c.
- Assign closed-loop eigenvalues λ_k and eigenvectors ϕ_{dk} for $k = 1, 2, \ldots, n$.
- Form matrix Γ_k for each eigenvalue λ_k, Eq. (8.43).
- Determine the SVD of Γ_k to obtain the basis vectors V_{ok}, Eq. (8.44).
- Solve for coefficient vector c_k by using the pseudoinverse, Eq. (8.55).
- Compute the control gain matrix $\mathcal{F}_c$, Eq. (8.52).
- Verify by solving the closed-loop eigenvalue problem to examine its degree of stability, Eq. (8.41).

The approaches introduced in this section are also applicable for deriving the pole-placement techniques for the discrete-time model. All formulations for the discrete-time model are identical to those shown in this section with the replacement of the continuous-time matrices A_c and B_c with the discrete-time matrices A and B. Of course, the closed-loop discrete-time model is stable if its eigenvalues are within the unit circle.

8.6 Optimal Control

In an automatic control problem, the primary concern of the control designer is to meet the performance requirements. In control design, it is practical to assume that the power available for control is limited. The user of a controller will be dissatisfied if the controller consumes too much power. The optimal theory sets a bound on performance in that the designer knows he or she can do no better than the optimal. The typical optimal control problem is concerned with optimizing the response of the controlled variable(s) to disturbances of the system to be controlled. Therefore the optimal control involves at least one cost index that must be minimized in order to quantify its optimal response of the controlled variables.

In structural systems, there are several obvious candidates that can be used to form a cost index. Good candidates to define the cost index include the kinetic energy, the potential energy, and the control energy. Let w be the displacement vector and u be the control vector of a system to be controlled. The kinetic energy T, the potential energy V, and the control energy U can be mathematically expressed by

$$\text{kinetic energy, } T = \frac{1}{2} \dot{w}^T M \dot{w};$$

$$\text{potential energy, } V = \frac{1}{2} w^T K w;$$

$$\text{control energy, } U = \frac{1}{2} u^T u.$$

Any combination of these energy terms may also be considered as a cost index. For example, the total energy is defined as

$$T + V = \frac{1}{2} [w^T \quad \dot{w}^T] \begin{bmatrix} K & 0 \\ 0 & M \end{bmatrix} \begin{bmatrix} w \\ \dot{w} \end{bmatrix}$$

$$= \frac{1}{2} x^T Q x. \tag{8.57}$$

For real structures, the stiffness matrix K and the mass matrix M are generally satisfied by

$$K \geq 0, \qquad M > 0$$

where $\geq$ means positive semidefinite and $>$ implies positive definite. The energy expression can be generalized to become

$$S = \frac{1}{2} x^T Q x = \frac{1}{2} [w^T \quad \dot{w}^T] \begin{bmatrix} Q_{11} & Q_{12} \\ Q_{12}^T & Q_{22} \end{bmatrix} \begin{bmatrix} w \\ \dot{w} \end{bmatrix}, \tag{8.58}$$

where Q is a constant matrix so that $S = T + V$ when $Q_{11} = K$, $Q_{22} = M$, and $Q_{12} = 0$. Adding the control energy term to Eq. (8.58) yields

$$S + U = \frac{1}{2} x^T Q x + \frac{1}{2} u^T R u. \tag{8.59}$$

The matrix Q is specified to weight the state vector x, whereas the matrix R is given to weight the control force u. Because x and u are time dependent, an integration over time leads to the following cost index:

$$J = \frac{1}{2} \int_0^\infty (x^T Q x + u^T R u) \, dt.$$

The control force that minimizes the cost index gives an optimal control. The first term $x^T Q x$ is related to the kinetic energy and the potential energy, whereas the second term $u^T R u$ is related to the control effort.

The linear quadratic (LQ) steady-state optimal control design problem can then be stated as follows:

Given a system described by the continuous-time model with $x(0)$ known,

$$\dot{x} = A_c x + B_c u, \tag{8.60}$$

find the control law for u,

$$u = \mathcal{F}_c x, \tag{8.61}$$

that minimizes the cost index

$$J = \frac{1}{2} \int_0^\infty (x^T Q x + u^T R u) \, dt, \tag{8.62}$$

where Q is positive semidefinite and R is positive definite.

The optimal control problem requires computing the control input that minimizes the cost index, Eq. (8.62), subject to constraint equations (8.60) and (8.61).

EXAMPLE 8.5

Consider the simple example

$$\dot{x} = ax + bu.$$

Assume full state feedback:

$$u = fx.$$

The state equation becomes

$$\dot{x} = (a + bf)x,$$

which has the solution

$$x(t) = e^{(a+bf)t}x(0).$$

Substituting the solution into the cost index yields

$$J = \frac{1}{2}\int_0^\infty (x^2 + ru^2)\,dt = \int_0^\infty (1 + rf^2)x^2 dt$$

$$= \int_0^\infty (1 + rf^2)x^2(0)e^{2(a+bf)t}\,dt$$

$$= (1 + rf^2)x^2(0)\left[\frac{e^{2(a+bf)t}}{2(a + bf)}\right]_0^\infty.$$

Now $a + bf$ must be negative in order to have a solution for J, i.e.,

$$a + bf < 0.$$

Hence, the cost index becomes

$$J = \frac{-(1 + rf^2)x^2(0)}{2(a + bf)}.$$

To minimize the cost index, first take the partial derivative with respect to f to yield

$$\frac{\partial J}{\partial f} = \frac{x^2(0)}{2}\left[\frac{-2(a + bf)rf + b(1 + rf^2)}{(a + bf)^2}\right]$$

$$= \frac{x^2(0)}{2}\left[\frac{-brf^2 - 2arf + b}{(a + bf)^2}\right].$$

Because $a + bf < 0$, the control gain f that minimizes the cost index J should satisfy the following equality,

$$brf^2 + 2arf - b = 0,$$

in order to make $\partial J/\partial f = 0$. Solving the roots of the polynomial equation thus produces

$$f = \frac{-a}{b} \pm \sqrt{\left(\frac{a}{b}\right)^2 + \frac{1}{r}}.$$

There are two solutions for the control gain f. Does that mean that the solution is not unique?

Let us assume that the open-loop system is stable, $a < 0$. For the case in which $r \to \infty$, the control gain f has two solutions

$$f = \frac{-2a}{b} \quad \text{and} \quad 0.$$

The first solution indicates that

$$2a + bf = 0 \implies a + bf = -a > 0 \quad \text{for} \quad a < 0.$$

Because of the assumption that $a < 0$ and the requirement that $a + bf < 0$, it becomes obvious that $f = -2a/b$ cannot be a solution. Hence, the only solution left is $f = 0$, implying that the control energy is zero, i.e., no input is needed.

As a result, the solution that minimizes the cost index J and satisfies the inequalities $a + bf < 0$ and $a < 0$ is

$$f = \frac{-a}{b} + \sqrt{\left(\frac{a}{b}\right)^2 + \frac{1}{r}}.$$

We eliminate the other possibility because it violates the requirement that $a + bf < 0$ in the special case $a < 0$ and $r \to \infty$ as shown above.

If $r \to 0$, f becomes

$$f \approx \sqrt{\frac{1}{r}}.$$

The gain f can be arbitrarily large, depending on the value of r. But we have to worry about saturation of the actuator.

8.6.1 Continuous-Time Technique

Now consider the multi-input multi-output case,

$$\dot{x} = (A_c + B_c \mathcal{F}_c)x, \qquad u = \mathcal{F}_c x, \tag{8.63}$$

which has the solution

$$x(t) = e^{(A_c + B_c \mathcal{F}_c)t} x(0). \tag{8.64}$$

Substituting the solution into the cost index yields

$$
\begin{aligned}
J &= \frac{1}{2}\int_0^\infty (x^T Q x + u^T R u)\, dt = \frac{1}{2}\int_0^\infty x^T\left(Q + \mathcal{F}_c^T R \mathcal{F}_c\right) x\, dt \\
&= \frac{1}{2}x^T(0)\left[\int_0^\infty e^{(A_c + B_c \mathcal{F}_c)^T t}\left(Q + \mathcal{F}_c^T R \mathcal{F}_c\right) e^{(A_c + B_c \mathcal{F}_c)t}\, dt\right] x(0). \tag{8.65}
\end{aligned}
$$

Now introduce a constant matrix P that satisfies the following equation,

$$-(A_c + B_c \mathcal{F}_c)^T P - P(A_c + B_c \mathcal{F}_c) = Q + \mathcal{F}_c^T R \mathcal{F}_c, \tag{8.66}$$

and note the following equality:

$$
\begin{aligned}
\frac{d}{dt}&\left\{e^{(A_c + B_c \mathcal{F}_c)^T t}(-P)e^{(A_c + B_c \mathcal{F}_c)t}\right\} \\
&= e^{(A_c + B_c \mathcal{F}_c)^T t}(A_c + B_c \mathcal{F}_c)^T(-P)e^{(A_c + B_c \mathcal{F}_c)t} \\
&\quad + e^{(A_c + B_c \mathcal{F}_c)^T t}(-P)(A_c + B_c \mathcal{F}_c)e^{(A_c + B_c \mathcal{F}_c)t} \\
&= e^{(A_c + B_c \mathcal{F}_c)^T t}[-(A_c + B_c \mathcal{F}_c)^T P - P(A_c + B_c \mathcal{F}_c)]e^{(A_c + B_c \mathcal{F}_c)t} \\
&= e^{(A_c + B_c \mathcal{F}_c)^T t}\left(Q + \mathcal{F}_c^T R \mathcal{F}_c\right)e^{(A_c + B_c \mathcal{F}_c)t}. \tag{8.67}
\end{aligned}
$$

Substituting Eq. (8.67) into Eq. (8.65) thus leads to

$$J = \frac{1}{2}x^T(0)\left[\int_0^\infty e^{(A_c+B_c\mathcal{F}_c)^T t}(Q + \mathcal{F}_c^T R\mathcal{F}_c)e^{(A_c+B_c\mathcal{F}_c)t}dt\right]x(0)$$

$$= \frac{1}{2}x^T(0)\left[\int_0^\infty \frac{d}{dt}\{e^{(A_c+B_c\mathcal{F}_c)^T t}(-P)e^{(A_c+B_c\mathcal{F}_c)t}\}dt\right]x(0)$$

$$= \frac{1}{2}x^T(0)\{e^{(A_c+B_c\mathcal{F}_c)^T t}(-P)e^{(A_c+B_c\mathcal{F}_c)t}\}_0^\infty x(0). \tag{8.68}$$

The matrix $A_c + B_c\mathcal{F}_c$ must be stable for the cost index J to have a solution. This means that the real part of the eigenvalues of $A_c + B_c\mathcal{F}_c$ must be negative. Hence Eq. (8.68) becomes

$$J = \frac{1}{2}x^T(0)Px(0). \tag{8.69}$$

With the size of x as $n \times 1$, where n is the number of states, the $n \times n$ matrix P must be symmetric and positive definite in order to make the cost index J positive. The question now is how to obtain a P such that J is minimum.

Let the control gain matrix $\mathcal{F}_c$ be explicitly expressed by

$$\mathcal{F}_c = \begin{bmatrix} f_{11} & f_{12} & \cdots & f_{1n} \\ f_{21} & f_{22} & \cdots & f_{2n} \\ \vdots & \vdots & \ddots & \vdots \\ f_{r1} & f_{r2} & \cdots & f_{rn} \end{bmatrix},$$

where n is the order of the system and r is the number of inputs. Choose any element f_{ij} for $i = 1, 2, \ldots, r$ and $j = 1, 2, \ldots, n$ and note the following equalities:

$$\frac{\partial}{\partial f_{ij}}(A_c + B_c\mathcal{F}_c)^T P = (A_c + B_c\mathcal{F}_c)^T \frac{\partial P}{\partial f_{ij}} + \frac{\partial \mathcal{F}_c^T}{\partial f_{ij}}B_c^T P \tag{8.70}$$

$$\frac{\partial}{\partial f_{ij}}P(A_c + B_c\mathcal{F}_c) = \frac{\partial P}{\partial f_{ij}}(A_c + B_c\mathcal{F}_c) + PB_c\frac{\partial \mathcal{F}_c}{\partial f_{ij}} \tag{8.71}$$

$$\frac{\partial}{\partial f_{ij}}(Q + \mathcal{F}_c^T R\mathcal{F}_c) = \frac{\partial \mathcal{F}_c^T}{\partial f_{ij}}R\mathcal{F}_c + \mathcal{F}_c^T R\frac{\partial \mathcal{F}_c}{\partial f_{ij}}. \tag{8.72}$$

With the aid of Eqs. (8.70)–(8.72), taking the partial derivative of Eq. (8.66) with respect to the gain element f_{ij} yields

$$-(A_c + B_c\mathcal{F}_c)^T \frac{\partial P}{\partial f_{ij}} - \frac{\partial P}{\partial f_{ij}}(A_c + B_c\mathcal{F}_c)$$

$$= \frac{\partial \mathcal{F}_c^T}{\partial f_{ij}}(R\mathcal{F}_c + B_c^T P) + (\mathcal{F}_c^T R + PB_c)\frac{\partial \mathcal{F}_c}{\partial f_{ij}} \tag{8.73}$$

for $i = 1, 2, \ldots, r$ and $j = 1, 2, \ldots, n$ with a total of rn equations. The matrix $\partial \mathcal{F}_c/\partial f_{ij}$ is an $r \times n$ zero matrix except with 1 at the ith row and the jth column. On the other hand, the matrix $\partial \mathcal{F}_c^T/\partial f_{ij}$ is an $n \times r$ zero matrix except with 1 at the jth row and the ith column.

Minimizing the cost index J shown in Eq. (8.69) requires making the partial derivative of P with respect to every element of $\mathcal{F}_c$ zero, i.e.,

$$\frac{\partial P}{\partial f_{ij}} = 0 \Longrightarrow \frac{\partial J}{\partial f_{ij}} = 0, \tag{8.74}$$

for any nonzero initial condition $x(0)$. The closed-loop state matrix $A_c + B_c\mathcal{F}_c$ must be full rank. Otherwise, it will have zero eigenvalues that conflict with the assumption that the real part of the eigenvalues for $A_c + B_c\mathcal{F}_c$ must be negative. If the matrix $A_c + B_c\mathcal{F}_c$ is full rank, inserting Eq. (8.74) into Eq. (8.73) produces one solution:

$$B_c^T P + R\mathcal{F}_c = 0$$

or

$$\mathcal{F}_c = -R^{-1}B_c^T P. \tag{8.75}$$

Equation (8.75) exists only when the symmetric matrix R is invertible. It has been shown earlier that the matrices Q and R must be at least positive semidefinite, i.e., $Q \geq 0$ and $R \geq 0$, in order to have a nonnegative cost index as defined in Eq. (8.62). Because R must be nonsingular for Eq. (8.75) to exist, R should be positive definite, i.e., $R > 0$. As a result, the weighting matrices Q and R must have the following properties,

$$Q \geq 0, \qquad R > 0, \tag{8.76}$$

for an optimal gain $\mathcal{F}_c$ to exist that minimizes a positive scalar cost index J defined in Eq. (8.62).

Substituting the result for computing the control gain $\mathcal{F}_c$ shown in Eq. (8.75) into Eq. (8.66) gives the following equation

$$-\left(A_c - B_cR^{-1}B_c^T P\right)^T P - P\left(A_c - B_cR^{-1}B_c^T P\right) = Q + PB_cR^{-1}B_c^T P$$

or

$$A_c^T P + PA_c - PB_cR^{-1}B_c^T P + Q = 0. \tag{8.77}$$

Equation (8.77) is the well-known steady-state Riccati equation, after Count Riccati, an Italian who investigated a scalar version of it in 1724. Equation (8.77) is used to solve for a symmetric and positive-definite matrix P to compute the gain matrix $\mathcal{F}_c$. The feedback control law for u becomes

$$u = \mathcal{F}_c x = -R^{-1}B_c^T Px. \tag{8.78}$$

The optimal control approach to a controller design requires full state feedback. Thus the number of sensors must be equal to the number of states so that the measurement matrix C is square. Otherwise, a state estimator (e.g., a Kalman filter) must be designed for state estimation. If R is positive definite and Q is positive semidefinite, then P is positive definite, which guarantees that the eigenvalues of

$$A_c + B_c\mathcal{F}_c = A_c - B_cR^{-1}B_c^T P$$

are stable. If $A_c + B_c\mathcal{F}_c$ has any eigenvalue with positive real part, J has no minimum.

As a summary, the typical procedure for an optimal controller design includes the following steps:

- Derive an analytical model or use a model identified from experimental data. The model provides the state matrix A_c and the input matrix B_c.
- Specify a positive-semidefinite Q and positive-definite R.
- Solve for the positive-definite matrix P from steady-state Riccati equation (8.77).
- Compute the control gain matrix $\mathcal{F}_c$.
- Verify by solving the closed-loop eigenvalue problem to examine its degree of stability.

There are some disadvantages of the optimal control approach. Closed-loop system usually has no mechanical analog. Design may lack stability robustness with respect to uncertainties in the system. Full state measurement is often not available and a state estimator, which in turn requires a good model of the system, is needed.

EXAMPLE 8.6

Consider the simple example

$$A_c = a, \quad B_c = b, \quad R = r, \quad Q = 1.$$

Riccati equation (8.77) becomes

$$2ap - \frac{b^2 p^2}{r} + 1 = 0.$$

The control gain from Eq. (8.75) is

$$f = -\frac{bp}{r} \implies p = -\frac{rf}{b}.$$

Substituting the expression for p into the Riccati equation yields

$$brf^2 + 2arf - b = 0,$$

which is the same equation as that given in Example 8.5. Because p must be positive, the only solution that minimizes the cost index J and satisfies the inequalities $a + bf < 0$ and $a < 0$ is

$$f = \frac{-a}{b} + \sqrt{\left(\frac{a}{b}\right)^2 + \frac{1}{r}}.$$

8.6.2 Discrete-Time Technique

Similar to the continuous-time model, the linear quadratic (LQ) steady-state optimal control design problem for a discrete-time model can be stated as follows:

Given a system described by the discrete-time model with $x(0)$ known,

$$x(k + 1) = Ax(k) + Bu(k), \tag{8.79}$$

find the control law for $u(k)$,

$$u(k) = \mathcal{F}x(k), \tag{8.80}$$

that minimizes the cost index

$$J = \sum_{k=0}^{\infty} x^T(k)Qx(k) + u^T(k)Ru(k), \tag{8.81}$$

where Q is positive semidefinite and R is positive definite.

The optimal control problem requires computing the control input that minimizes the cost index, Eq. (8.81), subject to constraint equations (8.79) and (8.80). The optimal control input $u(k)$ can be obtained such that J is minimized by the solution of the discrete-time steady-state Riccati equation,

$$P = A^T PA - A^T PB(R + B^T PB)^{-1} B^T PA + Q \tag{8.82}$$

with the optimal control gain given by

$$\mathcal{F} = (R + B^T PB)^{-1} B^T PA. \tag{8.83}$$

It is recommended that the reader prove Eq. (8.82) as an exercise problem.

8.6.3   Realistic Discrete-Time Implementation

The control $u(k)$ thus obtained is not implementable because both $x(k)$ and $u(k)$ are given at the same time k. There is no time to compute the input $u(k)$ to control the state $x(k)$. To allow time for computation, the following scheme can be used. Because the state equation gives

$$x(k) = Ax(k-1) + Bu(k-1), \tag{8.84}$$

the control force $u(k)$ can be written as follows,

$$u(k) = \mathcal{F}x(k) = \mathcal{F}Ax(k-1) + \mathcal{F}Bu(k-1), \tag{8.85}$$

or, with one time step ahead,

$$u(k+1) = \mathcal{F}Ax(k) + \mathcal{F}Bu(k). \tag{8.86}$$

As a result, the combined state and control equations become

$$\begin{bmatrix} x(k+1) \\ u(k+1) \end{bmatrix} = \begin{bmatrix} A & B \\ \mathcal{F}A & \mathcal{F}B \end{bmatrix} \begin{bmatrix} x(k) \\ u(k) \end{bmatrix}. \tag{8.87}$$

Now one sampling time period can be used for computing the control force $u(k)$.

8.7   Concluding Remarks

Several control techniques were introduced in this chapter. The practical implication of controllability and observability was studied. For a linear time-invariant controllable system, an introduction of state feedback can arbitrarily relocate the eigenvalues of the system. This technique is commonly referred to as pole-placement. It is accomplished by use of the SVD of a matrix formed from the desired closed-loop eigenvalues to generate the space wherein the closed-loop eigenvectors should reside. The desired eigenvectors should be chosen as orthogonal as possible to minimize the state-feedback gain. We also studied the notion of optimal control. Unlike the pole-placement technique, the optimal control technique requires the minimization of a cost index. The control gain is obtained by the solution of a Riccati equation. For state feedback, it is assumed that all the state variables are available. If not, a state estimator must be constructed. The combination of state feedback and state estimation will be introduced in the next chapter.

8.8 Problems

8.1 Assume that a linear discrete-time, time-invariant dynamical model, Eq. (8.1), of the order of n is controllable and observable. Prove that its corresponding continuous-time model, Eq. (8.29), is also controllable and observable, i.e., the following controllability and observability matrices,

$$[A_c^{n-1} B_c \quad \cdots \quad A_c B_c \quad B_c],$$

$$\begin{bmatrix} C \\ C A_c \\ \vdots \\ C A_c^{n-1} \end{bmatrix},$$

have a rank of n. Note that the state matrices for the discrete-time and continuous-time models share the same eigenvectors.

8.2 Prove that the continuous-time model represented by the system matrices

$$A_c = \begin{bmatrix} 0 & 1 & 0 & 0 \\ 0 & 0 & -1 & 0 \\ 0 & 0 & 0 & 1 \\ 0 & 0 & 11 & 0 \end{bmatrix}, \quad B_c = \begin{bmatrix} 0 \\ 1 \\ 0 \\ -1 \end{bmatrix}, \quad C = [1 \ 0 \ 0 \ 1]$$

is controllable and observable even though the system has two repeated zero eigenvalues.

8.3 Given the following discrete-time model,

$$x(k+1) = \begin{bmatrix} 0 & 1 \\ 1 & 1 \end{bmatrix} x(k) + \begin{bmatrix} 0 \\ 1 \end{bmatrix} u(k),$$

find the state-feedback gain matrices that will produce the closed-loop eigenvalues (poles) at

(a) $\lambda = 0$ and $\lambda = 0.6$,
(b) $\lambda = 0.1 \pm j0.4$.

8.4 Consider the discrete-time model

$$x(k+1) = \begin{bmatrix} 0.5 & 0.1 \\ 0.1 & 0.5 \end{bmatrix} x(k) + \begin{bmatrix} 1 & 1 \\ -1 & 1 \end{bmatrix} u(k);$$

find at least two feedback gain matrices that will both give the closed-loop eigenvalues at $\lambda = 0$ and $\lambda = 0.3$.

8.5 Let the state matrix A and the input influence matrix b be

$$A = \begin{bmatrix} 1 & 1 \\ 0 & 1 \end{bmatrix}, \quad b = \begin{bmatrix} 0 \\ 1 \end{bmatrix}.$$

(a) Find the values of $\mathcal{F} = [f_1 \quad f_2]$ such that the matrix $A + b\mathcal{F}$ has the distinct eigenvalues λ_1 and λ_2.
(b) What are the values of f_1 and f_2 that make $\lambda_1 = \lambda_2$?

8.6 Consider the state–space model

$$\dot{x} = \begin{bmatrix} -2 & 1 & 0 & 0 \\ 0 & -2 & 0 & 0 \\ 0 & 0 & 1 & 1 \\ 0 & 0 & 0 & 1 \end{bmatrix} x + \begin{bmatrix} 1 \\ 0 \\ 0 \\ 1 \end{bmatrix} u$$

that has an unstable eigenvalue 1.

(a) Is this system controllable?
(b) Is there any eigenvalue that is uncontrollable?
(c) Is it possible to stabilize the system by state feedback?
(d) If the answer to (c) is yes, find the gain matrix such that the closed-loop eigenvalues are $-1, -1, -2$, and -2.

8.7 Consider the state–space model

$$\dot{x} = \begin{bmatrix} 2 & 1 & 0 & 0 \\ 0 & 2 & 0 & 0 \\ 0 & 0 & -1 & 0 \\ 0 & 0 & 0 & -1 \end{bmatrix} x + \begin{bmatrix} 0 \\ 1 \\ 0 \\ 1 \end{bmatrix} u.$$

(a) Is this system controllable?
(b) Is there any eigenvalue that is uncontrollable?
(c) Is it possible to stabilize the system by state feedback?
(d) Is it possible to have the following closed-loop eigenvalues:
   (1) $-1, -1, -2$, and -2,
   (2) $-1, -2, -2$, and -2,
   (3) $-2, -2, -2$, and -2.

8.8 Consider the equation of motion

$$M\ddot{w} + Kw = B_2 u,$$

where

$$w = \begin{bmatrix} w_1 \\ w_2 \\ w_3 \end{bmatrix},$$

$$M = \text{diag}(1, 1, 1),$$

$$K = \begin{bmatrix} 10 & -5 & 0 \\ -5 & 25 & -20 \\ 0 & -20 & 20 \end{bmatrix},$$

$$B_2 = \begin{bmatrix} 1 & 0 \\ 0 & 0 \\ 0 & 1 \end{bmatrix},$$

$$u = \begin{bmatrix} u_1 \\ u_3 \end{bmatrix}.$$

(a) Form the first-order equation

$$\dot{x} = A_c x + B_c u; \quad x = \begin{bmatrix} w \\ \dot{w} \end{bmatrix}.$$

(b) Compute the open-loop eigenvalues and eigenvectors.
(c) Compute the closed-loop eigenvalues with the direct velocity feedback:

$$u_1 = -2\dot{w}_1, \qquad u_3 = -2\dot{w}_3.$$

(d) Use the closed-loop eigenvalues obtained from (c) as the desired eigenvalues and the open-loop eigenvectors as the desired eigenvectors. Find the gain matrix $\mathcal{F}_c$ such that $A_c + B_c \mathcal{F}_c$ has the desired eigenvalues.
(e) Repeat (d) but use the identity matrix of the 6×6 as the desired eigenvectors.
(f) Compute the condition number of $\mathcal{F}_c$ from (d) and (e) and compare with the condition number for the direct velocity feedback, i.e., $\sqrt{2^2 + 2^2}$.

8.9 Consider the undamped second-order scalar equation,

$$m\ddot{w} + kw = bu,$$

or, in the first-order state–space expression,

$$\dot{x} = A_c x + B_c u,$$

where

$$x = \begin{bmatrix} w \\ \dot{w} \end{bmatrix}, \qquad A_c = \begin{bmatrix} 0 & 1 \\ -\dfrac{k}{m} & 0 \end{bmatrix}, \qquad B_c = \begin{bmatrix} 0 \\ \dfrac{b}{m} \end{bmatrix}.$$

Define the following performance index:

$$J = \frac{1}{2} \int_0^\infty \left(x^T \begin{bmatrix} \alpha k & 0 \\ 0 & \beta m \end{bmatrix} x + \frac{b^2}{k} u^2 \right) dt,$$

where α and β are arbitrary positive constants. Solve the steady-state Riccati equation to obtain the P matrix and compute the feedback gain matrix.

8.10 The longitudinal perturbations of an airplane in horizontal cruising flight are reasonably well described by the following equations of motion,

$$\dot{\psi} = -\frac{1}{\tau}\psi + \phi$$

$$\dot{\phi} = -\omega_o^2 \psi + bu,$$

where ψ is the perturbation from the cruise angle of attack, ϕ is the perturbation of cruise pitch angular velocity, τ is the lifting time constant, ω_o is the undamped pitch natural frequency, b is the elevator influence constant, u is the elevator deflection. Determine the steady-state constant gains f_1 and f_2 such that

$$u = f_1 \psi + f_2 \phi$$

to minimize

$$J = \frac{1}{2} \int_0^\infty \left(\frac{\psi^2}{\psi_o^2} + \frac{\phi^2}{\phi_o^2} + \frac{u^2}{u_o^2} \right) dt,$$

where ψ_o^2, ϕ_o^2, and u_o^2 are weighting terms. Such a feedback controller (autopilot) will maintain small vertical acceleration but not horizontal flight.

8.11 With reference to Problem 8.10, consider the equations of motion

$$\dot{\psi} = -\frac{1}{\tau}\psi + \phi,$$

$$\dot{\phi} = -\omega_o^2 \psi + bu,$$

$$\dot{\theta} = \phi,$$

where an additional equation has been added to maintain small vertical flight. Derive the equations to determine the steady-state constant gains f_1, f_2, and f_3 such that

$$u = f_1\psi + f_2\phi + f_3\theta$$

to minimize

$$J = \frac{1}{2}\int_0^\infty \left(\frac{(\theta - \psi)^2}{\gamma_o^2} + \frac{u^2}{u_o^2}\right) dt,$$

where γ_o^2 and u_o^2 are weighting terms. Such a feedback controller will maintain nearly horizontal flight.

8.12 Prove that the Riccati equation, Eq. (8.82), minimizes the cost index, Eq. (8.81), subject to the constraint, Eq. (8.79), with the control force given by Eqs. (8.80) and (8.83).

BIBLIOGRAPHY

[1] Brogan, W. L., *Modern Control Theory*, Prentice-Hall, Englewood Cliffs, NJ, 1985.

[2] Bryson, A. E. and Ho, Y.-C., *Applied Optimal Control*, Wiley, New York, 1975.

[3] Chen, C,-T., *Linear System Theory and Design*, CBS College Publishing, New York, 1984.

[4] Fortman, T. E. and Hitz, K. L., *An Introduction to Linear Control Systems*, Marcel Dekker, Inc., New York, 1977.

[5] Meirovitch, L., *Introduction to Dynamics and Control*, Wiley, New York, 1985.

[6] Meirovitch, L., *Dynamics and Control of Structures*, Wiley-Interscience, New York, 1990.

[7] Ogata, K., *Discrete-Time Control Systems*, Prentice-Hall, Englewood Cliffs, NJ, 1987.

[8] Stengel, R. F., *Stochastic Optimal Control*, Wiley, New York, 1986.

[9] Juang, J. N., Turner, J. D., and Chun, H. M., "Closed-Form Solution of Feedback Control with Terminal Constraints," *J. Guid. Control Dyn.*, Vol. 1, pp. 39–43, 1985.

[10] Juang, J. N., Lim, K. B., and Junkins, J. L., "Robust Eigensystem Assignment for Flexible Structures," *J. Guid. Control Dyn.*, Vol. 12, pp. 381–387, 1989.

9

Dynamic Feedback Controller

9.1 Introduction

In a state–space model, the relationship between the input(s) and the output(s) of a system is described by means of an intermediate variable called the state. Recall that in a state-feedback-control system, the state information is required for computing the control input. In Chap. 8, we assume that the state of the system can be directly measured and used in the state-feedback-control law. This is certainly possible for a simple system (if enough sensors are available to measure all elements of the state vector), but usually not the case in practice. Fortunately, we can use a state estimator, otherwise known as an observer, to estimate the state from input and output measurements. The estimated state is then used in the state-feedback-control law as if it is the true state. Obviously it is desirable that the estimated state provided by an observer is as close as possible to the true state. Indeed, under ideal conditions, the estimated state can be the same as the true state (in theory at least).

The first part of this chapter addresses the problem of state estimation by use of an observer (Ref. [1–2]). The issue of integrating a state estimator into a state-feedback-control system and the stability analysis of the overall system will then be discussed. Under certain conditions, the determination of a state-feedback-control law and the design of an observer to be used can be treated as separate problems. Specifically, if the state-feedback-control law is stable with perfect knowledge of the state, it will also produce a stable closed-loop system with the state estimated by a stable observer. This fortunate result is quite important. It is known as the separation principle, which will be proved in this chapter.

Later in the chapter, a more general feedback scheme is introduced, namely the dynamic output-feedback controller, which is sometimes known as the dynamic compensator. We will focus on the generic form and the stability analysis of the dynamic output-feedback controller without dwelling on the general design aspects of these controllers. The static (or direct) output-feedback case will be examined first to better lead to the dynamic output-feedback case. Both the continuous-time and discrete-time cases will be discussed.

9.2 Continuous-Time State Estimation

The state–space model of a linear time-invariant dynamic system in continuous-time format is

$$\frac{dx}{dt} = A_c x(t) + B_c u(t), \tag{9.1}$$

$$y(t) = Cx(t) + Du(t). \tag{9.2}$$

The system is assumed to be known exactly (i.e., A_c, B_c, C, D are known), and so are the input $u(t)$ and the output $y(t)$. The initial conditions of the system, $x(0)$, however, are not known. We wish to find an estimate of $x(t)$ from the available information. Note that if $x(0)$ is known, then the state of the system at any time can be determined simply by integration of the state equation. This is certainly possible in theory but is not desirable in practice because such an approach is very sensitive to any errors in the system matrices A_c and B_c. For the present discussion, we ignore any issues associated with imperfect measurements (i.e., we assume noise-free data).

The basic approach of estimating the states is to simulate the state and the output measurement equations of the system on a computer with an assumed initial state vector. The actual system may be subjected to unmeasurable disturbances that cannot be used in the simulation but affect the output measurements. To make sure that the estimated state does not deviate too much from the actual state values, the difference between the actual output and the simulation output can be used as one of the driving inputs in the simulation equation.

A state estimator has the form

$$\frac{d\hat{x}}{dt} = A_c \hat{x}(t) + B_c u(t) - G_c [y(t) - \hat{y}(t)], \tag{9.3}$$

$$\hat{y}(t) = C\hat{x}(t) + Du(t), \tag{9.4}$$

where $\hat{x}(t)$ and $\hat{y}(t)$ denote the estimated state and the estimated output, respectively. The matrix G_c is called the observer gain matrix to be determined. At the moment, assume that G_c is some appropriately chosen constant matrix. Note that the difference between the true (measured) output and the estimated output is used to compensate for the unknown initial conditions. There is a rather convenient way to write the observer equation by combining Eqs. (9.3) and (9.4) as follows:

$$\begin{aligned}
\frac{d\hat{x}}{dt} &= A_c \hat{x}(t) + B_c u(t) - G_c [y(t) - \hat{y}(t)] \\
&= A_c \hat{x}(t) + B_c u(t) - G_c [y(t) - C\hat{x}(t) - Du(t)] \\
&= (A_c + G_c C)\hat{x}(t) + (B_c + G_c D)u(t) - G_c y(t).
\end{aligned} \tag{9.5}$$

Note that the last equality,

$$\frac{d\hat{x}}{dt} = (A_c + G_c C)\hat{x}(t) + (B_c + G_c D)u(t) - G_c y(t), \tag{9.6}$$

is simply a matrix differential equation driven by both input $u(t)$ and output $y(t)$. The solution of such a differential equation will yield $\hat{x}(t)$, which is the estimated state vector. How do we know that $\hat{x}(t)$ would converge to the true state? In fact, the estimated state will not converge to the true state if G_c is not properly chosen. Define the state-estimation error as

$$e(t) = x(t) - \hat{x}(t). \tag{9.7}$$

The goal is to choose a matrix G_c such that the estimation error becomes zero. The governing equation for $e(t)$ is

$$
\begin{aligned}
\frac{de}{dt} &= \frac{dx}{dt} - \frac{d\hat{x}}{dt} \\
&= A_c x(t) + B_c u(t) - [(A_c + G_c C)\hat{x}(t) + (B_c + G_c D) u(t) - G_c y(t)] \\
&= A_c x(t) + B_c u(t) - (A_c + G_c C)\hat{x}(t) \\
&\quad - (B_c + G_c D) u(t) + G_c [Cx(t) + Du(t)] \\
&= (A_c + G_c C) x(t) - (A_c + G_c C)\hat{x}(t) \\
&= (A_c + G_c C) [x(t) - \hat{x}(t)] \\
&= (A_c + G_c C) e(t).
\end{aligned}
\tag{9.8}
$$

Note that the terms involving $u(t)$ and $y(t)$ cancel themselves out to make the state-estimation error simply governed by

$$\frac{de}{dt} = (A_c + G_c C) e(t), \tag{9.9}$$

which is another first-order matrix differential equation. If G_c is chosen such that all the eigenvalues of $(A_c + G_c C)$ have negative real parts, then the solution for Eq. (9.9) that governs the state-estimation error will converge to zero as time tends to infinity, i.e.,

$$\lim_{t \to \infty} e(t) = 0 \tag{9.10}$$

for any initial error $e(0) = x(0) - \hat{x}(0)$. In other words, any initial estimated state $\hat{x}(0)$ can be assumed for the observer equation. The matrix $(A_c + G_c C)$ is called the observer state matrix. Convergence of $e(t)$ to zero implies that the estimated state converges to the true state as desired,

$$\lim_{t \to \infty} \hat{x}(t) = x(t). \tag{9.11}$$

To implement the observer, we must be able to integrate observer Eq. (9.6) in real time by using real-time measurements of both input $u(t)$ and output $y(t)$. Figure 9.1 gives the block diagram summarizing the state estimator.

When the estimated state $\hat{x}(t)$ of the system converges to the true state $x(t)$, what will happen to the estimated output $\hat{y}(t)$? The estimated output will also converge to the true output as well because

$$\hat{y}(t) = C\hat{x}(t) + Du(t). \tag{9.12}$$

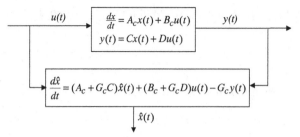

Figure 9.1. Continuous-time state estimator.

9.3 Discrete-Time State Estimation

In this section, the observer equation in the discrete-time domain will be derived. The discrete-time observer equations are very similar to the continuous-time ones. The original continuous-time system can be represented by a discrete-time model with a zero-order hold on the input as

$$x(k + 1) = Ax(k) + Bu(k), \tag{9.13}$$

$$y(k) = Cx(k) + Du(k). \tag{9.14}$$

Again, the system is assumed to be known exactly, i.e., A, B, C, and D are known and so are the input $u(k)$ and the output $y(k)$. The initial conditions of the system, $x(0)$, however, are not known. A discrete-time observer has the form

$$\hat{x}(k + 1) = A\hat{x}(k) + Bu(k) - G\left[y(k) - \hat{y}(k)\right], \tag{9.15}$$

$$\hat{y}(k) = C\hat{x}(k) + Du(k), \tag{9.16}$$

where $\hat{x}(k)$ and $\hat{y}(k)$ denote the estimated state and output, respectively. The matrix G is called the (discrete-time) observer gain matrix to be determined. The difference between the true (measured) output and the estimated output is used to compensate for the unknown initial conditions. The observer equation can be rewritten in another way, as follows:

$$\begin{aligned}
\hat{x}(k + 1) &= A\hat{x}(k) + Bu(k) - G\left[y(k) - \hat{y}(k)\right] \\
&= A\hat{x}(k) + Bu(k) - G\left[y(k) - C\hat{x}(k) - Du(k)\right] \\
&= (A + GC)\hat{x}(k) + (B + GD)u(k) - Gy(k). \tag{9.17}
\end{aligned}$$

Equation (9.17) is a matrix difference equation driven by both the input $u(k)$ and the output $y(k)$. The solution of such a difference equation will yield the estimated state $\hat{x}(k)$. To show that the estimated state $\hat{x}(k)$ will indeed converge to the true state $x(k)$ if the observer gain matrix G is properly chosen, let the state-estimation error be defined by

$$e(k) = x(k) - \hat{x}(k). \tag{9.18}$$

The equation that governs $e(k)$ is derived as follows:

$$e(k+1) = x(k+1) - \hat{x}(k+1)$$
$$= Ax(k) + Bu(k) - [(A+GC)\hat{x}(k) + (B+GD)u(k) - Gy(k)]$$
$$= Ax(k) + Bu(k) - (A+GC)\hat{x}(k) - (B+GD)u(k) + G[Cx(k) + Du(k)]$$
$$= (A+GC)x(k) - (A+GC)\hat{x}(k)$$
$$= (A+GC)[x(k) - \hat{x}(k)]$$
$$= (A+GC)e(k). \tag{9.19}$$

The last equality is the governing equation of the state-estimation error,

$$e(k+1) = (A+GC)e(k), \tag{9.20}$$

which is a first-order matrix difference equation. Thus, if G is chosen such that the magnitudes of all the eigenvalues of $(A+GC)$ are less than one then the solution for Eq. (9.20) that governs the state-estimation error will converge to zero as time tends to infinity, i.e.,

$$\lim_{k \to \infty} e(k) = 0, \tag{9.21}$$

for any initial error $e(0) = x(0) - \hat{x}(0)$. Thus any initial estimated state $\hat{x}(0)$ can be assumed for the observer equation. As in the continuous-time case, convergence of $e(k)$ to zero means that the estimated state converges to the true state, as desired:

$$\lim_{k \to \infty} \hat{x}(k) = x(k). \tag{9.22}$$

To implement the observer, we recursively solve the observer equation in real time to estimate the state at the next time step $\hat{x}(k+1)$ by using the measurements of input $u(k)$ and output $y(k)$ at the current time step k. With a digital computer, the computation can be done rather easily because it involves only addition, subtraction, and multiplication. The system and the state-estimation equation are summarized in Fig. 9.2.

If the estimated state $\hat{x}(k)$ of the system converges to the true state $x(k)$, then the estimated output $\hat{y}(k)$ will also converge to the true output $y(k)$ as well because $\hat{y}(k) = C\hat{x}(k) + Du(k)$. Again, to be able to design an observer, the system state–space model must be known.

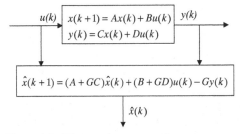

Figure 9.2. Discrete-time state estimator.

9.4 Placement of Observer Eigenvalues

In both the continuous-time and the discrete-time cases, the observer gain to be used in a state estimator may not be unique. The behavior of the observer (say, how fast the estimated state converges to the true state) is dependent on the location of the observer eigenvalues (or poles).

9.4.1 Definition Based on the First-Order Model

In the case of an observable system, any real or complex-conjugate pairs of observer eigenvalues in the complex plane may be achieved. This result is mathematically stated as follows.

Consider the state–space model

$$\frac{dx}{dt} = A_c x(t) + B_c u(t),$$

$$y(t) = Cx(t) + Du(t),$$

with any prespecified real or complex-conjugate pairs of the observer eigenvalues (poles) in the complex plane. There exists an observer of the form

$$\frac{d\hat{x}}{dt} = (A_c + G_c C)\,\hat{x}(t) + (B_c + G_c D)\,u(t) - G_c y(t),$$

such that the eigenvalues of the observer state matrix $(A_c + G_c C)$ lie in the specified locations if and only if the system is observable.

This statement is very similar to the one given in Section 8.5 of Chap. 8 for the placement of closed-loop eigenvalues. For a single-output system, such an observer gain matrix is unique for a particular prespecified set of the observer eigenvalues. For a multiple-output system, however, such an observer gain is not unique. We may use this extra freedom to find an observer gain that possesses further useful properties such as robustness with respect to parameter uncertainty or measurement noise.

In the discrete-time case, the same result applies without alteration. Recall that, in the discrete-time case, the observer poles must be such that they all have magnitudes less than one for stability (as opposed to the continuous-time case, in which they must all have negative real parts). In the discrete-time case, if all the observer eigenvalues are placed at the origin, we have a deadbeat observer. A deadbeat observer has the special property that the estimated state will converge to the true state exactly in (at most) n steps, where n is the order of the state–space model. It is the fastest observer that is theoretically possible. A deadbeat observer is significant from the theoretical point of view but not so from a practical point of view because it is very sensitive to system uncertainties and measurement noise. If the statistical structure of the uncertainties is known, then we can make use of this information to design an optimal observer that minimizes the mean-square error of the estimated state. This observer has exactly the same structure as the one studied here except that the observer gain is time-varying (but soon reaches a steady-state value). It is computed based on the known model of the system as well as the statistics of the uncertainties. This optimal observer has a special name that is the famous Kalman filter.

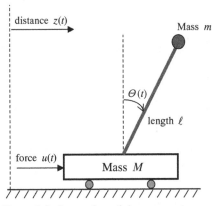

distance $z(t)$

Mass m

$\Theta(t)$

length ℓ

force $u(t)$

Mass M

Figure 9.3. Inverted pendulum.

EXAMPLE 9.1

A classic example is given to illustrate the state-estimation problem. An inverted pendulum on a moving cart is shown in Fig. 9.3 (see Ref. [3]). The linearized equation of motion of the system (for small-angle motion of the inverted pendulum) is

$$\frac{d}{dt}\begin{bmatrix} z(t) \\ \dot{z}(t) \\ \theta(t) \\ \dot{\theta}(t) \end{bmatrix} = \begin{bmatrix} 0 & 1 & 0 & 0 \\ 0 & 0 & -\dfrac{mg}{M} & 0 \\ 0 & 0 & 0 & 1 \\ 0 & 0 & \dfrac{(M+m)g}{M\ell} & 0 \end{bmatrix}\begin{bmatrix} z(t) \\ \dot{z}(t) \\ \theta(t) \\ \dot{\theta}(t) \end{bmatrix} + \begin{bmatrix} 0 \\ \dfrac{1}{M} \\ 0 \\ \dfrac{1}{M\ell} \end{bmatrix} u(t).$$

Suppose that the system parameters are

$$M = 1\,\text{Kg}, \quad m = 0.1\,\text{Kg}, \quad \ell = 1\,\text{m},$$

and, with $g = 10$ m/s^2, the system state–space model becomes

$$\frac{d}{dt}\begin{bmatrix} z(t) \\ \dot{z}(t) \\ \theta(t) \\ \dot{\theta}(t) \end{bmatrix} = \begin{bmatrix} 0 & 1 & 0 & 0 \\ 0 & 0 & -1 & 0 \\ 0 & 0 & 0 & 1 \\ 0 & 0 & 11 & 0 \end{bmatrix}\begin{bmatrix} z(t) \\ \dot{z}(t) \\ \theta(t) \\ \dot{\theta}(t) \end{bmatrix} + \begin{bmatrix} 0 \\ 1 \\ 0 \\ -1 \end{bmatrix} u(t).$$

Furthermore, we assume that only the position of the cart can be directly measured,

$$y(t) = z(t) = \begin{bmatrix} 1 & 0 & 0 & 0 \end{bmatrix}\begin{bmatrix} z(t) \\ \dot{z}(t) \\ \theta(t) \\ \dot{\theta}(t) \end{bmatrix},$$

and it is necessary to design an observer to estimate the remaining elements of the state vector. The system matrices are

$$
A_c = \begin{bmatrix} 0 & 1 & 0 & 0 \\ 0 & 0 & -1 & 0 \\ 0 & 0 & 0 & 1 \\ 0 & 0 & 11 & 0 \end{bmatrix}, \qquad B_c = \begin{bmatrix} 0 \\ 1 \\ 0 \\ -1 \end{bmatrix}, \qquad C = [1 \ \ 0 \ \ 0 \ \ 0].
$$

The observability matrix (formed from A and C) is full rank, which implies that the system is observable. Thus the entire state of the system can be estimated from the output $y(t) = z(t)$ alone by an observer of the form

$$
\frac{d\hat{x}}{dt} = (A_c + G_c C)\hat{x}(t) + (B_c + G_c D)u(t) - G_c y(t),
$$

where

$$
\hat{x}(t) = \begin{bmatrix} \hat{z}(t) \\ \hat{\dot{z}}(t) \\ \hat{\theta}(t) \\ \hat{\dot{\theta}}(t) \end{bmatrix}.
$$

Actually, it is not needed to estimate z because it happens to be the output y. For the sake of simplicity, however, we choose to ignore this fact because our theory does not care whether y is one element of the state vector x or not. With the elements of G_c written out as

$$
G_c = \begin{bmatrix} g_{c1} \\ g_{c2} \\ g_{c3} \\ g_{c4} \end{bmatrix},
$$

the observer thus has the form

$$
\frac{d}{dt} \begin{bmatrix} \hat{z}(t) \\ \hat{\dot{z}}(t) \\ \hat{\theta}(t) \\ \hat{\dot{\theta}}(t) \end{bmatrix} = \begin{bmatrix} g_{c1} & 1 & 0 & 0 \\ g_{c2} & 0 & -1 & 0 \\ g_{c3} & 0 & 0 & 1 \\ g_{c4} & 0 & 11 & 0 \end{bmatrix} \begin{bmatrix} \hat{z}(t) \\ \hat{\dot{z}}(t) \\ \hat{\theta}(t) \\ \hat{\dot{\theta}}(t) \end{bmatrix} + \begin{bmatrix} 0 \\ 1 \\ 0 \\ -1 \end{bmatrix} u(t) - \begin{bmatrix} g_{c1} \\ g_{c2} \\ g_{c3} \\ g_{c4} \end{bmatrix} y(t).
$$

To select G_c, we must prescribe a set of stable pole configurations. Let us choose the desired observer eigenvalues as $-2, -3, -2 \pm i$, which give the following desired characteristic equation for $(A + G_c C)$:

$$
\begin{aligned}
p(\lambda) &= (\lambda + 2)(\lambda + 3)(\lambda + 2 + i)(\lambda + 2 - i) \\
&= \lambda^4 + 9\lambda^3 + 31\lambda^2 + 49\lambda + 30 = 0.
\end{aligned}
$$

The characteristic equation of $(A + G_c C)$ is obtained from

$$
|A + G_c C - \lambda I| = 0,
$$

yielding the following polynomial equation:

$$\lambda^4 - g_{c1}\lambda^3 - (g_{c2} + 11)\lambda^2 + (11g_{c1} + g_{c3})\lambda + (11g_{c2} + g_{c4}) = 0.$$

Setting the characteristic equation to be equal to the desired characteristic equation by equating like powers of λ, the elements of the observer gain matrix G_c can be computed by

$$g_{c1} = -9,$$

$$g_{c2} = -31 - 11 = -42,$$

$$g_{c3} = 49 - 11g_{c1} = 148,$$

$$g_{c4} = 30 - 11g_{c2} = 492.$$

Using these values for G_c, we now have an observer that continuously estimates the state vector $\hat{x}(t)$ that will converge to the true state $x(t)$. The state-estimation error is governed by the matrix differential equation:

$$\frac{de}{dt} = (A_c + G_c C)\,e(t),$$

where

$$A_c + G_c C = \begin{bmatrix} -9 & 1 & 0 & 0 \\ -42 & 0 & -1 & 0 \\ 148 & 0 & 0 & 1 \\ 492 & 0 & 11 & 0 \end{bmatrix}.$$

The observer eigenvalues are placed to determine the rate at which the state-estimation error converges to zero. To integrate the observer equation, any initial condition $\hat{x}(0)$ can be used. Of course, it is preferred to use an $\hat{x}(0)$ that is close to the actual initial state, but this is not always possible. In such cases, for the lack of a better choice, zero initial conditions are generally chosen for the estimated state vector.

9.4.2 Second-Order Formulation

In Example 9.1, a technique has been shown to place the eigenvalues of the observer used in a state estimator. Although the technique is conceptually simple, it is very difficult, if not impossible, to be used for multi-input and multi-output systems with more than one degree of freedom. In the following, another technique is introduced with the approach similar to the one used in Section 8.5 of Chap. 8 for the placement of closed-loop eigenvalues (Ref. [4–5]). Instead of using the first-order model, we will use the second-order model in this section to derive a somewhat different pole-placement technique (Ref. [6]).

In the analysis and design of the dynamics and vibration control of flexible structures, two sets of second-order linear, constant coefficient, ordinary differential equations are frequently used:

$$M\ddot{w}(t) + \Xi\dot{w}(t) + Kw(t) = B_f u(t),\tag{9.23}$$

$$y(t) = C_v\dot{w}(t) + C_d w(t).\tag{9.24}$$

Equation (9.23) is the system dynamic equation, having w as the position vector of dimension n, and M, Ξ, and K of dimension $n \times n$ as the mass, damping, and stiffness matrices, respectively, which generally are symmetric and sparse. The $n \times r$ influence matrix B_f describes the actuator locations for the $r \times 1$ control force vector u. Equation (9.24) is the measurement equation, having y as the measurement vector of length m, C_v as the $m \times n$ velocity influence matrix, and C_d as the $m \times n$ displacement influence matrix. The measurement equation may be used either directly or indirectly for feedback-controller design for control of flexible structures.

Equation (9.23) can be rewritten in a first-order form:

$$\begin{bmatrix} I_n & 0_n \\ 0_n & M \end{bmatrix}\begin{bmatrix} \dot{w}(t) \\ \ddot{w}(t) \end{bmatrix} + \begin{bmatrix} 0_n & -I_n \\ K & \Xi \end{bmatrix}\begin{bmatrix} w(t) \\ \dot{w}(t) \end{bmatrix} = \begin{bmatrix} 0_{n\times r} \\ B_f \end{bmatrix}u(t),\tag{9.25}$$

where I_n is an $n \times n$ identity matrix, 0_n is an $n \times n$ zero matrix and $0_{n\times r}$ is an $n \times r$ zero matrix. Correspondingly, the state-estimation equation becomes

$$\begin{bmatrix} I_n & 0_n \\ 0_n & M \end{bmatrix}\begin{bmatrix} \dot{\hat{w}}(t) \\ \ddot{\hat{w}}(t) \end{bmatrix} + \begin{bmatrix} 0_n & -I_n \\ K & \Xi \end{bmatrix}\begin{bmatrix} \hat{w}(t) \\ \dot{\hat{w}}(t) \end{bmatrix}$$
$$= \begin{bmatrix} 0_{n\times r} \\ B_f \end{bmatrix}u(t) - \begin{bmatrix} G_v \\ G_d \end{bmatrix}[C_d \quad C_v]\begin{bmatrix} w(t) - \hat{w}(t) \\ \dot{w}(t) - \dot{\hat{w}}(t) \end{bmatrix}.\tag{9.26}$$

Note that G_v and G_d are $n \times m$ gain matrices. To determine the matrices G_v and G_d, the state-estimation error is defined by

$$e = \begin{bmatrix} w - \hat{w} \\ \dot{w} - \dot{\hat{w}} \end{bmatrix}.\tag{9.27}$$

Subtracting Eq. (9.26) from Eq. (9.25) yields the error equation

$$\begin{bmatrix} I_n & 0_n \\ 0_n & M \end{bmatrix}\dot{e} + \begin{bmatrix} 0_n & -I_n \\ K & \Xi \end{bmatrix}e - \begin{bmatrix} G_v \\ G_d \end{bmatrix}[C_d \quad C_v]e = 0_{2n\times 1}.\tag{9.28}$$

There are two ways for solving the eigenvalue problem of Eq. (9.28), i.e., a right-hand eigenvalue problem and a left-hand eigenvalue problem. Because the gain matrices G_v and G_d are both on the left-hand side of their corresponding terms in Eq. (9.28), we consider the left-hand eigenvalue problem so that they can be lumped together with the observer left-hand eigenvectors. The left-hand eigenvectors defined by

$$\phi_k = \begin{bmatrix} \phi_{kv} \\ \phi_{kd} \end{bmatrix}, \quad k = 1, \ldots, n\tag{9.29}$$

and the corresponding eigenvalues λ_k for $k = 1, \ldots, n$ for the system given by Eq. (9.28) produce the following left-hand eigenvalue problem:

$$\begin{bmatrix} \phi_{kv}^T & \phi_{kd}^T \end{bmatrix} \left\{ \begin{bmatrix} I_n & 0_n \\ 0_n & M \end{bmatrix} \lambda_k + \begin{bmatrix} 0_n & -I_n \\ K & \Xi \end{bmatrix} - \begin{bmatrix} G_v \\ G_d \end{bmatrix} \begin{bmatrix} C_d & C_v \end{bmatrix} \right\} = 0_{1 \times 2n}, \qquad (9.30)$$

where the subscript k refers to the eigenvalue number. The transpose of Eq. (9.30) is

$$\left\{ \begin{bmatrix} I_n & 0_n \\ 0_n & M^T \end{bmatrix} \lambda_k + \begin{bmatrix} 0_n & K^T \\ -I_n & \Xi^T \end{bmatrix} - \begin{bmatrix} C_d^T \\ C_v^T \end{bmatrix} \begin{bmatrix} G_v^T & G_d^T \end{bmatrix} \right\} \begin{bmatrix} \phi_{kv} \\ \phi_{kd} \end{bmatrix} = 0_{2n \times 1}. \qquad (9.31)$$

For generality, the matrices M, Ξ, and K are not necessarily assumed to be symmetric. Even though most mechanical systems possess symmetric mass, damping, and stiffness matrices, there exist some exceptional cases. For example, the inverted-pendulum problem used in Example 9.1 has a stiffness matrix that is not symmetric.

Equation (9.31) is in the first-order matrix form. We may use the same approach as that given in Subsection 8.5.1 of Chap. 8 to solve for the gain matrices G_v and G_d. It is quite easy to prove (see Problem 9.1) that the formulations such as Eq. (8.49) derived from the null-space technique in Subsection 8.5.1 can be readily used for the observer pole-placement problem if A_c, B_c, and $\mathcal{F}_c$ are replaced with A_c^T, C^T, and G^T, respectively. This is really based on the so-called duality property of the full-state-feedback controller and full-state observer.

On the other hand, Eq. (9.31) can be decomposed into two parts:

$$\lambda_k \phi_{kv} + K^T \phi_{kd} - C_d^T \left[G_v^T \phi_{kv} + G_d^T \phi_{kd} \right] = 0_{n \times 1}, \qquad (9.32)$$

$$\lambda_k M^T \phi_{kd} - \phi_{kv} + \Xi^T \phi_{kd} - C_v^T \left[G_v^T \phi_{kv} + G_d^T \phi_{kd} \right] = 0_{n \times 1}. \qquad (9.33)$$

Premultiplying Eq. (9.33) by λ_k and adding the resulting equation to Eq. (9.32) yields

$$\left[\lambda_k^2 M^T + \lambda_k \Xi^T + K^T \right] \phi_{kd} - \left[\lambda_k C_v^T + C_d^T \right] \left[G_v^T \phi_{kv} + G_d^T \phi_{kd} \right] = 0_{n \times 1} \qquad (9.34)$$

or, in a compact form,

$$\Gamma_k \psi_k = 0_{n \times 1}, \qquad (9.35)$$

where

$$\Gamma_k = \left[\lambda_k^2 M^T + \lambda_k \Xi^T + K^T \quad -\lambda_k C_v^T - C_d^T \right], \qquad (9.36)$$

$$\psi_k = \begin{bmatrix} \phi_{kd} \\ G_v^T \phi_{kv} + G_d^T \phi_{kd} \end{bmatrix}. \qquad (9.37)$$

The dimension of Γ_k is $n \times (n + m)$ and of ψ_k is $(n + m) \times 1$. We have reduced the number of equations by half in comparison with the first-order approach. If the closed-loop eigenvalues $\lambda_k (k = 1, \ldots, n)$ are assigned, including their complex conjugate for a complex eigenvalue, Eq. (9.35) can be used to determine the gain matrices G_v and G_d. Because the vector ψ_k is in the null space of the matrix Γ_k, it is necessary to compute the null spaces of the matrices Γ_k $(k = 1, \ldots, n)$ corresponding to the eigenvalues λ_k $(k = 1, \ldots, n)$.

To obtain the nontrivial solution space of the homogeneous Eq. (9.35), the SVD is applied to the matrix Γ_k, yielding

$$\Gamma_k = U_k \Sigma_k V_k^* = U_k \begin{bmatrix} \sigma_k & 0_{q \times (n+m-q)} \\ 0_{(n-q) \times q} & 0_{(n-q) \times (n+m-q)} \end{bmatrix} \begin{bmatrix} V_{\sigma k}^* \\ V_{ok}^* \end{bmatrix}. \tag{9.38}$$

Because λ_k in Γ_k is a complex value, all the quantities are complex except the $q \times q$ diagonal matrix σ_k, which contains q nonzero and positive singular values. Here the superscript $*$ means complex conjugate transpose. In general, the integer q is equal to n unless the desired observer eigenvalue λ_k is chosen to be one of the system eigenvalues. For this case, the rank of the matrix $\lambda_k^2 M^T + \lambda_k \Xi^T + K^T$ is less than n, implying that the matrix Γ_k is rank deficient.

From Eq. (9.38), it is clear that

$$\Gamma_k [V_{\sigma k} \quad V_{ok}] = U_k \begin{bmatrix} \sigma_k & 0_{q \times (n+m-q)} \\ 0_{(n-q) \times q} & 0_{(n-q) \times (n+m-q)} \end{bmatrix}, \tag{9.39}$$

where the orthonormal equality $V_k^* V_k = I_n$ has been used to derive this equation. Equation (9.39) implies that

$$\Gamma_k V_{ok} = 0_{n \times (n+m-q)}. \tag{9.40}$$

The matrix V_{ok} represents a set of orthogonal basis vectors spanning the null space of the matrix Γ_k. A comparison of Eq. (9.35) and Eq. (9.40) thus suggests that

$$\Gamma_k \psi_k = \Gamma_k V_{ok} c_k = 0_{n \times 1}, \tag{9.41}$$

where c_k is an arbitrary $(n + m - q) \times 1$ column vector.

Instead of using the SVD to decompose the matrix Γ_k, another way of solving for ψ_k is to use Eq. (9.35) directly. If the matrix $[\lambda_k^2 M^T + \lambda_k \Xi^T + K^T]$ is invertible, the vector

$$\psi_k = \begin{bmatrix} \bar{\psi}_k \\ \hat{\psi}_k \end{bmatrix}, \tag{9.42}$$

where $\hat{\psi}_k$ is an arbitrary vector of length m and

$$\bar{\psi}_k = [\lambda_k^2 M^T + \lambda_k \Xi^T + K^T]^{-1} [\lambda_k C_v^T + C_d^T] \hat{\psi}_k \tag{9.43}$$

should be in the null space of the matrix Γ_k.

Choose a particular set of vectors $\psi_1, \psi_2, \ldots, \psi_n$ satisfying Eq. (9.41) and partition it into two components, as shown in Eq. (9.42). Equation (9.35) implies that

$$\phi_{kd} = \bar{\psi}_k, \tag{9.44}$$

$$G_v^T \phi_{kv} + G_d^T \phi_{kd} = \hat{\psi}_k, \tag{9.45}$$

where the matrix ϕ_{kv} can be solved from Eqs. (9.32) with the aid of Eqs. (9.44) and (9.45):

$$\phi_{kv} = -\frac{1}{\lambda_k}\left(K^T\phi_{kd} - C_d^T\hat{\psi}_k\right) \tag{9.46}$$

The observer eigenvalue λ_k cannot be zero, because it implies that the observer is not asymptotically stable. Recall that the eigenvalue must be assigned to have a negative real part to guarantee the observer stability. Both Eqs. (9.45) and (9.46) are $n \times 1$ vector equations. Now, let

$$\Phi_d = [\phi_{1d} \quad \phi_{2d} \quad \cdots \quad \phi_{nd}], \tag{9.47a}$$

$$\Phi_v = [\phi_{1v} \quad \phi_{2v} \quad \cdots \quad \phi_{nv}], \tag{9.47b}$$

$$\hat{\Psi} = [\hat{\psi}_1 \quad \hat{\psi}_2 \quad \cdots \quad \hat{\psi}_n]. \tag{9.47c}$$

For a pair of complex eigenvalues λ_k, and $\lambda_{k+1} = \lambda_k^*$, only the eigenvalue λ_k is needed for use in computing real and imaginary parts of ϕ_{kd}, ϕ_{kv}, and $\hat{\psi}_k$, because the complex-conjugate eigenvalue λ_k^* will produce identical real and imaginary quantities. Therefore the number of columns for Φ_d, Φ_v, and $\hat{\Psi}$ may be less than n (the number of states).

Based on the definition of Eqs. (9.47) for Φ_d, Φ_v, and $\hat{\Psi}$, Eqs. (9.45) gives

$$G_v^T\Phi_{vr} + G_d^T\Phi_{dr} = \hat{\Psi}_r, \tag{9.48}$$

$$G_v^T\Phi_{vi} + G_d^T\Phi_{di} = \hat{\Psi}_i, \tag{9.49}$$

or, in matrix form,

$$\begin{bmatrix} G_v^T & G_d^T \end{bmatrix}\begin{bmatrix} \Phi_{vr} & \Phi_{vi} \\ \Phi_{dr} & \Phi_{di} \end{bmatrix} = [\hat{\Psi}_r \quad \hat{\Psi}_i], \tag{9.50}$$

where the subscript r and i refer to real and imaginary parts of the associated quantities, respectively. The gain matrices G_v and G_d can then be solved with

$$\begin{bmatrix} G_v^T & G_d^T \end{bmatrix} = [\hat{\Psi}_r \quad \hat{\Psi}_i]\begin{bmatrix} \Phi_{vr} & \Phi_{vi} \\ \Phi_{dr} & \Phi_{di} \end{bmatrix}^{-1}. \tag{9.51}$$

A matrix inversion is required in the computation of the gain matrices G_v and G_d in Eq. (9.51).

The pole-placement technique presented in this section is somewhat different from the conventional one, which uses the first-order model to build a state estimator. From the computational point of view, the second-order dynamics models may be more attractive for use in designing the state estimators because the dimension of the mathematical models remains unchanged rather than increases by the factor of two for the first-order models. Furthermore, the fundamental structure of the mathematical models such as the symmetry and sparsity of the mass, damping, and stiffness matrices is maintained in the second-order approach.

EXAMPLE 9.2

Let us consider a simple rigid-body problem described by

$$\ddot{w} = u,$$

$$y = w.$$

The model has two zero eigenvalues. This is a single-output case in which the only output measurement is the displacement w. Observation of Eqs. (9.23) and (9.24) reveals that

$$M = 1, \quad \Xi = 0, \quad K = 0, \quad B_f = 1, \quad C_v = 0, \quad C_d = 1.$$

Single-output case: Substituting these quantities into Eq. (9.36) yields

$$\Gamma_k = \begin{bmatrix} \lambda_k^2 & -1 \end{bmatrix}.$$

The matrix V_{ok} shown in Eq. (9.40) has only one column vector, i.e.,

$$V_{ok} = \frac{1}{\sqrt{1 + \lambda_k^2}} \begin{bmatrix} 1 \\ \lambda_k^2 \end{bmatrix}.$$

It is clear that $\Gamma_k V_{ok} = 0$. Let us choose c_k to be

$$c_k = \sqrt{1 + \lambda_k^2}$$

Using Eqs. (9.41) and (9.42) yields

$$\psi_k = V_{ok} c_k = \begin{bmatrix} \bar{\psi}_k \\ \hat{\psi}_k \end{bmatrix} = \begin{bmatrix} 1 \\ \lambda_k^2 \end{bmatrix}.$$

From Eq. (9.44), it gives

$$\phi_{kd} = \bar{\psi}_k = 1 \quad \text{and} \quad \hat{\psi}_k = \lambda_k^2.$$

Application of Eq. (9.46) produces

$$\phi_{kv} = -\frac{1}{\lambda_k}(K^T \phi_{kd} - C_d^T \hat{\psi}_k) = \lambda_k.$$

Only two eigenvalues can be placed.

The gain constants G_d and G_v can be computed by

$$\begin{bmatrix} G_v^T & G_d^T \end{bmatrix} = \begin{bmatrix} \hat{\psi}_1 & \hat{\psi}_2 \end{bmatrix} \begin{bmatrix} \phi_{1v} & \phi_{2v} \\ \phi_{1d} & \phi_{2d} \end{bmatrix}^{-1}$$

$$= \begin{bmatrix} \lambda_1^2 & \lambda_2^2 \end{bmatrix} \begin{bmatrix} \lambda_1 & \lambda_2 \\ 1 & 1 \end{bmatrix}^{-1}$$

$$= \begin{bmatrix} \lambda_1 + \lambda_2 & -\lambda_1 \lambda_2 \end{bmatrix}.$$

It is clear that λ_1 and λ_2 cannot be assigned to be identical, or the matrix inverse in the above equation does not exist, i.e., no solution for the observer gains G_v and G_d. The observer must not possess repeated eigenvalues.

If both eigenvalues are a pair of complex values $\lambda_{1,2} = \lambda_r \pm i\,\lambda_i$ with a negative real part, another way of computing the observer gains with complex arithmetic is

$$
\begin{bmatrix} G_v^T & G_d^T \end{bmatrix} = \begin{bmatrix} \hat{\psi}_{1r} & \hat{\psi}_{1i} \end{bmatrix} \begin{bmatrix} \phi_{1vr} & \phi_{1vi} \\ \phi_{1dr} & \phi_{1di} \end{bmatrix}^{-1}
$$

$$
= \begin{bmatrix} \lambda_r^2 - \lambda_i^2 & 2\lambda_r\lambda_i \end{bmatrix} \begin{bmatrix} \lambda_r & \lambda_i \\ 1 & 0 \end{bmatrix}^{-1}
$$

$$
= \begin{bmatrix} 2\lambda_r & -\lambda_r^2 - \lambda_i^2 \end{bmatrix}.
$$

Note that the equality $\lambda_1^2 = (\lambda_r^2 - \lambda_i^2) + 2\,i\,\lambda_r\lambda_i$ gives the real and the imaginary parts of λ_1^2. For the pair of complex eigenvalues λ_1 and $\lambda_2 = \lambda_1^*$, only λ_1 is used to compute real and imaginary parts of ϕ_{1d}, ϕ_{1v}, and $\hat{\psi}_1$, because the complex-conjugate eigenvalue λ_1^* will produce identical real and imaginary quantities.

Two-output case: In addition to the displacement measurement, assume that the velocity measurement is also available. Actually, we may not need to estimate the state which consists of the displacement and velocity because both quantities are available from measurements. To demonstrate the design of an observer, we choose to ignore this fact. The measurement equation for this case becomes

$$
y = \begin{bmatrix} w \\ \dot{w} \end{bmatrix} = \begin{bmatrix} 1 \\ 0 \end{bmatrix} w + \begin{bmatrix} 0 \\ 1 \end{bmatrix} \dot{w}.
$$

It gives

$$
C_d = \begin{bmatrix} 1 \\ 0 \end{bmatrix}, \quad C_v = \begin{bmatrix} 0 \\ 1 \end{bmatrix}.
$$

Substituting these quantities into Eq. (9.36) yields

$$
\Gamma_k = \begin{bmatrix} \lambda_k^2 & -1 & -\lambda_k \end{bmatrix}.
$$

The matrix V_{ok} satisfying $\Gamma_k V_{ok} = 0_{1\times 2}$ is

$$
V_{ok} = \begin{bmatrix} 0 & \dfrac{1 + \lambda_k^2}{\lambda_k^2} \\ -\lambda_k & 1 \\ 1 & \lambda_k \end{bmatrix}.
$$

The two columns in V_{ok} are orthogonal but not orthonormal, i.e., $V_{ok}^* V_{ok} \neq I_2$, where I_2 is a 2×2 identity matrix. For simplicity, we ignore the normalization of the matrix V_{ok}.

For the eigenvalues λ_1 and λ_2, let us arbitrarily choose different coefficient vectors c_1 and c_2 to be

$$
c_1 = \begin{bmatrix} 1 \\ 0 \end{bmatrix}, \quad c_2 = \begin{bmatrix} 1 \\ \lambda_2 \end{bmatrix}.
$$

Application of Eqs. (9.41) and (9.42) produces

$$\psi_1 = V_{o1}c_1 = \begin{bmatrix} \bar{\psi}_1 \\ \hat{\psi}_1 \end{bmatrix} = \begin{bmatrix} 0 \\ -\lambda_1 \\ 1 \end{bmatrix},$$

$$\psi_2 = V_{o2}c_2 = \begin{bmatrix} \bar{\psi}_2 \\ \hat{\psi}_2 \end{bmatrix} = \begin{bmatrix} \dfrac{1+\lambda_2^2}{\lambda_2} \\ 0 \\ 1+\lambda_2^2 \end{bmatrix}.$$

From Eq. (9.44), it gives

$$\phi_{1d} = \bar{\psi}_1 = 0, \quad \hat{\psi}_1 = \begin{bmatrix} -\lambda_1 \\ 1 \end{bmatrix},$$

$$\phi_{2d} = \bar{\psi}_2 = \frac{1+\lambda_2^2}{\lambda_2}, \quad \hat{\psi}_2 = \begin{bmatrix} 0 \\ 1+\lambda_2^2 \end{bmatrix}.$$

Using Eq. (9.46) yields

$$\phi_{1v} = -\frac{1}{\lambda_1}\left(K^T \phi_{1d} - C_d^T \hat{\psi}_1\right) = -1,$$

$$\phi_{2v} = -\frac{1}{\lambda_2}\left(K^T \phi_{2d} - C_d^T \hat{\psi}_2\right) = 0.$$

The gain matrices G_d and G_v can then be computed by

$$\begin{bmatrix} G_v^T & G_d^T \end{bmatrix} = \begin{bmatrix} \hat{\psi}_1 & \hat{\psi}_2 \end{bmatrix} \begin{bmatrix} \phi_{1v} & \phi_{2v} \\ \phi_{1d} & \phi_{2d} \end{bmatrix}^{-1}$$

$$= \begin{bmatrix} -\lambda_1 & 0 \\ 1 & 1+\lambda_2^2 \end{bmatrix} \begin{bmatrix} -1 & 0 \\ 0 & \dfrac{1+\lambda_2^2}{\lambda_2} \end{bmatrix}^{-1}$$

$$= \begin{bmatrix} \lambda_1 & 0 \\ -1 & \lambda_2 \end{bmatrix}.$$

It can be easily proved from Eq. (9.28) by use of the gain matrices G_d and G_v just computed that λ_1 and λ_2 are indeed the observer eigenvalues.

9.5 Continuous-Time Observer-Based State Feedback

If the output measurements are used directly for a feedback-control design, an output-feedback controller is obtained. The output-feedback control is generally attractive because it is simple and easy for real-time implementation. However, a stable and robust output-feedback controller may require either too many measurements that are not practical or some measurement devices that are not yet available and need to be developed. On the other hand, a state-feedback-control law assumes that all states are measurable. In many practical control designs for dynamical systems, it is physically or economically impractical to install all the sensors that would be necessary to measure the entire state vector. For such cases, a state estimator is needed to estimate the states from the measurement outputs and provide enough freedom for a stable feedback-controller design.

The state–space model of a linear time-invariant dynamic system in the continuous-time domain is

$$\frac{dx}{dt} = A_c x(t) + B_c u(t), \tag{9.52}$$

$$y(t) = Cx(t) + Du(t). \tag{9.53}$$

From Chap. 8, we know that if the state is available for feedback then a control law of the form

$$u(t) = \mathcal{F}_c x(t) \tag{9.54}$$

can always be found to make the controlled (closed-loop) system stable, provided that the system is controllable. The controller gain $\mathcal{F}_c$ can be found to place the poles of the closed-loop system,

$$\frac{dx}{dt} = A_c x(t) + B_c \mathcal{F}_c x(t)$$
$$= (A_c + B_c \mathcal{F}_c) x(t), \tag{9.55}$$

in any desired (symmetric) configuration in a complex plane. When it is not possible to measure the state directly, we can use an observer of the form

$$\frac{d\hat{x}}{dt} = (A_c + G_c C)\hat{x}(t) + (B_c + G_c D)u(t) - G_c y(t) \tag{9.56}$$

to obtain an estimation of the state from input and output measurements, provided that the system is observable. The observer gain G_c can be found to place the poles of the observer system in any desired (symmetric) configuration in a complex plane. When implemented with an observer, the control law will then be

$$u(t) = \mathcal{F}_c \hat{x}(t), \tag{9.57}$$

where $\hat{x}(t)$ is provided by the observer. This configuration is shown in Fig. 9.4.

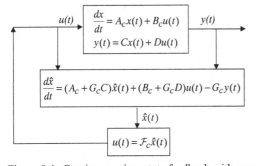

Figure 9.4. Continuous-time state feedback with an estimator.

We are interested in obtaining an equation that describes the dynamics of the overall system. First, because the state-feedback-control law is not implemented with the true state of the system but with its estimated value, the desired closed-loop system dynamics described by Eq. (9.55) can exist in the steady state only after the estimated state $\hat{x}(t)$ has converged to the true state $x(t)$. To have a truly complete picture, the overall dynamics must involve the observer equation in the transient. From Eq. (9.57), the overall dynamics in terms of both the estimated state $\hat{x}(t)$ and the true state $x(t)$ becomes

$$\frac{dx}{dt} = A_c x(t) + B_c u(t)$$
$$= A_c x(t) + B_c \mathcal{F}_c \hat{x}(t) \tag{9.58}$$

and the observer equation is

$$\frac{d\hat{x}}{dt} = (A_c + G_c C)\,\hat{x}(t) + (B_c + G_c D)\, u(t) - G_c y(t)$$
$$= (A_c + G_c C)\,\hat{x}(t) + (B_c + G_c D)\,\mathcal{F}_c \hat{x}(t) - G_c\,[Cx(t) + Du(t)]$$
$$= (A_c + G_c C)\,\hat{x}(t) + B_c \mathcal{F}_c \hat{x}(t) + G_c D \mathcal{F}_c \hat{x}(t) - G_c Cx(t) - G_c D \mathcal{F}_c \hat{x}(t)$$
$$= -G_c Cx(t) + (A_c + G_c C + B_c \mathcal{F}_c)\,\hat{x}(t). \tag{9.59}$$

Both Eqs. (9.58) and (9.59) can be combined to yield

$$\frac{d}{dt}\begin{bmatrix} x(t) \\ \hat{x}(t) \end{bmatrix} = \begin{bmatrix} A_c & B_c \mathcal{F}_c \\ -G_c C & A_c + G_c C + B_c \mathcal{F}_c \end{bmatrix} \begin{bmatrix} x(t) \\ \hat{x}(t) \end{bmatrix}. \tag{9.60}$$

This is a first-order matrix differential equation. It describes the dynamics of the overall closed-loop system in which the state of the overall system consists of the actual system state and its estimated state. To determine if the overall closed-loop system is stable, we simply check to see if all the eigenvalues of

$$\bar{A}_c = \begin{bmatrix} A_c & B_c \mathcal{F}_c \\ -G_c C & A_c + G_c C + B_c \mathcal{F}_c \end{bmatrix} \tag{9.61}$$

have negative real parts. Note that this matrix involves the system matrices A_c, B_c, C, the observer gain G_c, and the controller gain $\mathcal{F}_c$, as expected. The solution of the above equation describes how the actual state of the system $x(t)$ and the estimated state $\hat{x}(t)$ change as a function of time. As for the output of the system, it can be written in terms of the overall state vector as

$$y(t) = [\,C \quad D\mathcal{F}_c\,]\begin{bmatrix} x(t) \\ \hat{x}(t) \end{bmatrix} \tag{9.62}$$

because

$$y(t) = Cx(t) + Du(t)$$
$$= Cx(t) + D\mathcal{F}_c \hat{x}(t). \tag{9.63}$$

Although the matrix $\bar{A}_c$ describes correctly the dynamics of the overall closed-loop system, it does not reveal explicitly whether or not the closed-loop system is stable if the controller gain $\mathcal{F}_c$ is designed such that $(A_c + B_c\mathcal{F}_c)$ is stable and the observer gain G_c is designed such that $(A_c + G_cC)$ is stable. Indeed, based on our analysis earlier and in Chap. 8, it is quite counterintuitive to design $\mathcal{F}_c$ and G_c otherwise. In other words, we ask the following mathematical question: If $\mathcal{F}_c$ and G_c are designed such that $(A_c + B_c\mathcal{F}_c)$ and $(A_c + G_cC)$ are stable, then does it follow that

$$\bar{A}_c = \begin{bmatrix} A_c & B_c\mathcal{F}_c \\ -G_cC & A_c + G_cC + B_c\mathcal{F}_c \end{bmatrix}$$

is also stable? The answer is yes, although it is quite impossible to see this result directly from the above matrix. This is a mathematical statement of what is known as the separation principle. Formally, we can show the separation principle by demonstrating that $\bar{A}_c$ is related to the following matrix,

$$\tilde{A}_c = \begin{bmatrix} A_c + B_c\mathcal{F}_c & -B_c\mathcal{F}_c \\ 0 & A_c + G_cC \end{bmatrix}, \tag{9.64}$$

by a coordinate transformation that transforms the state vector

$$\begin{bmatrix} x(t) \\ \hat{x}(t) \end{bmatrix}$$

into the state and error vector

$$\begin{bmatrix} x(t) \\ e(t) \end{bmatrix} = \begin{bmatrix} x(t) \\ x(t) - \hat{x}(t) \end{bmatrix}$$

by means of the transformation

$$\begin{bmatrix} x(t) \\ e(t) \end{bmatrix} = \begin{bmatrix} I & 0 \\ I & -I \end{bmatrix} \begin{bmatrix} x(t) \\ \hat{x}(t) \end{bmatrix}.$$

Indeed, we can verify rather easily that

$$\begin{bmatrix} A_c + B_c\mathcal{F}_c & -B_c\mathcal{F}_c \\ 0 & A_c + G_cC \end{bmatrix}$$

$$= \begin{bmatrix} I & 0 \\ I & -I \end{bmatrix} \begin{bmatrix} A_c & B_c\mathcal{F}_c \\ -G_cC & A_c + G_cC + B_c\mathcal{F}_c \end{bmatrix} \begin{bmatrix} I & 0 \\ I & -I \end{bmatrix}^{-1} \tag{9.65}$$

by using an interesting property of the particular transformation matrix, i.e., its inverse is the same as itself;

$$\begin{bmatrix} I & 0 \\ I & -I \end{bmatrix}^{-1} = \begin{bmatrix} I & 0 \\ I & -I \end{bmatrix},$$

where I is an identity matrix. Mathematically, a coordinate transformation is a similarity transformation that does not affect the matrix eigenvalues. Therefore the eigenvalues of $\bar{A}_c$ are exactly the same as the eigenvalues of the transformed matrix $\tilde{A}_c$. Because $\tilde{A}_c$ is an upper-block triangular matrix, its eigenvalues are made up of the eigenvalues of the matrices on the main-block diagonal (the statement holds for any upper- or lower-block triangular matrix). In other words, the eigenvalues of $\bar{A}_c$ are made up of the eigenvalues of $(A_c + B_c \mathcal{F}_c)$ and $(A_c + G_c C)$. Consequently, if the controller gain $\mathcal{F}_c$ is such that $(A_c + B \mathcal{F}_c)$ is stable and the observer gain G_c is such that $(A_c + G_c C)$ is stable, then the overall closed-loop system is stable. The separation principle is thus proved.

Rather than using the transformation matrix, it is perhaps simpler to derive the closed-loop equations in terms of $x(t)$ and $e(t)$ directly. First, the equation for $e(t)$ has been derived earlier for the observer,

$$\frac{de}{dt} = (A_c + G_c C) e(t).$$

The system state equation can be expressed in terms of $x(t)$ and $e(t)$ by

$$\begin{aligned} \frac{dx}{dt} &= A_c x(t) + B_c u(t) \\ &= A_c x(t) + B_c \mathcal{F}_c \hat{x}(t) \\ &= A_c x(t) + B_c \mathcal{F}_c [x(t) - e(t)] \\ &= (A_c + B_c \mathcal{F}_c) x(t) - B_c \mathcal{F}_c e(t). \end{aligned} \tag{9.66}$$

Thus, in terms of $x(t)$ and $e(t)$, the overall closed-loop equation is

$$\frac{d}{dt} \begin{bmatrix} x(t) \\ e(t) \end{bmatrix} = \begin{bmatrix} A_c + B_c \mathcal{F}_c & -B_c \mathcal{F}_c \\ 0 & A_c + G_c C \end{bmatrix} \begin{bmatrix} x(t) \\ e(t) \end{bmatrix}. \tag{9.67}$$

Note that the overall closed-loop system matrix is identical to $\tilde{A}_c$, shown in Eq. (9.64).

There is yet another way to view the result shown in Eq. (9.67). Recall that, in the ideal case in which the state of the system is known exactly, there is no need to use an observer and the closed-loop system is governed by the equation

$$\frac{dx}{dt} = (A_c + B_c \mathcal{F}_c) x(t). \tag{9.68}$$

Now if the state of the system can not be directly measured and an observer is used to provide state-estimation, then the closed-loop system is governed instead by

$$\frac{dx}{dt} = (A_c + B_c \mathcal{F}_c) x(t) - B_c \mathcal{F}_c e(t), \tag{9.69}$$

where $e(t)$ is the state-estimation error. Comparing Eqs. (9.68) and (9.69) reveals that the presence of the term $-B_c \mathcal{F}_c e(t)$ as a forcing function makes the closed-loop system response deviate from the ideal behavior. The behavior of $e(t)$ itself is completely determined by the observer system matrix $A_c + G_c C$, which is independent of $x(t)$, B_c, and $\mathcal{F}_c$. In the steady state as the state-estimation error $e(t)$ converges to zero, then the behavior of the closed-loop system will approach the ideal behavior, as we would intuitively expect.

EXAMPLE 9.3

In this example, we revisit the case of stabilizing an inverted pendulum on a moving cart, as described in Example 9.1. The system state–space model is

$$
\frac{d}{dt}
\begin{bmatrix} z(t) \\ \dot{z}(t) \\ \theta(t) \\ \dot{\theta}(t) \end{bmatrix}
=
\begin{bmatrix} 0 & 1 & 0 & 0 \\ 0 & 0 & -1 & 0 \\ 0 & 0 & 0 & 1 \\ 0 & 0 & 11 & 0 \end{bmatrix}
\begin{bmatrix} z(t) \\ \dot{z}(t) \\ \theta(t) \\ \dot{\theta}(t) \end{bmatrix}
+
\begin{bmatrix} 0 \\ 1 \\ 0 \\ -1 \end{bmatrix}
u(t).
$$

Assume that only the linear position of the cart can be directly measured. Thus, the output equation is

$$
y(t) = z(t) = \begin{bmatrix} 1 & 0 & 0 & 0 \end{bmatrix}
\begin{bmatrix} z(t) \\ \dot{z}(t) \\ \theta(t) \\ \dot{\theta}(t) \end{bmatrix}.
$$

From Chap. 8, a state-feedback controller was designed to stabilize the system

$$
u(t) = \begin{bmatrix} 0.4 & 1 & 21.4 & 6 \end{bmatrix}
\begin{bmatrix} z(t) \\ \dot{z}(t) \\ \theta(t) \\ \dot{\theta}(t) \end{bmatrix},
$$

where the controller gain was chosen to place the closed-loop poles at -1, -2, and $-1 \pm i$. To implement this state-feedback controller, complete knowledge of the state vector is required. In our present problem, however, only the linear position of the cart is available; thus the state-estimation is needed. In Example 9.1, an observer was designed to estimate the entire state vector from the linear position of the cart alone:

$$
\frac{d}{dt}
\begin{bmatrix} \hat{z}(t) \\ \hat{\dot{z}}(t) \\ \hat{\theta}(t) \\ \hat{\dot{\theta}}(t) \end{bmatrix}
=
\begin{bmatrix} -9 & 1 & 0 & 0 \\ -42 & 0 & -1 & 0 \\ 148 & 0 & 0 & 1 \\ 492 & 0 & 11 & 0 \end{bmatrix}
\begin{bmatrix} \hat{z}(t) \\ \hat{\dot{z}}(t) \\ \hat{\theta}(t) \\ \hat{\dot{\theta}}(t) \end{bmatrix}
+
\begin{bmatrix} 0 \\ 1 \\ 0 \\ -1 \end{bmatrix}
u(t) -
\begin{bmatrix} -9 \\ -42 \\ 148 \\ 492 \end{bmatrix}
y(t),
$$

where the observer poles were chosen to be at -2, -3, $-2 \pm i$. Now this observer will be combined with the state-feedback controller to form an observer-based state-feedback-control system. Our observer-based state feedback controller now is

$$
u(t) = \begin{bmatrix} 0.4 & 1 & 21.4 & 6 \end{bmatrix}
\begin{bmatrix} \hat{z}(t) \\ \hat{\dot{z}}(t) \\ \hat{\theta}(t) \\ \hat{\dot{\theta}}(t) \end{bmatrix},
$$

where the estimated state is provided by the observer. The system closed-loop behavior is governed by

$$
\frac{d}{dt}\begin{bmatrix} z \\ \dot{z} \\ \theta \\ \dot{\theta} \\ \hat{z} \\ \dot{\hat{z}} \\ \hat{\theta} \\ \dot{\hat{\theta}} \end{bmatrix} = \begin{bmatrix} 0 & 1 & 0 & 0 & 0 & 0 & 0 & 0 \\ 0 & 0 & -1 & 0 & 0.4 & 1 & 21.4 & 6 \\ 0 & 0 & 0 & 1 & 0 & 0 & 0 & 0 \\ 0 & 0 & 11 & 0 & -0.4 & -1 & -21.4 & -6 \\ 9 & 0 & 0 & 0 & -9 & 1 & 0 & 0 \\ 42 & 0 & 0 & 0 & -41.6 & 1 & 20.4 & 6 \\ -148 & 0 & 0 & 0 & 148 & 0 & 0 & 1 \\ -492 & 0 & 0 & 0 & 491.6 & -1 & -10.4 & -6 \end{bmatrix}\begin{bmatrix} z \\ \dot{z} \\ \theta \\ \dot{\theta} \\ \hat{z} \\ \dot{\hat{z}} \\ \hat{\theta} \\ \dot{\hat{\theta}} \end{bmatrix},
$$

or, equivalently,

$$
\frac{d}{dt}\begin{bmatrix} z \\ \dot{z} \\ \theta \\ \dot{\theta} \\ e_z \\ e_{\dot{z}} \\ e_\theta \\ e_{\dot{\theta}} \end{bmatrix} = \begin{bmatrix} 0 & 1 & 0 & 0 & 0 & 0 & 0 & 0 \\ 0.4 & 1 & 20.4 & 6 & -0.4 & -1 & -21.4 & -6 \\ 0 & 0 & 0 & 1 & 0 & 0 & 0 & 0 \\ -0.4 & -1 & -10.4 & -6 & 0.4 & 1 & 21.4 & 6 \\ 0 & 0 & 0 & 0 & -9 & 1 & 0 & 0 \\ 0 & 0 & 0 & 0 & -42 & 0 & -1 & 0 \\ 0 & 0 & 0 & 0 & 148 & 0 & 0 & 1 \\ 0 & 0 & 0 & 0 & 492 & 0 & 11 & 0 \end{bmatrix}\begin{bmatrix} z \\ \dot{z} \\ \theta \\ \dot{\theta} \\ e_z \\ e_{\dot{z}} \\ e_\theta \\ e_{\dot{\theta}} \end{bmatrix}.
$$

It can be verified that the eigenvalues describing the overall closed-loop dynamics in both cases are $-1, -2, -2, -3, -1 \pm i$, and $-2 \pm i$, which are made up of those of the controller and of the observer, as expected.

9.6 Discrete-Time Observer-Based State Feedback

We now derive the closed-loop equations for observer-based state feedback in the discrete-time domain. The discussion that follows is quite similar to that of the continuous-time case. The state–space model of a linear time-invariant dynamic system in discrete-time format is

$$x(k+1) = Ax(k) + Bu(k), \tag{9.70}$$

$$y(k) = Cx(k) + Du(k). \tag{9.71}$$

If the state is available for feedback, then a control law of the form

$$u(k) = \mathcal{F}x(k) \tag{9.72}$$

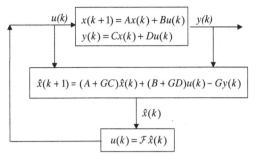

Figure 9.5. Discrete-time state feedback with an estimator.

can always be found to place the poles of the closed-loop system,

$$x(k + 1) = Ax(k) + B\mathcal{F}x(k)$$
$$= (A + B\mathcal{F})x(k), \tag{9.73}$$

in any desired (symmetric) configuration in a complex plane, provided that the system is controllable. When the state is not available for direct measurements, then an observer of the form

$$\hat{x}(k + 1) = (A + GC)\hat{x}(k) + (B + GD)u(k) - Gy(k) \tag{9.74}$$

can be used to estimate the state from input and output measurements, provided that the system is observable. The poles of the observer system can be placed in any desired (symmetric) configuration. When implemented with an observer, the control law will then be

$$u(k) = \mathcal{F}\hat{x}(k), \tag{9.75}$$

where $\hat{x}(k)$ is provided by an observer. This control design is shown in the block diagram of Fig. 9.5.

We are interested in obtaining an equation that describes the dynamics of the overall closed-loop system. For the system equation, we have

$$x(k + 1) = Ax(k) + Bu(k)$$
$$= Ax(k) + B\mathcal{F}\hat{x}(k), \tag{9.76}$$

and for the observer equation we have

$$\hat{x}(k + 1) = (A + GC)\hat{x}(k) + (B + GD)u(k) - Gy(k)$$
$$= (A + GC)\hat{x}(k) + (B + GD)\mathcal{F}\hat{x}(k) - G[Cx(k) + Du(k)]$$
$$= (A + GC)\hat{x}(k) + B\mathcal{F}\hat{x}(k) + GD\mathcal{F}\hat{x}(k) - GCx(k) - GD\mathcal{F}\hat{x}(k)$$
$$= -GCx(k) + (A + GC + B\mathcal{F})\hat{x}(k). \tag{9.77}$$

Therefore the overall closed-loop system dynamics is described by

$$\begin{bmatrix} x(k + 1) \\ \hat{x}(k + 1) \end{bmatrix} = \begin{bmatrix} A & B\mathcal{F} \\ -GC & A + GC + B\mathcal{F} \end{bmatrix} \begin{bmatrix} x(k) \\ \hat{x}(k) \end{bmatrix}. \tag{9.78}$$

Equation (9.78) is an algebraic matrix equation, in contrast to the differential matrix equation, Eq. (9.60), in the continuous-time domain. To ensure overall closed-loop stability, the magnitudes of all the eigenvalues of

$$\bar{A} = \begin{bmatrix} A & B\mathcal{F} \\ -GC & A + GC + B\mathcal{F} \end{bmatrix} \tag{9.79}$$

must be less than one. The solution of Eq. (9.78) describes how the actual state of the system $x(t)$ and the estimated state $\hat{x}(t)$ evolve as a function of time. Indeed, the solution can be expressed as

$$\begin{bmatrix} x(k) \\ \hat{x}(k) \end{bmatrix} = \begin{bmatrix} A & B\mathcal{F} \\ -GC & A + GC + B\mathcal{F} \end{bmatrix}^k \begin{bmatrix} x(0) \\ \hat{x}(0) \end{bmatrix}. \tag{9.80}$$

The system output can be written in terms of the overall state vector as

$$y(k) = \begin{bmatrix} C & D\mathcal{F} \end{bmatrix} \begin{bmatrix} x(k) \\ \hat{x}(k) \end{bmatrix}. \tag{9.81}$$

as the measurement equation is

$$y(k) = Cx(k) + Du(k)$$
$$= Cx(k) + D\mathcal{F}\hat{x}(k). \tag{9.82}$$

Again, it can be shown that the eigenvalues of $\bar{A}$ consists of those of $(A + B\mathcal{F})$ and $(A + GC)$ by a coordinate transformation. Similar to the continuous-time case, it is also possible to develop the closed-loop equations in terms of $x(k)$ and $e(k)$. The equation for $e(k)$ has been derived earlier in the chapter for the discrete-time observer as

$$e(k + 1) = (A + GC)\, e(k). \tag{9.83}$$

The system state equation can be expressed in terms of $x(k)$ and $e(k)$ as follows:

$$x(k + 1) = Ax(k) + Bu(k)$$
$$= Ax(k) + B\mathcal{F}\hat{x}(k)$$
$$= Ax(k) + B\mathcal{F}[x(k) - e(k)]$$
$$= (A + B\mathcal{F})x(k) - B\mathcal{F}e(k). \tag{9.84}$$

Thus, in terms of $x(k)$ and $e(k)$, the overall closed-loop equation is

$$\begin{bmatrix} x(k + 1) \\ e(k + 1) \end{bmatrix} = \begin{bmatrix} A + B\mathcal{F} & -B\mathcal{F} \\ 0 & A + GC \end{bmatrix} \begin{bmatrix} x(k) \\ e(k) \end{bmatrix}. \tag{9.85}$$

Note that the matrix describing the overall closed-loop system dynamics is an upper-block triangular matrix. Therefore its eigenvalues consist of the eigenvalues of $(A + B\mathcal{F})$ and $(A + GC)$. Consequently, if the controller gain $\mathcal{F}$ is such that $(A + B\mathcal{F})$ is stable and the observer gain G is such that $(A + GC)$ is stable then the overall closed-loop system is also stable. The separation principle thus applies in the discrete-time domain in exactly the same way it does in the continuous-time domain.

EXAMPLE 9.4

Consider a simple mass–spring system with its discrete-time model described by

$$x(k+1) = Ax(k) + Bu(k),$$

$$y(k) = Cx(k),$$

where

$$A = \begin{bmatrix} 0 & 1 \\ -1 & 0 \end{bmatrix}, \qquad B = \begin{bmatrix} 1 \\ 1 \end{bmatrix}, \qquad C = [1 \quad 0].$$

The open-loop system eigenvalues are $\pm i$ on the unit circle.

Controller Design: Let the desired closed-loop eigenvalues be λ_{ck}. Construct the matrix Γ_{ck} shown in Eq. (8.43) for the discrete-time model

$$\Gamma_{ck} = [A - \lambda_{ck}I \quad B] = \begin{bmatrix} -\lambda_{ck} & 1 & 1 \\ -1 & -\lambda_{ck} & 1 \end{bmatrix}.$$

By inspection, the following vector,

$$V_{ok} = \begin{bmatrix} \lambda_{ck} + 1 \\ \lambda_{ck} - 1 \\ \lambda_{ck}^2 + 1 \end{bmatrix},$$

represents a basis vector spanning the null space of the matrix Γ_{ck}, i.e.,

$$\Gamma_{ck} V_{ok} c_{ck} = 0,$$

where c_{ck} can be any complex number. In practice, we should use the SVD shown in Eq. (8.43) to compute the basis vectors spanning the null space of the matrix Γ_{ck}.

For the desired eigenvalues λ_{c1} and λ_{c2}, let us now choose and partition a particular set of vectors ϕ_{ck} ($k = 1, 2$) satisfying Eq. (8.47), corresponding to the choice $c_{c1} = c_{c2} = 1$, i.e.,

$$\phi_{ck} = \begin{bmatrix} \bar{\phi}_{ck} \\ \hat{\phi}_{ck} \end{bmatrix} = V_{ok} = \begin{bmatrix} \lambda_{ck} + 1 \\ \lambda_{ck} - 1 \\ \lambda_{ck}^2 + 1 \end{bmatrix} \implies \Gamma_{ck} \phi_{ck} = 0, \quad k = 1, 2.$$

The control gain matrix $\mathcal{F}$ can then be solved with Eq. (8.51),

$$\mathcal{F}[\bar{\phi}_{c1} \quad \bar{\phi}_{c2}] = [\hat{\phi}_{c1} \quad \hat{\phi}_{c2}],$$

or

$$\mathcal{F} \begin{bmatrix} \lambda_{c1} + 1 & \lambda_{c2} + 1 \\ \lambda_{c1} - 1 & \lambda_{c2} - 1 \end{bmatrix} = [\lambda_{c1}^2 + 1 \quad \lambda_{c2}^2 + 1].$$

Assuming that $\lambda_{c1} = 0.5 + 0.6i$ and $\lambda_{c1} = 0.5 - 0.6i$ and substituting these values into the above equation, we find that the gain matrix $\mathcal{F}$ is

$$\mathcal{F} = [0.575 \quad 0.025].$$

The closed-loop state matrix becomes

$$A + B\mathcal{F} = \begin{bmatrix} 0 & 1 \\ -1 & 0 \end{bmatrix} + \begin{bmatrix} 1 \\ 1 \end{bmatrix} [\, 0.575 \quad 0.025 \,]$$

$$= \begin{bmatrix} 0.575 & 1.025 \\ -0.425 & 0.025 \end{bmatrix}.$$

Observer Design: For the state-estimation, let the desired observer eigenvalues be λ_{ok} $(k = 1, 2)$ and construct the matrix Γ_{ok} for the discrete-time model

$$\Gamma_{ok} = [\, A^T - \lambda_{ok}I \quad C^T \,] = \begin{bmatrix} -\lambda_{ok} & -1 & 1 \\ 1 & -\lambda_{ok} & 0 \end{bmatrix}.$$

This matrix is identical to the matrix Γ_{ck} shown earlier for the controller pole placement with the replacement of A and B with A^T and C^T, respectively. It is easy to show that the following vector,

$$V_{ok} = \begin{bmatrix} \lambda_{ok} \\ 1 \\ \lambda_{ok}^2 + 1 \end{bmatrix},$$

represents a basis vector spanning the null space of the matrix Γ_{ok}, i.e.,

$$\Gamma_{ok} V_{ok} c_{ok} = 0,$$

where c_{ok} can be any complex number.

For the desired eigenvalues λ_{o1} and λ_{o2}, a particular set of vectors ϕ_{ok} $(k = 1, 2)$ can be selected and partitioned corresponding to the choice $c_{o1} = c_{o2} = 1$, such that

$$\phi_{ok} = \begin{bmatrix} \bar{\phi}_{ok} \\ \hat{\phi}_{ok} \end{bmatrix} = V_{ok} = \begin{bmatrix} \lambda_{ok} \\ 1 \\ \lambda_{ok}^2 + 1 \end{bmatrix} \implies \Gamma_{ok} \phi_{ok} = 0, \quad k = 1, 2.$$

The observer gain matrix G can then be solved by

$$G^T [\, \bar{\phi}_{o1} \quad \bar{\phi}_{o2} \,] = [\, \hat{\phi}_{o1} \quad \hat{\phi}_{o2} \,]$$

or

$$G^T \begin{bmatrix} \lambda_{o1} & \lambda_{o2} \\ 1 & 1 \end{bmatrix} = [\, \lambda_{o1}^2 + 1 \quad \lambda_{o2}^2 + 1 \,].$$

Let us choose $\lambda_{o1} = 0.1$ and $\lambda_{o1} = 0.2$. The observer gain matrix G becomes

$$G^T = [\, 0.3 \quad 0.98 \,]$$

and the observer state matrix becomes

$$
A + GC = \begin{bmatrix} 0 & 1 \\ -1 & 0 \end{bmatrix} + \begin{bmatrix} 0.3 \\ 0.98 \end{bmatrix} \begin{bmatrix} 1 & 0 \end{bmatrix}
$$

$$
= \begin{bmatrix} 0.3 & 1 \\ -0.02 & 0 \end{bmatrix}.
$$

In terms of $x(k)$ and $e(k)$ (the difference between the true state and the estimated state), the overall closed-loop state matrix shown in Eq. (9.85) is

$$
\begin{bmatrix} A + BF & -BF \\ 0 & A + GC \end{bmatrix} = \begin{bmatrix} 0.575 & 1.025 & -0.575 & -0.575 \\ -0.425 & 0.025 & -0.025 & -0.025 \\ 0 & 0 & 0.3 & 1 \\ 0 & 0 & -0.02 & 0 \end{bmatrix}.
$$

9.7 Static Output Feedback

Consider the case in which there is no direct-transmission term. The state–space model of a linear time-invariant dynamic system in continuous-time format is

$$
\frac{dx}{dt} = A_c x(t) + B_c u(t), \tag{9.86}
$$

$$
y(t) = Cx(t). \tag{9.87}
$$

A direct output-feedback controller takes the form

$$
u(t) = \mathcal{F}_c y(t), \tag{9.88}
$$

where $\mathcal{F}_c$ is some constant gain matrix to be determined. The closed-loop system is governed by

$$
\frac{dx}{dt} = A_c x(t) + B_c \mathcal{F}_c y(t)
$$

$$
= (A_c + B_c \mathcal{F}_c C) x(t). \tag{9.89}
$$

Thus, stability of the closed-loop system is determined by the closed-loop system matrix $(A_c + B_c \mathcal{F}_c C)$. Unlike the direct state-feedback-control case, it is not always possible to place the closed-loop poles in any desired closed-loop eigenvalues. In fact, finding an output-feedback gain matrix $\mathcal{F}_c$ just to stabilize the closed-loop system can be difficult.

If the direct-transmission term is present, the resultant closed-loop equations become slightly more complicated. In the continuous-time case, the output-feedback controller becomes

$$
u(t) = \mathcal{F}_c y(t)
$$

$$
= \mathcal{F}_c [Cx(t) + Du(t)]
$$

$$
= \mathcal{F}_c Cx(t) + \mathcal{F}_c Du(t). \tag{9.90}
$$

The control input is related to the state of the system by

$$
u(t) = (I - \mathcal{F}_c D)^{-1} \mathcal{F}_c Cx(t). \tag{9.91}
$$

Thus the closed-loop system dynamics is described by

$$\frac{dx}{dt} = A_c x(t) + B_c u(t)$$

$$= [A_c + B_c (I - \mathcal{F}_c D)^{-1} \mathcal{F}_c C] x(t). \tag{9.92}$$

The closed-loop system matrix now is $A_c + B_c (I - \mathcal{F}_c D)^{-1} \mathcal{F}_c C$ rather than $A_c + B_c \mathcal{F}_c C$ as before. Thus, the problem of stabilizing a system by direct output feedback in the presence of a direct-transmission term can be quite complicated.

We now derive the closed-loop equations for the static output-feedback-control system in the discrete-time domain. Again, for simplicity, we first consider the case in which there is no direct-transmission term. The state–space model of a linear time-invariant dynamic system in discrete-time format is

$$x(k+1) = Ax(k) + Bu(k), \tag{9.93}$$

$$y(k) = Cx(k). \tag{9.94}$$

If the output is used directly for feedback with a control law of the form

$$u(k) = \mathcal{F}y(k), \tag{9.95}$$

then the closed-loop system is governed by

$$x(k+1) = Ax(k) + B\mathcal{F}y(k)$$

$$= (A + B\mathcal{F}C)x(k). \tag{9.96}$$

Thus, stability of the closed-loop system is determined by the closed-loop system matrix $(A + B\mathcal{F}C)$. Analogous statements regarding the existence of an output-feedback-gain matrix to make the discrete-time closed-loop system stable apply in this case.

When the direct-transmission term is present, an output-feedback-control law has the form

$$u(k) = \mathcal{F}y(k)$$

$$= \mathcal{F}[Cx(k) + Du(k)]$$

$$= \mathcal{F}Cx(k) + \mathcal{F}Du(k), \tag{9.97}$$

from which the control input is related to the state of the system by

$$u(k) = (I - \mathcal{F}D)^{-1} \mathcal{F}Cx(k). \tag{9.98}$$

Thus, the closed-loop system is now described by

$$x(k+1) = Ax(k) + Bu(k)$$

$$= [A + B (I - \mathcal{F}D)^{-1} \mathcal{F}C]x(k). \tag{9.99}$$

The discrete-time closed-loop system matrix is $A + B (I - \mathcal{F}D)^{-1} \mathcal{F}C$.

9.8 Dynamic Output Feedback

One way to view a dynamic controller is that it is some dynamic system that takes the system output as its input and produces an output that becomes the input to the

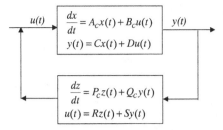

Figure 9.6. Continuous-time dynamic output feedback.

controlled system. In other words, the system output becomes the controller input and the controller output becomes the system input.

CONTINUOUS-TIME DYNAMIC OUTPUT FEEDBACK

Consider a linear dynamic system in state–space format of the form

$$\frac{dx}{dt} = A_c x(t) + B_c u(t), \tag{9.100}$$

$$y(t) = Cx(t) + Du(t). \tag{9.101}$$

A (linear) dynamic controller then assumes the general form

$$\frac{dz}{dt} = P_c z(t) + Q_c y(t), \tag{9.102}$$

$$u(t) = Rz(t) + Sy(t), \tag{9.103}$$

where the dynamics of the controller is described by the state–space model P_c, Q_c, R, S. The subscript c is used here to signify the correspondence between A_c and P_c and B_c and Q_c. The vector $z(t)$ denotes the controller state, which does not necessarily have the same dimensions as the system state $x(t)$. Note that, in this controller, the output $y(t)$ of the system is the controller input and the controller output becomes the system input $u(t)$. A block diagram of this feedback-control mechanism is given in Fig. 9.6.

We are interested in determining the conditions for which the dynamic feedback produces a stable closed-loop system. First, we need to form a set of equations describing the closed-loop dynamics. Let us consider the case in which the overall closed-loop dynamics is described by a state vector, which consists of the system state $x(t)$ and the controller state $z(t)$. We begin with describing the system input and output in terms of $x(t)$ and $z(t)$. For the system input, we have

$$\begin{aligned} u(t) &= Rz(t) + Sy(t) \\ &= Rz(t) + S[Cx(t) + Du(t)] \\ &= Rz(t) + SCx(t) + SDu(t), \end{aligned} \tag{9.104}$$

which yields

$$u(t) = (I - SD)^{-1} Rz(t) + (I - SD)^{-1} SCx(t). \tag{9.105}$$

Substituting Eq. (9.105) into the system output equation gives

$$
\begin{aligned}
y(t) &= Cx(t) + Du(t) \\
&= Cx(t) + D[(I - SD)^{-1} Rz(t) + (I - SD)^{-1} SCx(t)] \\
&= [C + D(I - SD)^{-1} SC]x(t) + D(I - SD)^{-1} Rz(t).
\end{aligned}
\tag{9.106}
$$

Having expressed both the input and the output of the system in terms of $x(t)$ and $z(t)$, we can write the system state equation as

$$
\begin{aligned}
\frac{dx}{dt} &= A_c x(t) + B_c u(t) \\
&= A_c x(t) + B_c[(I - SD)^{-1} Rz(t) + (I - SD)^{-1} SCx(t)] \\
&= [A_c + B_c (I - SD)^{-1} SC]x(t) + B_c (I - SD)^{-1} Rz(t).
\end{aligned}
\tag{9.107}
$$

Similarly, for the controller state equation,

$$
\begin{aligned}
\frac{dz}{dt} &= P_c z(t) + Q_c y(t) \\
&= P_c z(t) + Q_c\{[C + D(I - SD)^{-1} SC]x(t) + D(I - SD)^{-1} Rz(t)\} \\
&= [Q_c C + Q_c D(I - SD)^{-1} SC]x(t) + [P_c + Q_c D(I - SD)^{-1} R]z(t).
\end{aligned}
\tag{9.108}
$$

We can now write the overall closed-loop system in the form

$$
\begin{aligned}
&\frac{d}{dt}\begin{bmatrix} x(t) \\ z(t) \end{bmatrix} \\
&= \begin{bmatrix} A_c + B_c (I - SD)^{-1} SC & B_c (I - SD)^{-1} R \\ Q_c C + Q_c D(I - SD)^{-1} SC & P_c + Q_c D(I - SD)^{-1} R \end{bmatrix} \begin{bmatrix} x(t) \\ z(t) \end{bmatrix},
\end{aligned}
\tag{9.109}
$$

which describes the closed-loop dynamics. Thus, the behavior of the closed-loop system depends on the eigenvalues of the closed-loop system matrix

$$
\bar{A}_c = \begin{bmatrix} A_c + B_c (I - SD)^{-1} SC & B_c (I - SD)^{-1} R \\ Q_c C + Q_c D(I - SD)^{-1} SC & P_c + Q_c D(I - SD)^{-1} R \end{bmatrix},
\tag{9.110}
$$

whose eigenvalues must have negative real parts for closed-loop asymptotic stability. One typical objective of feedback control is to add damping to the open-loop system (i.e., to make it more stable). In this case, it is desirable that the eigenvalues of the closed-loop system matrix $\bar{A}_c$ have more damping than those of the system matrix A_c.

DISCRETE-TIME DYNAMIC OUTPUT FEEDBACK

The discrete-time dynamic output feedback is very similar to the continuous-time dynamic output feedback. A linear discrete-time dynamic system in the state–space format has the form

$$
x(k + 1) = Ax(k) + Bu(k),
\tag{9.111}
$$

$$
y(k) = Cx(k) + Du(k).
\tag{9.112}
$$

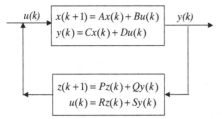

Figure 9.7. Discrete-time dynamic output feedback.

A discrete-time (linear) dynamic controller assumes the general form

$$z(k + 1) = Pz(k) + Qy(k), \tag{9.113}$$

$$u(k) = Rz(k) + Sy(k), \tag{9.114}$$

where the dynamics of the controller is determined by the state–space model P, Q, R, and S. Again, the controller state vector $z(k)$ does not necessarily have the same dimensions as the system state $x(k)$. A block diagram of this feedback-control mechanism is given in Fig. 9.7.

The overall closed-loop dynamics can be described in terms of the system state $x(k)$ and the controller state $z(k)$. We must first derive the expressions for the system input $u(k)$ and output $y(k)$ in terms of $x(k)$ and $z(k)$. For the system input, we have

$$u(k) = Rz(k) + Sy(k)$$

$$= Rz(k) + S[Cx(k) + Du(k)]$$

$$= Rz(k) + SCx(k) + SDu(k). \tag{9.115}$$

Thus,

$$u(k) = (I - SD)^{-1} Rz(k) + (I - SD)^{-1} SCx(k). \tag{9.116}$$

Substituting Eq. (9.116) into the system output equation produces

$$y(k) = Cx(k) + Du(k)$$

$$= Cx(k) + D[(I - SD)^{-1} Rz(k) + (I - SD)^{-1} SCx(k)]$$

$$= [C + D(I - SD)^{-1} SC]x(k) + D(I - SD)^{-1} Rz(k). \tag{9.117}$$

Having expressed both the input and output of the system in terms of $x(k)$ and $z(k)$, we can write the system state equation as

$$x(k + 1) = Ax(k) + Bu(k)$$

$$= Ax(k) + B[(I - SD)^{-1} Rz(k) + (I - SD)^{-1} SCx(k)]$$

$$= [A + B(I - SD)^{-1} SC]x(k) + B(I - SD)^{-1} Rz(k). \tag{9.118}$$

Similarly, the controller state equation can be written as

$$z(k + 1) = Pz(k) + Qy(k)$$
$$= Pz(k) + Q\{[C + D(I - SD)^{-1} SC]x(k) + D(I - SD)^{-1} Rz(k)\}$$
$$= [QC + QD(I - SD)^{-1} SC]x(k) + [P + QD(I - SD)^{-1} R]z(k). \quad (9.119)$$

The overall discrete-time closed-loop system is then governed by

$$\begin{bmatrix} x(k + 1) \\ z(k + 1) \end{bmatrix}$$
$$= \begin{bmatrix} A + B(I - SD)^{-1} SC & B(I - SD)^{-1} R \\ QC + QD(I - SD)^{-1} SC & P + QD(I - SD)^{-1} R \end{bmatrix} \begin{bmatrix} x(k) \\ z(k) \end{bmatrix}. \quad (9.120)$$

The behavior of the closed-loop system depends on the eigenvalues of the closed-loop system matrix

$$\bar{A} = \begin{bmatrix} A + B(I - SD)^{-1} SC & B(I - SD)^{-1} R \\ QC + QD(I - SD)^{-1} SC & P + QD(I - SD)^{-1} R \end{bmatrix}, \quad (9.121)$$

whose eigenvalues must have magnitudes less than one for closed-loop asymptotic stability.

A SPECIAL CASE: OBSERVER-BASED STATE-FEEDBACK CONTROL

It will be shown that the observer-based state-feedback control is a special case of the dynamic output feedback. We will make this connection in both the continuous-time and the discrete-time domain. Recall that a continuous-time observer has the form

$$\frac{d\hat{x}}{dt} = (A_c + G_c C)\hat{x}(t) + (B_c + G_c D)u(t) - G_c y(t)$$

and the controller is computed from

$$u(t) = \mathcal{F}_c \hat{x}(t).$$

Combining the two equations yields an equivalent dynamic feedback controller of the form

$$\frac{d\hat{x}}{dt} = [A_c + G_c C + (B_c + G_c D)\mathcal{F}_c]\hat{x}(t) - G_c y(t), \quad (9.122)$$

$$u(t) = \mathcal{F}_c \hat{x}(t). \quad (9.123)$$

Comparing this set of equations with those of the generic dynamic output-feedback controller given in Eq. (9.109) produces

$$z(t) = \hat{x}(t),$$
$$P_c = A_c + G_c C + (B_c + G_c D)\mathcal{F}_c,$$
$$Q_c = -G_c,$$
$$R = \mathcal{F}_c,$$
$$S = 0. \quad (9.124)$$

Thus it is clear that an observer-based state-feedback controller is simply a special dynamic output-feedback controller.

In the discrete-time domain, we can easily arrive at the same conclusions. Recall that the observer equation in the discrete-time domain is

$$\hat{x}(k+1) = (A + GC)\,\hat{x}(k) + (B + GD)\,u(k) - Gy(k)$$

and the control input is computed from

$$u(k) = \mathcal{F}\hat{x}(k).$$

Combining the two equations produces a dynamic output-feedback-controller equivalent:

$$\hat{x}(k+1) = [A + GC + (B + GD)\,\mathcal{F}]\,\hat{x}(k) - Gy(k), \tag{9.125}$$

$$u(k) = \mathcal{F}\hat{x}(k). \tag{9.126}$$

Thus, the equivalent discrete-time dynamic feedback controller is given by

$$z(k) = \hat{x}(k),$$

$$P = A + GC + (B + GD)\,\mathcal{F},$$

$$Q = -G,$$

$$R = \mathcal{F},$$

$$S = 0. \tag{9.127}$$

Note that, in both cases, the dimensions of the controller state are the same as those of the system state. The general form of a dynamic output-feedback controller certainly allows this possibility, but this need not be the case in general.

EXAMPLE 9.5

We will show how one may design a dynamic output-feedback controller for a simple second-order spring–mass system (see Fig. 9.8) with no damping:

$$m\ddot{w}(t) + kw(t) = u(t),$$

where $w(t)$ is the linear position of the mass m and $u(t)$ is the control (force) input. Only the position of the mass is available as an output measurement:

$$y(t) = w(t).$$

A state–space representation of the above system is

$$\frac{d}{dt}\begin{bmatrix} w \\ \dot{w} \end{bmatrix} = \begin{bmatrix} 0 & 1 \\ -k/m & 0 \end{bmatrix}\begin{bmatrix} w \\ \dot{w} \end{bmatrix} + \begin{bmatrix} 0 \\ 1/m \end{bmatrix}u(t)$$

$$y(t) = \begin{bmatrix} 1 & 0 \end{bmatrix}\begin{bmatrix} w \\ \dot{w} \end{bmatrix},$$

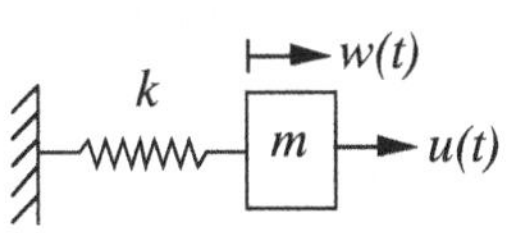

Figure 9.8. A mass–spring system.

with the following system state and state–space matrices:

$$x(t) = \begin{bmatrix} w \\ \dot{w} \end{bmatrix}, \quad A = \begin{bmatrix} 0 & 1 \\ -k/m & 0 \end{bmatrix}, \quad B = \begin{bmatrix} 0 \\ 1/m \end{bmatrix}, \quad C = [1 \quad 0], \quad D = 0.$$

The open-loop system has no damping, and it is desirable to add damping to the system by feedback control. We certainly know how to do this with an observer-based state-feedback controller. Let us consider a dynamic output-feedback controller instead. A dynamic output-feedback controller is of the form

$$m_c \ddot{w}_c(t) + \varsigma_c \dot{w}_c(t) + k_c w_c(t) = k_c y(t),$$

$$u(t) = k_c w_c(t) - k_c y(t),$$

which, in state–space format, is

$$\frac{d}{dt} \begin{bmatrix} w_c \\ \dot{w}_c \end{bmatrix} = \begin{bmatrix} 0 & 1 \\ -k_c/m_c & -\varsigma_c/m_c \end{bmatrix} \begin{bmatrix} w_c \\ \dot{w}_c \end{bmatrix} + \begin{bmatrix} 0 \\ k_c/m_c \end{bmatrix} y(t),$$

$$u(t) = [k_c \quad 0] \begin{bmatrix} w_c \\ \dot{w}_c \end{bmatrix} - k_c y(t),$$

where w_c is the virtual displacement of the controller mass m_c, ς_c, and k_c are virtual damping and stiffness constants, respectively.

The controller state and its corresponding state–space matrices are

$$z(t) = \begin{bmatrix} w_c(t) \\ \dot{w}_c(t) \end{bmatrix},$$

$$P_c = \begin{bmatrix} 0 & 1 \\ -k_c/m_c & -\varsigma_c/m_c \end{bmatrix},$$

$$Q_c = \begin{bmatrix} 0 \\ k_c/m_c \end{bmatrix},$$

$$R = [k_c \quad 0],$$

$$S = -k_c.$$

How do we know this controller will work? To see this, we must derive the closed-loop dynamics equations. For the physical system, we have

$$m\ddot{w}(t) + kw(t) = u(t)$$
$$= k_c w_c(t) - k_c y(t)$$
$$= k_c w_c(t) - k_c w(t),$$

which yields

$$m\ddot{w}(t) + (k + k_c)\,w(t) - k_c w_c(t) = 0.$$

For the controller system, we have

$$m_c \ddot{w}_c(t) + \varsigma_c \dot{w}_c(t) + k_c w_c(t) = k_c y(t) = k_c w(t),$$

which yields

$$m_c \ddot{w}_c(t) + \varsigma_c \dot{w}_c(t) + k_c w_c(t) - k_c w(t) = 0.$$

Writing the system equation and the controller equation together in (second-order) matrix form, we obtain

$$\begin{bmatrix} m & 0 \\ 0 & m_c \end{bmatrix} \begin{bmatrix} \ddot{w}(t) \\ \ddot{w}_c(t) \end{bmatrix} + \begin{bmatrix} 0 & 0 \\ 0 & \varsigma_c \end{bmatrix} \begin{bmatrix} \dot{w}(t) \\ \dot{w}_c(t) \end{bmatrix}$$

$$+ \begin{bmatrix} k + k_c & -k_c \\ -k_c & k_c \end{bmatrix} \begin{bmatrix} w(t) \\ w_c(t) \end{bmatrix} = \begin{bmatrix} 0 \\ 0 \end{bmatrix}$$

or, equivalently, in first-order matrix form,

$$\frac{d}{dt} \begin{bmatrix} w(t) \\ w_c(t) \\ \dot{w}(t) \\ \dot{w}_c(t) \end{bmatrix} = \begin{bmatrix} 0 & 0 & 1 & 0 \\ 0 & 0 & 0 & 1 \\ -(k + k_c)/m & k_c/m & 0 & 0 \\ k_c/m_c & -k_c/m_c & 0 & -\varsigma_c/m_c \end{bmatrix} \begin{bmatrix} w(t) \\ w_c(t) \\ \dot{w}(t) \\ \dot{w}_c(t) \end{bmatrix}.$$

Does this set of closed-loop equations represent a stable system? Certainly it does as long as m_c, ς_c, and k_c are positive. It represents the dynamic system shown in Fig. 9.9.

We can make several statements about this controller. First, the overall closed-loop system is always stable regardless of the choices of m_c, ς_c, and k_c (as long as m_c, ς_c, and k_c are positive) and regardless of the actual system parameters m and k. In this sense, the design is extremely robust. Second, the mechanism by which the controller performs the vibration suppression function is clearly displayed. The presence of the damper in the controller will dissipate energy from the physical system to bring it to rest if it is disturbed from the equilibrium position. Third, the parameters m_c, ς_c, and k_c that make up the controller matrices have physical meanings that facilitate optimal tuning of these parameters. This feature stands in contrast to many modern control design methods in which the controller gains do not necessarily have any physical meaning. Note that the above diagram is an interpretation of the controller only. There is no physical spring, mass, or damper attached to the system to be controlled at all. The dynamic feedback controller will generate the control input as if it comes from an attached mass–spring–damper system, which in reality is a virtual system, as discussed in Chapter 6. Indeed, we can attach a physical spring–mass–damper to the system to suppress its vibration

Dynamic Output Feedback Controller

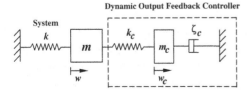

Figure 9.9. Single-degree-of-freedom system with a single-degree-of-freedom dynamic controller.

(like a shock absorber in an automobile) rather than use a controller. This concept is called passive stabilization. For these reasons, this controller is called a virtual passive controller. Thus, virtual passive control is one special type of dynamic output-feedback controller that possesses many interesting and useful properties. We have shown here a simple example on how this design applies to a simple spring–mass system.

9.9 Concluding Remarks

State feedback and state-estimation are the dual theories that provide the foundation for the field of modern control. State feedback is designed under the unrealistic assumption that all state variables are available for feedback. In practical control systems, the measurement of all state variables is generally impossible. It becomes necessary to estimate the unmeasurable state variables to implement a design based on state feedback. In this chapter, we have introduced the design methods for a state observer that estimates all state variables required for feedback to complete a control design. The fundamental design method based on pole placement was discussed in detail because it is one of the fundamental design methods available to control engineers for use in practice. It is demonstrated that the pole-placement design process is mathematically identical in form for both feedback and observer designs. In other words, the computer program written for the feedback design can be also readily adapted for use in the observer design.

In the observer-based controller design, the state feedback is accomplished by use of the estimated state variables rather than actual state variables that are not available for direct measurement. The separation principle has proved that stability of the overall closed-loop system is governed by the stability of the observer and controller separately as if all state variables are available. The combination of state feedback and state observer is a special case of the dynamic output feedback in which the dimension of the controller states is the same as that of the system states. Another special form of dynamic output feedback was also discussed, i.e., the virtual passive controller.

9.10 Problems

9.1 From Eq. (9.23), we may form a first-order matrix equation as

$$
\begin{bmatrix} \dot{w} \\ \ddot{w} \end{bmatrix} = \begin{bmatrix} 0_n & I_n \\ -M^{-1}K & -M^{-1}\Xi \end{bmatrix} \begin{bmatrix} w \\ \dot{w} \end{bmatrix} + \begin{bmatrix} 0_{n \times r} \\ B_f \end{bmatrix},
$$

$$
y = [C_d \quad C_v] \begin{bmatrix} w \\ \dot{w} \end{bmatrix}.
$$

In view of Eq. (8.37) for closed-loop state response and Eq. (9.9) for state-estimation error, both equations show significant similarity. The pole-placement technique introduced in Subsection 8.5.1 is based on the right-hand eigenvalue problem whereas the pole-placement technique presented in Subsection 9.4.2 uses the left-hand eigenvalue problem. Prove that the pole-placement formulations derived for state feedback are identical to those for state observation with the replacement of A_c, B_c, and $\mathcal{F}_c$ with A_c^T, C^T, and G^T, respectively.

9.2 Consider a simple mass–spring problem described by

$$\ddot{w} + w = u,$$

$$y = \begin{bmatrix} w \\ \dot{w} \end{bmatrix} = \begin{bmatrix} 1 \\ 0 \end{bmatrix} w + \begin{bmatrix} 0 \\ 1 \end{bmatrix} \dot{w}.$$

Compute a set of observer gain matrices G_v and G_d in terms of the desired observer gain eigenvalues λ_1 and λ_2. Is this set of gain matrices unique? If not, derive another one and prove that both sets are correct.

9.3 Consider the inverted pendulum on a moving cart described in Example 9.1. The linearized equation of motion of the system (for small-angle motion of the inverted pendulum) is

$$\frac{d}{dt} \begin{bmatrix} z \\ \dot{z} \\ \theta \\ \dot{\theta} \end{bmatrix} = \begin{bmatrix} 0 & 1 & 0 & 0 \\ 0 & 0 & -1 & 0 \\ 0 & 0 & 0 & 1 \\ 0 & 0 & 11 & 0 \end{bmatrix} \begin{bmatrix} z \\ \dot{z} \\ \theta \\ \dot{\theta} \end{bmatrix} + \begin{bmatrix} 0 \\ 1 \\ 0 \\ -1 \end{bmatrix} u(t).$$

In addition, assume that the position of the cart and the angle of the pendulum can be directly measured:

$$y(t) = \begin{bmatrix} z \\ \theta \end{bmatrix} = \begin{bmatrix} 1 & 0 & 0 & 0 \\ 0 & 0 & 1 & 0 \end{bmatrix} \begin{bmatrix} z \\ \dot{z} \\ \theta \\ \dot{\theta} \end{bmatrix}.$$

Choose the desired observer eigenvalues as -2, -3, and $-2 \pm i$. Find an observer gain matrix $G_c = [\, G_d \; G_v \,]$. Is this gain matrix unique? Prove that the gain matrix is a correct one.

9.4 Let the system matrices A and C be

$$A = \begin{bmatrix} 1 & 1 \\ 0 & 1 \end{bmatrix}, \qquad C = [0 \;\; 1].$$

(a) Find the values of $G^T = [g_1 \;\; g_2\,]$ such that the matrix $A + GC$ has the distinct eigenvalues λ_1 and λ_2.

(b) What are the values of g_1 and g_2 that make $\lambda_1 = \lambda_2$?

9.5 Let the state matrix A and the output matrix C be

$$A = \begin{bmatrix} -2 & 1 & 0 & 0 \\ 0 & -2 & 0 & 0 \\ 0 & 0 & 1 & 1 \\ 0 & 0 & 0 & 1 \end{bmatrix}, \qquad C = [1 \;\; 0 \;\; 0 \;\; 1],$$

which has an unstable eigenvalue 1.

(a) Is this system observable?

(b) Is there any eigenvalue that is unobservable?

(c) Is it possible to design a stable observable?

(d) If the answer to (c) is yes, find the observable gain matrix such that the eigenvalues are $-1, -1, -2$, and -2.

9.6 Consider the state–space model described by the system matrices

$$A = \begin{bmatrix} 2 & 1 & 0 & 0 \\ 0 & 2 & 0 & 0 \\ 0 & 0 & -1 & 0 \\ 0 & 0 & 0 & -1 \end{bmatrix}, \qquad C = [0 \quad 1 \quad 0 \quad 1].$$

(a) Is this system observable?

(b) Is there any eigenvalue that is unobservable?

(c) Is it possible to design a stable observer?

(d) Is it possible to have the observer eigenvalues?

    (1) $-1, -1, -2$, and -2.

    (2) $-1, -2, -2$, and -2.

    (3) $-2, -2, -2$, and -2.

9.7 Given the following discrete-time model,

$$x(k+1) = \begin{bmatrix} 0 & 1 \\ 1 & 1 \end{bmatrix} x(k) + \begin{bmatrix} 0 \\ 1 \end{bmatrix} u(k),$$

$$y(k) = [0 \quad 1] x(k) + u(k),$$

find the state-feedback and observer gain matrices that will produce both closed-loop and observer eigenvalues (poles) at

(a) $\lambda = 0$ and $\lambda = 0.6$,

(b) $\lambda = 0.1 \pm j0.4$.

9.8 Consider the discrete-time model

$$x(k+1) = \begin{bmatrix} \cos\gamma & \sin\gamma \\ -\sin\gamma & \cos\gamma \end{bmatrix} x(k) + \begin{bmatrix} 1 - \cos\gamma \\ \sin\gamma \end{bmatrix} u(k),$$

$$y(k) = \begin{bmatrix} 0 & 1 \\ 1 & 0 \end{bmatrix} x(k).$$

Set $\gamma = \pi/3$.

(a) Find a feedback gain matrix with the closed-loop eigenvalues at $\lambda = 0.2$ and $\lambda = 0.3$

(b) Compute two different observer gain matrices with observer eigenvalues at $\lambda = 0$ and $\lambda = 0.1$ and choose the one with minimum norm.

(c) Form the overall closed-loop state matrix as shown in Eq. (9.85).

9.9 Consider the equation of motion

$$M\ddot{w} + Kw = B_2 u,$$

$$y = Cw,$$

where

$$M = \text{diag}(1, 1, 1),$$

$$K = \begin{bmatrix} 10 & -5 & 0 \\ -5 & 25 & -20 \\ 0 & -20 & 20 \end{bmatrix},$$

$$C = \begin{bmatrix} 1 & 0 & 0 \\ 0 & 0 & 1 \end{bmatrix}.$$

(a) Form a first-order equation.

(b) Compute the open-loop eigenvalues and eigenvectors.

(c) Use the eigenvalues $\lambda_1 = -2 + i$, $\lambda_2 = -2 - i$ $\lambda_3 = -1$, and $\lambda_4 = -2$ as the desired eigenvalues and the open-loop eigenvectors as the desired eigenvectors. Find the observer gain matrix G_c such that the observer has the desired eigenvalues.

(d) Repeat (d) but use a 6×6 identity matrix as the desired eigenvectors.

(e) Compute the norm of G_c from (d) and (e).

9.10 Consider the state error equation

$$\frac{de}{dt} = [A_c + G_c C]e(t),$$

where

$$A_c = \begin{bmatrix} 0 & 1 \\ -1 & 0 \end{bmatrix}, \qquad C = [1 \quad 0].$$

Define the following performance index:

$$J = \frac{1}{2} \int_0^\infty e^T(t)e(t)\,dt.$$

Solve for the gain matrix $G_c^T = [\,g_1 \quad g_2\,]$ that minimizes the index. This example is given to demonstrate that the observer problem may also be formulated as an optimal estimation problem similar to the optimal control problem.

BIBLIOGRAPHY

[1] Lewis, F. L., *Optimal Estimation*, Wiley, New York, 1986.

[2] Ogata, K., *Discrete-Time Control Systems*, Prentice-Hall, Upper Saddle River, NJ, 1995.

[3] Fortman, T. E. and Hitz, K. L., *An Introduction to Linear Control Systems*, Marcel Dekker, Inc., New York, 1977.

[4] Maghami, P. G. and Juang, J.-N., "Efficient Eigenvalue Assignment for Large Space Structures," *J. Guid. Control Dyn.*, Vol. 13, pp. 1033–1039, 1990.

[5] Maghami, P. G., Juang, J.-N., and Lim, K. B., "Eigensystem Assignment with Output Feedback," *J. Guid. Control Dyn.*, Vol. 15, pp. 531–536, 1992.

[6] Juang, J.-N. and Maghami, P. G., "Robust Eigensystem Assignment for State Estimators Using Second-Order Models," *J. Guid. Control Dyn.*, Vol. 15, pp. 920–927, 1992.

10

System Identification

10.1 Introduction

Identification is the process of developing a mathematical model for a physical system by use of experimental data (Ref. [1–3]). This chapter starts with introducing the relationship between the input and output data in terms of a time-domain finite-difference model, known as the autoregressive model with exogeneous input (ARX). Formulations are derived to compute the coefficient of the finite-difference model from input and output data. From the coefficient, the output-prediction model is derived, giving the relationship between future input–output data and past input–output data. The system Markov parameters are introduced and their relationship with the pulse-response time history is discussed. The system Markov parameters may be computed from the coefficient of the finite-difference model in a recursive fashion. The connection between the state–space model and the finite-difference model of a linear system is also shown and discussed. Finally, identification of a state–space observer model is derived, giving formulation for computing the observer gain for state estimation.

10.2 Finite-Difference Model

The input–output relationship of a linear system, even a nonlinear system, is commonly described by a finite-difference model. Given input and output time histories $u(k)$ and $y(k)$, a general finite-difference model or ARX model in the time domain may be written as

$$y(k) + \alpha_1 y(k-1) + \cdots + \alpha_p y(k-p)$$

$$= \beta_0 u(k) + \beta_1 u(k-1) + \cdots + \beta_p u(k-p), \tag{10.1}$$

where p is an integer indicating the order of the model. The coefficient matrices $\alpha_1, \ldots, \alpha_p$ of dimension $m \times m$ are associated with the $m \times 1$ output vector y at time steps $k-1, \ldots, k-p$, respectively. The coefficient matrices $\beta_0, \beta_1, \ldots, \beta_p$ of dimension $m \times r$ correspond to the $r \times 1$ input vector u at the time step $k, k-1, \ldots, k-p$, respectively. Equation (10.1) can be rearranged to be

$$y(k) = -\alpha_1 y(k-1) - \cdots - \alpha_p y(k-p)$$

$$+ \beta_0 u(k) + \beta_1 u(k-1) + \cdots + \beta_p u(k-p). \tag{10.2}$$

This equation means that the current output $y(k)$ at the time step k may be expressed in terms of p previous output and input measurements, $y(k-1), \ldots, y(k-p)$ and $u(k-1), \ldots, u(k-p)$, and the current input measurement $u(k)$. Predicting the current measurement from previous data is a forward-in-time prediction.

EXAMPLE 10.1

Consider the finite-difference model, i.e., the ARX model,

$$y(k)+\frac{3}{2}y(k-1) + \frac{5}{2}y(k-2) + \frac{3}{2}y(k-3) + y(k-4)$$

$$=u(k) - \frac{7}{6}u(k-1) + u(k-2) - \frac{4}{3}u(k-3) + \frac{1}{2}u(k-4).$$

Assume that the system is initially at rest, i.e., $y(0) = y(-1) = \cdots = y(-\infty) = 0$ and $u(0) = u(-1) = \cdots = u(-\infty) = 0$.

Basic Recursive Generation of Output: Given a time history for the input $u(1), u(2), \ldots, u(N)$, where N is an arbitrary integer, the output time history may be recursively computed as

$$y(1)=u(1),$$

$$y(2)=-\frac{3}{2}y(1) + u(2) - \frac{7}{6}u(1),$$

$$y(3)=-\frac{3}{2}y(2) - \frac{5}{2}y(1) + u(3) - \frac{7}{6}u(2) + u(1),$$

$$\vdots$$

$$y(N)=-\frac{3}{2}y(N-1) - \frac{5}{2}y(N-2) - \frac{3}{2}y(N-3) - y(N-4)$$

$$+u(N) - \frac{7}{6}u(N-1) + u(N-2) - \frac{4}{3}u(N-3) + \frac{1}{2}u(N-4).$$

Choose the time history

$$u(k) = \sin(2\pi\omega k),$$

for $u(1), u(2) \ldots, u(40)$ and $u(41) = u(42) = \cdots = u(80) = 0$ with $\omega = 0.3$ [see Fig. 10.1(a)]. The output time history $y(1), y(2), \ldots, y(80)$ can then be generated and is shown in Fig. 10.1(b).

Measurement Noise Problem: In practice, the output time history y is a measured quantity subject to measurement noises, and the system is not necessarily at rest initially. To illustrate, assume that the output noise is random as shown in Fig. 10.1(c). The measured output [see Fig. 10.1(d)] becomes the sum of the true output [Fig. 10.1(b)] and the noise [Fig. 10.1(c)]. Because the true output time history is generally unknown, the finite-difference model that describes the relationship between the input and the output time histories will not be perfectly satisfied. Thus, the equation used to estimate the output time history must be

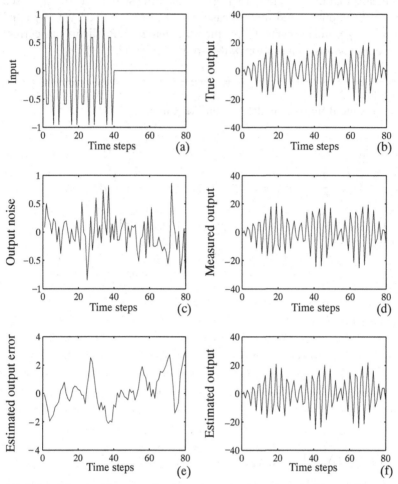

Figure 10.1. Measured output and estimated output with a sinusoidal input of frequency 0.3 Hz.

modified as follows:

$$\hat{y}(1)=u(1),$$

$$\hat{y}(2)=-\frac{3}{2}y(1)+u(2)-\frac{7}{6}u(1),$$

$$\hat{y}(3)=-\frac{3}{2}y(2)-\frac{5}{2}y(1)+u(3)-\frac{7}{6}u(2)+u(1),$$

$$\vdots$$

$$\hat{y}(N)=-\frac{3}{2}y(N-1)-\frac{5}{2}y(N-2)-\frac{3}{2}y(N-3)-y(N-4)$$

$$+u(N)-\frac{7}{6}u(N-1)+u(N-2)-\frac{4}{3}u(N-3)+\frac{1}{2}u(N-4),$$

where $y(1), y(2), \ldots, y(N)$ on the right-hand side are the measured quantities and $\hat{y}(1), \hat{y}(2), \ldots, \hat{y}(N)$ on the left-hand side are estimated quantities. Both measured and estimated quantities can be significantly different. The estimated output $\hat{y}$ approaches the true output when the signal-to-noise ratio, i.e., the ratio of the maximum amplitudes of the true output and the noise, becomes sufficiently large. The true output [Fig. 10.1(b)], the measured output [Fig. 10.1(d)], and the estimated output [Fig. 10.1(f)] time histories are not considerably different because the signal-to-noise ratio is sufficiently high:

$$\frac{\text{Maximum amplitude of the true output time history}}{\text{Maximum amplitude of the noise signal}} \approx \frac{25}{1} = 25.$$

For comparison, let the input frequency be $\omega = 0.1$ and assume the same noise signal. Figure 10.2 shows that the true output [Fig. 10.2(b)], the measured output [Fig. 10.2(d)], and the estimated output [Fig. 10.2(f)] time histories are considerably different because the signal-to-noise ratio is 1.7 only, that is ~15 times less than that of the previous case. In other words, the output time history estimated from the measured output is not accurate. To improve the output estimation, we must properly choose the input frequency and amplitude to sufficiently excite the system to obtain a larger output response.

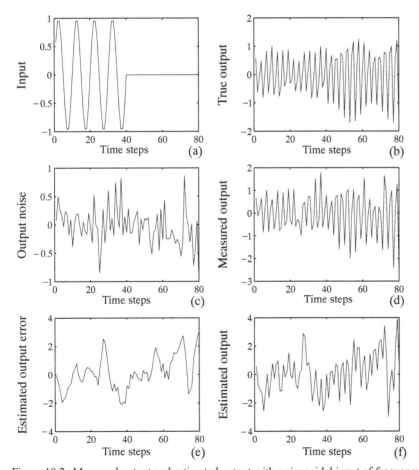

Figure 10.2. Measured output and estimated output with a sinusoidal input of frequency 0.1 Hz.

In this example, it has been shown that the current output can be estimated if past input and output time histories are given and the ARX model coefficients are known. In practice, the exact model coefficients are unknown and also need to be estimated from input and output signals.

Let the $(m + r) \times 1$ vector $v(k)$ be defined as

$$v(k) = \begin{bmatrix} y(k) \\ u(k) \end{bmatrix}, \quad k = 1, 2, \ldots, \ell, \tag{10.3}$$

where ℓ is the length of the data. Equation (10.1) with k from 1 to ℓ produces the following matrix equation:

$$[y_0 \quad y] = \theta[V_0 \quad V], \tag{10.4}$$

where the quantities y_0, y, θ, V_0, and V are defined by

$$y_0 = [y(1) \quad y(2) \quad \cdots \quad y(p)], \tag{10.5a}$$

$$y = [y(p+1) \quad y(p+2) \quad \cdots \quad y(\ell)], \tag{10.5b}$$

$$\theta = [\beta_0 \quad (-\alpha_1 \ \beta_1) \quad \cdots \quad (-\alpha_{p-1} \ \beta_{p-1}) \quad (-\alpha_p \ \beta_p)], \tag{10.5c}$$

$$V_0 = \begin{bmatrix} u(1) & u(2) & \cdots & u(p) \\ v(0) & v(1) & \cdots & v(p-1) \\ \vdots & \vdots & \ddots & \vdots \\ v(2-p) & v(3-p) & \cdots & v(1) \\ v(1-p) & v(2-p) & \cdots & v(0) \end{bmatrix}, \tag{10.5d}$$

$$V = \begin{bmatrix} u(p+1) & u(p+2) & \cdots & u(\ell) \\ v(p) & v(p+1) & \cdots & v(\ell-1) \\ \vdots & \vdots & \ddots & \vdots \\ v(2) & v(3) & \cdots & v(\ell-p+1) \\ v(1) & v(2) & \cdots & v(\ell-p) \end{bmatrix}. \tag{10.5e}$$

For the case in which $y(0) = y(-1) = \cdots = y(-\infty) = 0$ and $u(0) = u(-1) = \cdots = u(-\infty) = 0$, i.e., no initial transient response, the matrix V_0 becomes

$$V_0 = \begin{bmatrix} u(1) & u(2) & \cdots & u(p) \\ 0 & v(1) & \cdots & v(p-1) \\ \vdots & \vdots & \ddots & \vdots \\ 0 & 0 & \cdots & v(1) \\ 0 & 0 & \cdots & 0 \end{bmatrix}.$$

Equation (10.4) can be used to solve for the ARX parameter matrix θ by the least-squares method, yielding

$$\theta = [y_0 \quad y][V_0 \quad V]^\dagger, \tag{10.6}$$

where $\dagger$ means pseudoinverse. The matrix $[V_0 \quad V]$ has the dimension $[p(m+r)+r] \times \ell$, where r is the number of inputs, m is the number of outputs, p are forward-in-time steps, and ℓ is the data length. For the case in which $p = 20$, $r = 1$, $m = 10$, and $\ell = 10,000$, the matrix $[V_0 \quad V]$ has the size $221 \times 10,000$. Computing the pseudo-inverse of such a large matrix becomes numerically cumbersome and very time consuming. When the matrix $[V_0 \quad V]$ is full rank, its pseudoinverse becomes

$$[V_0 \quad V]^\dagger = \begin{bmatrix} V_0^T \\ V^T \end{bmatrix} [V_0 V_0^T + V V^T]^{-1}. \tag{10.7}$$

The matrix $[V_0 V_0^T + V V^T]$ is symmetric and of dimension $[p(m+r)+r] \times [p(m+r)+r]$, regardless of how long the data length is. When $[V_0 V_0^T + V V^T]$ is not a full-rank matrix, Eq. (10.7) is still valid, with the replacement of the matrix inverse with the matrix pseudoinverse. An efficient way of computing the matrix $V V^T$ is shown in the Appendix at the end of this chapter. The matrix $V V^T$ is known as the data correlation matrix or information matrix.

There is a disadvantage to using Eq. (10.7) for computing the pseudoinverse of $[V_0 \quad V]$ in comparison with the direct application of the SVD to $[V_0 \quad V]$. The matrix computation $[V_0 V_0^T + V V^T]$ squares its condition number, i.e., its singular values. As a result, the matrix inverse $[V_0 \quad V]^\dagger$ may lose some numerical accuracy. Scaling the input and the output data such that the maximum value does not exceed unity may help reduce the numerical inaccuracy caused by the increase of the matrix condition number.

When there is an initial transient response, V_0 does not exist because the data before the starting time step $k = 1$ are not available. For tests starting with some transient response, y_0 and V_0 are deleted (see Ref. [2]). For the input and the output data time histories that include transient responses, Eq. (10.6) reduces to

$$\theta = y V^\dagger. \tag{10.8}$$

Note that this equation is also applicable for the case without transient responses.

EXAMPLE 10.2

Given a set of input data $u(k)$ and output data $y(k)$ for $k = 1, 2, \ldots, 7$,

$$\begin{bmatrix} u(1) \\ u(2) \\ u(3) \\ u(4) \\ u(5) \\ u(6) \\ u(7) \end{bmatrix} = \begin{bmatrix} -1 \\ -2 \\ -3 \\ 0 \\ 1 \\ 4 \\ 5 \end{bmatrix}, \quad \begin{bmatrix} y(1) \\ y(2) \\ y(3) \\ y(4) \\ y(5) \\ y(6) \\ y(7) \end{bmatrix} = \begin{bmatrix} -1 \\ -1.5 \\ -0.5 \\ 7 \\ 14 \\ 22.5 \\ 26.5 \end{bmatrix}.$$

Assume that the system is initially at rest and the ARX order is $p = 2$. With the data length $\ell = 7$, y_0 and y on the left-hand side of Eq. (10.4) have the form

$$y_0 = [-1 \quad -1.5],$$
$$y = [-0.5 \quad 7 \quad 14 \quad 22.5 \quad 26.5]$$

and matrices V_0, and V on the right-hand side of Eq. (10.4) have the form

$$V_0 = \begin{bmatrix} -1 & -2 \\ \begin{pmatrix} 0 \\ 0 \end{pmatrix} & \begin{pmatrix} -1 \\ -1 \end{pmatrix} \\ \begin{pmatrix} 0 \\ 0 \end{pmatrix} & \begin{pmatrix} 0 \\ 0 \end{pmatrix} \end{bmatrix},$$

$$V = \begin{bmatrix} -3 & 0 & 1 & 4 & 5 \\ \begin{pmatrix} -1.5 \\ -2 \end{pmatrix} & \begin{pmatrix} -0.5 \\ -3 \end{pmatrix} & \begin{pmatrix} 7 \\ 0 \end{pmatrix} & \begin{pmatrix} 14 \\ 1 \end{pmatrix} & \begin{pmatrix} 22.5 \\ 4 \end{pmatrix} \\ \begin{pmatrix} -1 \\ -1 \end{pmatrix} & \begin{pmatrix} -1.5 \\ -2 \end{pmatrix} & \begin{pmatrix} -0.5 \\ -3 \end{pmatrix} & \begin{pmatrix} 7 \\ 0 \end{pmatrix} & \begin{pmatrix} 14 \\ 1 \end{pmatrix} \end{bmatrix}.$$

Application of Eq. (10.6) thus gives

$$\theta = [y_0 \quad y][V_0 \quad V]^{\dagger}$$
$$= [1 \quad 2 \quad -2.5 \quad -1 \quad 0.5].$$

For the case in which the system is not initially at rest, the matrix V_0 cannot be formed because $y(-1), y(-2), \ldots, y(-\infty)$ are unknown. We can use Eq. (10.8) to obtain

$$\theta = yV^{\dagger}$$
$$= [1 \quad 2 \quad -2.5 \quad -1 \quad 0.5].$$

The result is identical. In practice, it is sufficient to use only y and V to solve for the parameter matrix θ.

Let us now rewrite Eq. (10.4) as

$$[-I \quad \theta]\begin{bmatrix} y \\ V \end{bmatrix} = 0, \tag{10.9}$$

where y_0 and V_0 are not included, because the initial condition is generally unknown. Define the following quantities:

$$\tilde{\theta} = [-I \quad \theta], \tag{10.10}$$

$$\tilde{V} = \begin{bmatrix} y \\ V \end{bmatrix} = \begin{bmatrix} v(p+1) & v(p+2) & \cdots & v(\ell) \\ v(p) & v(p-1) & \cdots & v(\ell-1) \\ \vdots & \vdots & \ddots & \vdots \\ v(1) & v(2) & \cdots & v(\ell-p) \end{bmatrix}. \tag{10.11}$$

The matrix $\tilde{V}$ is identical to V except that V has an additional $m \times (\ell - p)$ output matrix,

$$[y(p+1) \quad y(p+2) \quad \cdots \quad y(\ell)],$$

at its top. From Eqs. (10.9)–(10.11), an alternative way of solving for the parameter matrix θ defined in Eq. (10.4) can thus be formulated by

$$\tilde{\theta} \tilde{V} = 0. \tag{10.12}$$

Equation (10.12) implies that the parameter matrix $\tilde{\theta}$ is in the column null space of the matrix $\tilde{V}$. We may use the SVD to factor $\tilde{V}$ to find a set of m basis vectors that are orthogonal to the columns of $\tilde{V}$ and use them to form the parameter matrix $\tilde{\theta}$. In practice, because of system and measurement uncertainties, there may not exist any orthogonal basis vectors available for use. In this case, we may use the left singular vectors corresponding to the m smallest singular values. The smaller the singular values are, the better the identified parameter matrix should be.

EXAMPLE 10.3
For the purpose of comparison, let us use the same data as shown in Example 10.2. First, form the matrix

$$\tilde{V} = \begin{bmatrix} \begin{pmatrix} -0.5 \\ -3 \end{pmatrix} & \begin{pmatrix} 7 \\ 0 \end{pmatrix} & \begin{pmatrix} 14 \\ 1 \end{pmatrix} & \begin{pmatrix} 22.5 \\ 4 \end{pmatrix} & \begin{pmatrix} 26.5 \\ 5 \end{pmatrix} \\ \begin{pmatrix} -1.5 \\ -2 \end{pmatrix} & \begin{pmatrix} -0.5 \\ -3 \end{pmatrix} & \begin{pmatrix} 7 \\ 0 \end{pmatrix} & \begin{pmatrix} 14 \\ 1 \end{pmatrix} & \begin{pmatrix} 22.5 \\ 4 \end{pmatrix} \\ \begin{pmatrix} -1 \\ -1 \end{pmatrix} & \begin{pmatrix} -1.5 \\ -2 \end{pmatrix} & \begin{pmatrix} -0.5 \\ -3 \end{pmatrix} & \begin{pmatrix} 7 \\ 0 \end{pmatrix} & \begin{pmatrix} 14 \\ 1 \end{pmatrix} \end{bmatrix}$$

by inserting y into the first row of V. Second, use the SVD to factor $\tilde{V}$ to obtain a triple product $\tilde{V} = P \Sigma Q^T$. Next, choose the last column of P, i.e.,

$$P(:, 5) = \begin{bmatrix} 0.27217 \\ -0.27217 \\ -0.54433 \\ 0.68041 \\ 0.27217 \\ -0.13608 \end{bmatrix}.$$

Note that the column vector $P(:, 5)$ corresponds to the last singular value of Σ. It is easy to prove that $[P(:, 5)]^T \tilde{V} = 0$. Now divide the column vector $P(:, 5)$ by the negative value of the first element to yield

$$\tilde{\theta}^T = \frac{-1}{0.27217} \begin{bmatrix} 0.27217 \\ -0.27217 \\ -0.54433 \\ 0.68041 \\ 0.27217 \\ -0.13608 \end{bmatrix} = \begin{bmatrix} -1 \\ 1 \\ 2 \\ -2.5 \\ -1 \\ 0.5 \end{bmatrix}.$$

The last five elements of $\tilde{\theta}$ consists of the parameter vector θ, i.e.,

$$\theta = [1 \quad 2 \quad -2.5 \quad -1 \quad 0.5].$$

It is worth emphasizing that this result is identical to the result from Example 10.2.

10.3 Multistep Output Prediction

There are many different ways of rewriting the finite-difference model. One of the models that is very useful for the control design is the output-prediction model. The basic idea rests simply on the relationship between the current output and the previous input and output signals [see Eq. (10.2)]. Given a system with r inputs and m outputs, the finite-difference equation for the $r \times 1$ input $u(k+1)$ and the $m \times 1$ output $y(k+1)$ at time $k+1$ is

$$y(k+1) = -\alpha_1 y(k) - \alpha_2 y(k-1) - \cdots - \alpha_p y(k-p+1)$$
$$+ \beta_0 u(k+1) + \beta_1 u(k) + \cdots + \beta_p u(k-p+1). \tag{10.13}$$

This is identical to Eq. (10.2) except that the time index has been shifted one step ahead. Define the following quantities:

$$\alpha_1^{(1)} = \alpha_2 - \alpha_1 \alpha_1,$$

$$\alpha_2^{(1)} = \alpha_3 - \alpha_1 \alpha_2,$$

$$\vdots \quad \vdots \quad \vdots$$

$$\alpha_{p-1}^{(1)} = \alpha - \alpha_1 \alpha_{p-1},$$

$$\alpha_p^{(1)} = -\alpha_1 \alpha_p; \tag{10.14}$$

$$\beta_1^{(1)} = \beta_2 - \alpha_1 \beta_1,$$

$$\beta_2^{(1)} = \beta_3 - \alpha_1 \beta_2,$$

$$\vdots \quad \vdots \quad \vdots$$

$$\beta_{p-1}^{(1)} = \beta_p - \alpha_1 \beta_{p-1},$$

$$\beta_p^{(1)} = -\alpha_1 \beta_p; \tag{10.15}$$

$$\beta_0^{(1)} = \beta_1 - \alpha_1 \beta_0. \tag{10.16}$$

Substituting $y(k)$ from Eq. (10.2) into Eq. (10.13) yields

$$y(k+1) = -\alpha_1^{(1)} y(k-1) - \alpha_2^{(1)} y(k-2) - \cdots - \alpha_p^{(1)} y(k-p)$$
$$+ \beta_0 u(k+1) + \beta_0^{(1)} u(k)$$
$$+ \beta_1^{(1)} u(k-1) + \beta_2^{(1)} u(k-2) + \cdots + \beta_p^{(1)} u(k-p). \tag{10.17}$$

The output measurement at time step $k+1$ can be expressed as the weighted sum of current and future input signals and past input and output data without the output measurement at time step k. Note that the future input signals are given a priori.

EXAMPLE 10.4

Consider the finite-difference model shown in Example 10.1:

$$y(k) = -\frac{3}{2} y(k-1) - \frac{5}{2} y(k-2) - \frac{3}{2} y(k-3) - y(k-4)$$
$$+ u(k) - \frac{7}{6} u(k-1) + u(k-2) - \frac{4}{3} u(k-3) + \frac{1}{2} u(k-4).$$

By shifting one time step, the model becomes

$$y(k+1) = -\frac{3}{2} y(k) - \frac{5}{2} y(k-1) - \frac{3}{2} y(k-2) - y(k-3)$$
$$+ u(k+1) - \frac{7}{6} u(k) + u(k-1) - \frac{4}{3} u(k-2) + \frac{1}{2} u(k-3).$$

Inserting $y(k)$ from the first equation in this example into the second equation thus yields

$$y(k+1) = -\frac{1}{4} y(k-1) + \frac{9}{4} y(k-2) + \frac{5}{4} y(k-3) + \frac{3}{2} y(k-4)$$
$$+ u(k+1) - \frac{16}{6} u(k)$$
$$+ \frac{11}{4} u(k-1) - \frac{17}{6} u(k-2) + \frac{5}{2} u(k-3) - \frac{3}{4} u(k-4).$$

If the current output $y(k)$ is not known, the future output $y(k+1)$ can be predicted from the past outputs $y(k-1), y(k-2), y(k-3), y(k-4)$ and the past inputs $u(k-1), u(k-2), u(k-3), u(k-4)$, once the current input $u(k)$ and the future input $u(k+1)$ are given.

If the process that derives Eq. (10.17) is applied several times, we may express the output measurement at the time step $k+j$ in terms of the past outputs $y(k-1)$,

$y(k-1), \ldots, y(k-p)$ and the inputs $u(k+j), u(k+j-1), \ldots, u(k-p)$ by

$$y(k+j) = -\alpha_1^{(j)} y(k-1) - \alpha_2^{(j)} y(k-2) - \cdots - \alpha_p^{(j)} y(k-p)$$
$$+ \beta_0 u(k+j) + \beta_0^{(1)} u(k+j-1) + \cdots + \beta_0^{(j)} u(k)$$
$$+ \beta_1^{(j)} u(k-1) + \beta_2^{(j)} u(k-2) + \cdots + \beta_p^{(j)} u(k-p), \tag{10.18}$$

where

$$\alpha_1^{(j)} = \alpha_2^{(j-1)} - \alpha_1^{(j-1)} \alpha_1,$$
$$\alpha_2^{(j)} = \alpha_3^{(j-1)} - \alpha_1^{(j-1)} \alpha_2,$$
$$\vdots \quad \vdots \quad \vdots$$
$$\alpha_{p-1}^{(j)} = \alpha_p^{(j-1)} - \alpha_1^{(j-1)} \alpha_{p-1},$$
$$\alpha_p^{(j)} = -\alpha_1^{(j-1)} \alpha_p; \tag{10.19}$$

$$\beta_1^{(j)} = \beta_2^{(j-1)} - \alpha_1^{(j-1)} \beta_1,$$
$$\beta_2^{(j)} = \beta_3^{(j-1)} - \alpha_1^{(j-1)} \beta_2$$
$$\vdots \quad \vdots \quad \vdots$$
$$\beta_{p-1}^{(j)} = \beta_p^{(j-1)} - \alpha_1^{(j-1)} \beta_{p-1},$$
$$\beta_p^{(j)} = -\alpha_1^{(j-1)} \beta_p; \tag{10.20}$$

$$\beta_0^{(j)} = \beta_1^{(j-1)} - \alpha_1^{(j-1)} \beta_0. \tag{10.21}$$

Note that $\beta_0^{(0)} = \beta_0$, and $\alpha_i^{(0)} = \alpha_i$ and $\beta_i^{(0)} = \beta_i$ for $i = 1, 2, \ldots, p$. After some algebraic operation, Eq. (10.21) can also be expressed by

$$\beta_0^{(0)} = \beta_0,$$
$$\beta_0^{(k)} = \beta_k - \sum_{i=1}^{k} \alpha_i \beta_0^{(k-i)} \quad \text{for} \quad k = 1, \ldots, p,$$
$$\beta_0^{(k)} = -\sum_{i=1}^{p} \alpha_i \beta_0^{(k-i)} \quad \text{for} \quad k = p+1, \ldots, \infty. \tag{10.22}$$

Similar to Eqs. (10.22), $\alpha_1^{(j)} = \alpha_1^{(j-1)} \alpha_1 + \alpha_2^{(j-1)}$ can also be written as

$$\alpha_1^{(0)} = \alpha_1,$$
$$\alpha_1^{(k)} = \alpha_{k+1} - \sum_{i=1}^{k} \alpha_i \alpha_1^{(k-i)}, \quad k = 1, \ldots, p-1,$$
$$\alpha_1^{(k)} = -\sum_{i=1}^{p} \alpha_i \alpha_1^{(k-i)}, \quad k = p, \ldots, \infty. \tag{10.23}$$

Close observation of Eqs. (10.22) and (10.23) reveals that $\beta_0^{(j)}$ and $\alpha_1^{(j)}$ for $j > p$ are a linear combination of their respective past p parameters weighted by the parameters $\alpha_1, \alpha_2, \ldots, \alpha_p$. This property is very useful in developing predictive control designs, which will be discussed in Chapter 11. The quantities $\beta_0^{(i)}$ $(i = 0, 1, \ldots)$ are, in fact, the pulse-response sequence, which will be shown later. On the other hand, the quantities $\alpha_1^{(i)}$ $(i = 0, 1, \ldots)$ are the observer gain Markov parameters, which can be used to compute an observer for state estimation.

EXAMPLE 10.5

Consider the same example as that given in Example 10.4,

$$y(k) = -\frac{3}{2}y(k-1) - \frac{5}{2}y(k-2) - \frac{3}{2}y(k-3) - y(k-4)$$
$$+ u(k) - \frac{7}{6}u(k-1) + u(k-2) - \frac{4}{3}u(k-3) + \frac{1}{2}u(k-4),$$

which gives

$$\alpha_1^{(0)} = \alpha_1 = \frac{3}{2},$$

$$\alpha_2^{(0)} = \alpha_2 = \frac{5}{2},$$

$$\alpha_3^{(0)} = \alpha_3 = \frac{3}{2},$$

$$\alpha_4^{(0)} = \alpha_4 = 1;$$

$$\beta_0^{(0)} = \beta_0 = 1,$$

$$\beta_1^{(0)} = \beta_1 = -\frac{7}{6},$$

$$\beta_2^{(0)} = \beta_2 = 1,$$

$$\beta_3^{(0)} = \beta_3 = -\frac{4}{3},$$

$$\beta_4^{(0)} = \beta_4 = \frac{1}{2}.$$

Application of Eqs. (10.19) for $j = 1$ yields

$$\alpha_1^{(1)} = \alpha_2^{(0)} - \alpha_1^{(0)}\alpha_1 = \frac{5}{2} - \left(\frac{3}{2}\right)^2 = \frac{1}{4},$$

$$\alpha_2^{(1)} = \alpha_3^{(0)} - \alpha_1^{(0)}\alpha_2 = \frac{3}{2} - \left(\frac{3}{2}\right)\left(\frac{5}{2}\right) = -\frac{9}{4},$$

$$\alpha_3^{(1)} = \alpha_4^{(0)} - \alpha_1^{(0)}\alpha_3 = 1 - \left(\frac{3}{2}\right)\left(\frac{3}{2}\right) = -\frac{5}{4},$$

$$\alpha_4^{(1)} = -\alpha_1^{(0)}\alpha_4 = -\frac{3}{2}.$$

From Eqs. (10.20), we obtain

$$\beta_1^{(1)} = \beta_2^{(0)} - \alpha_1^{(0)}\beta_1 = 1 - \left(\frac{3}{2}\right)\left(-\frac{7}{6}\right) = \frac{11}{4},$$

$$\beta_2^{(1)} = \beta_3^{(0)} - \alpha_1^{(0)}\beta_2 = -\frac{4}{3} - \left(\frac{3}{2}\right) = -\frac{17}{6},$$

$$\beta_3^{(1)} = \beta_4^{(0)} - \alpha_1^{(0)}\beta_3 = \frac{1}{2} - \left(\frac{3}{2}\right)\left(-\frac{4}{3}\right) = \frac{5}{2},$$

$$\beta_4^{(1)} = -\alpha_1^{(0)}\beta_4 = -\left(\frac{3}{2}\right)\left(\frac{1}{2}\right) = -\frac{3}{4},$$

and, similarly, from Eq. (10.21),

$$\beta_0^{(1)} = \beta_1^{(0)} - \alpha_1^{(0)}\beta_0 = -\left(\frac{7}{6}\right) - \left(\frac{3}{2}\right) = -\left(\frac{8}{3}\right).$$

The output $y(k+1)$ can then be expressed by (see Example 10.4)

$$y(k+1) = -\frac{1}{4}y(k-1) + \frac{9}{4}y(k-2) + \frac{5}{4}y(k-3) + \frac{3}{2}y(k-4)$$

$$+ u(k+1) - \frac{8}{3}u(k) + \frac{11}{4}u(k-1) - \frac{17}{6}u(k-2)$$

$$+ \frac{5}{2}u(k-3) - \frac{3}{4}u(k-4).$$

Repeat the above process for $j = 2$ from Eqs. (10.19) to produce

$$\alpha_1^{(2)} = \alpha_2^{(1)} - \alpha_1^{(1)}\alpha_1 = -\frac{9}{4} - \left(\frac{1}{4}\right)\left(\frac{3}{2}\right) = -\frac{21}{8},$$

$$\alpha_2^{(2)} = \alpha_3^{(1)} - \alpha_1^{(1)}\alpha_2 = -\frac{5}{4} - \left(\frac{1}{4}\right)\left(\frac{5}{2}\right) = -\frac{15}{8},$$

$$\alpha_3^{(2)} = \alpha_4^{(1)} - \alpha_1^{(1)}\alpha_3 = -\frac{3}{2} - \left(\frac{1}{4}\right)\left(\frac{3}{2}\right) = -\frac{15}{8},$$

$$\alpha_4^{(2)} = -\alpha_1^{(1)}\alpha_4 = -\frac{1}{4};$$

$$\beta_1^{(2)} = \beta_2^{(1)} - \alpha_1^{(1)}\beta_1 = -\left(\frac{17}{6}\right) - \left(\frac{1}{4}\right)\left(-\frac{7}{6}\right) = -\frac{61}{24},$$

$$\beta_2^{(2)} = \beta_3^{(1)} - \alpha_1^{(1)}\beta_2 = \frac{5}{2} - \frac{1}{4} = \frac{9}{4},$$

$$\beta_3^{(2)} = \beta_4^{(1)} - \alpha_1^{(1)}\beta_3 = -\frac{3}{4} - \left(\frac{1}{4}\right)\left(-\frac{4}{3}\right) = -\frac{5}{12},$$

$$\beta_4^{(2)} = -\alpha_1^{(1)}\beta_4 = -\left(\frac{1}{4}\right)\left(\frac{1}{2}\right) = -\frac{1}{8};$$

$$\beta_0^{(2)} = \beta_1^{(1)} - \alpha_1^{(1)}\beta_0 = \left(\frac{11}{4}\right) - \left(\frac{1}{4}\right) = \left(\frac{5}{2}\right).$$

The output $y(k+2)$ has the following expression:

$$y(k+2) = \frac{21}{8}y(k-1) + \frac{15}{8}y(k-2) + \frac{15}{8}y(k-3) + \frac{1}{4}y(k-4)$$

$$+ u(k+2) - \frac{8}{3}u(k+1) + \frac{5}{2}u(k)$$

$$- \frac{61}{24}u(k-1) + \frac{9}{4}u(k-2) - \frac{5}{12}u(k-3) - \frac{1}{8}u(k-4).$$

Without the output measurements $y(k)$ and $y(k+1)$, the future output $y(k+2)$ can be computed from the past outputs $y(k-1)$, $y(k-2)$, $y(k-3)$, and $y(k-4)$ and the past inputs $u(k-1)$, $u(k-2)$, $u(k-3)$, and $u(k-4)$ if the current input $u(k)$ and the future inputs $u(k+1)$ and $u(k+2)$ are given a priori. This property can be used to shape the future input to obtain the desired output.

Let the index j be $j = 1, 2, \ldots, s-1$. Equation (10.18) produces the following matrix equation:

$$y_s(k) = \mathcal{T}u_s(k) + \mathcal{B}u_p(k-p) - \mathcal{A}y_p(k-p), \tag{10.24}$$

where constant matrices $\mathcal{T}$, $\mathcal{B}$, and $\mathcal{A}$ are computed by

$$\mathcal{T} = \begin{bmatrix} \beta_0 & & & \\ \beta_0^{(1)} & \beta_0 & & \\ \vdots & \vdots & \ddots & \\ \beta_0^{(s-1)} & \beta_0^{(s-2)} & \cdots & \beta_0 \end{bmatrix}, \tag{10.25a}$$

$$\mathcal{B} = \begin{bmatrix} \beta_p & \beta_{p-1} & \cdots & \beta_1 \\ \beta_p^{(1)} & \beta_{p-1}^{(1)} & \cdots & \beta_1^{(1)} \\ \vdots & \vdots & \ddots & \vdots \\ \beta_p^{(s-1)} & \beta_{p-1}^{(s-1)} & \cdots & \beta_1^{(s-1)} \end{bmatrix}, \tag{10.25b}$$

$$\mathcal{A} = \begin{bmatrix} \alpha_p & \alpha_{p-1} & \cdots & \alpha_1 \\ \alpha_p^{(1)} & \alpha_{p-1}^{(1)} & \cdots & \alpha_1^{(1)} \\ \vdots & \vdots & \ddots & \vdots \\ \alpha_p^{(s-1)} & \alpha_{p-1}^{(s-1)} & \cdots & \alpha_1^{(s-1)} \end{bmatrix}, \tag{10.25c}$$

$$y_s(k) = \begin{bmatrix} y(k) \\ y(k+1) \\ \vdots \\ y(k+s-1) \end{bmatrix}, \tag{10.26a}$$

$$u_s(k) = \begin{bmatrix} u(k) \\ u(k+1) \\ \vdots \\ u(k+s-1) \end{bmatrix}. \tag{10.26b}$$

Similar to $y_s(k)$ and $u_s(k)$, the vectors $y_p(k-p)$ and $u_p(k-p)$ are defined by the replacement of s with p and k with $k-p$. The quantity $y_s(k)$ represents the output vector with a total of s points for each sensor from the time step k to $k+s-1$, whereas $y_p(k-p)$ includes the p data from $k-p$ to $k-1$. Similarly, $u_s(k)$ has s input points starting from the time step k and $u_p(k-p)$ has p input data points from $k-p$. A matrix of the form shown in Eq. (10.25a) is commonly called a Toeplitz matrix. The entries of $\mathcal{T}$ are constant down the diagonals parallel to the main diagonal.

The explicit expression of the multi-step output prediction equation is shown in Fig. 10.3, where q is an arbitrary integer and $q \le s-1$. Figure 10.3 is given to show the interrelationships among the parameter matrices. The quantities $\alpha^{(q)}_{p-1}$, $\beta^{(q)}_{p-1}$, and

Multistep Output Prediction Equation

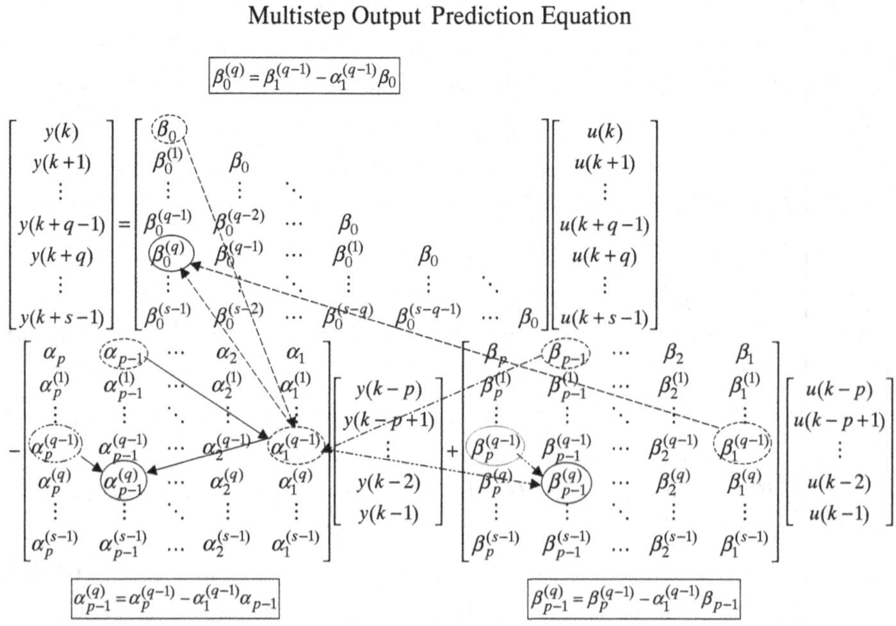

Figure 10.3. Explicit expression of the multistep output-prediction equation.

$\beta_0^{(q)}$ can be calculated with Eqs. (10.19), (10.20), and (10.21), respectively:

$$\alpha_{p-1}^{(q)} = \alpha_p^{(q-1)} - \alpha_1^{(q-1)}\alpha_{p-1},$$

$$\beta_{p-1}^{(q)} = \beta_p^{(q-1)} - \alpha_1^{(q-1)}\beta_{p-1},$$

$$\beta_0^{(q)} = \beta_1^{(q-1)} - \alpha_1^{(q-1)}\beta_0.$$

The solid lines in Fig. (10.3) show how the quantity $\alpha_{p-1}^{(q)}$ at the current step q is recursively calculated with the quantities $\alpha_p^{(q-1)}$, $\alpha_1^{(q-1)}$, and α_{p-1} from the previous step $q-1$ and the initial step. The dashed lines give the path of how $\beta_{p-1}^{(q)}$ at the current step q can be recursively determined from $\beta_p^{(q-1)}$, $\alpha_1^{(q-1)}$, and β_{p-1} at the previous step $q-1$ and the initial step. The quantity $\beta_0^{(q)}$ is computed similarly to $\beta_{p-1}^{(q)}$.

The expression for $y_s(k)$ in Eq. (10.24) consists of three terms. The first term is the input vector $u_s(k)$, including inputs from time step k to $k+s-1$. Relative to the same time k, the second and third terms, $u_p(k-p)$ and $y_p(k-p)$, are input and output vectors, respectively, with past known quantities from $k-p$ to $k-1$. For feedback-controller designs, the input vector $u_s(k)$ is to be determined to generate the feedback signal if $y_s(k)$ is specified.

EXAMPLE 10.6

Consider the same example as that given in Example 10.5:

$$y(k) = -\frac{3}{2}y(k-1) - \frac{5}{2}y(k-2) - \frac{3}{2}y(k-3) - y(k-4)$$

$$+ u(k) - \frac{7}{6}u(k-1) + u(k-2) - \frac{4}{3}u(k-3) + \frac{1}{2}u(k-4).$$

Let $s = 4$ and $p = 4$, shown as subscripts in Eq. (10.24). Equation (10.24) becomes

$$y_4(k) = \mathcal{T}u_4(k) + \mathcal{B}u_4(k-4) - \mathcal{A}y_4(k-4),$$

where

$$\mathcal{T} = \begin{bmatrix} \beta_0 & 0 & 0 & 0 \\ \beta_0^{(1)} & \beta_0 & 0 & 0 \\ \beta_0^{(2)} & \beta_0^{(1)} & \beta_0 & 0 \\ \beta_0^{(3)} & \beta_0^{(2)} & \beta_0^{(1)} & \beta_0 \end{bmatrix} = \begin{bmatrix} 1 & 0 & 0 & 0 \\ -\frac{8}{3} & 1 & 0 & 0 \\ \frac{5}{2} & -\frac{8}{3} & 1 & 0 \\ \frac{1}{12} & \frac{5}{2} & -\frac{8}{3} & 1 \end{bmatrix},$$

$$\mathcal{B} = \begin{bmatrix} \beta_4 & \beta_3 & \beta_2 & \beta_1 \\ \beta_4^{(1)} & \beta_3^{(1)} & \beta_2^{(1)} & \beta_1^{(1)} \\ \beta_4^{(2)} & \beta_3^{(2)} & \beta_2^{(2)} & \beta_1^{(2)} \\ \beta_4^{(3)} & \beta_3^{(3)} & \beta_2^{(3)} & \beta_1^{(3)} \end{bmatrix} = \begin{bmatrix} \frac{1}{2} & -\frac{4}{3} & 1 & -\frac{7}{6} \\ -\frac{3}{4} & \frac{5}{2} & -\frac{17}{6} & \frac{11}{4} \\ -\frac{1}{8} & -\frac{5}{12} & \frac{9}{4} & -\frac{61}{24} \\ \frac{21}{16} & -\frac{29}{8} & \frac{53}{24} & -\frac{13}{16} \end{bmatrix},$$

$$
A = \begin{bmatrix} \alpha_4 & \alpha_3 & \alpha_2 & \alpha_1 \\ \alpha_4^{(1)} & \alpha_3^{(1)} & \alpha_2^{(1)} & \alpha_1^{(1)} \\ \alpha_4^{(2)} & \alpha_3^{(2)} & \alpha_2^{(2)} & \alpha_1^{(2)} \\ \alpha_4^{(3)} & \alpha_3^{(3)} & \alpha_2^{(3)} & \alpha_1^{(3)} \end{bmatrix} = \begin{bmatrix} 1 & \frac{3}{2} & \frac{5}{2} & \frac{3}{2} \\ -\frac{3}{2} & -\frac{5}{4} & -\frac{9}{4} & \frac{1}{4} \\ -\frac{1}{4} & -\frac{15}{8} & -\frac{15}{8} & -\frac{21}{8} \\ \frac{21}{8} & \frac{59}{16} & \frac{75}{16} & \frac{33}{16} \end{bmatrix},
$$

$$
y_4(k) = \begin{bmatrix} y(k) \\ y(k+1) \\ y(k+2) \\ y(k+3) \end{bmatrix}, \qquad u_4(k) = \begin{bmatrix} u(k) \\ u(k+1) \\ u(k+2) \\ u(k+3) \end{bmatrix},
$$

$$
y_4(k-4) = \begin{bmatrix} y(k-4) \\ y(k-3) \\ y(k-2) \\ y(k-1) \end{bmatrix}, \qquad u_4(k-4) = \begin{bmatrix} u(k-4) \\ u(k-3) \\ u(k-2) \\ u(k-1) \end{bmatrix}.
$$

Note that the first column of $\mathcal{T}$ is the output of $y_4(k)$ by setting $u(k) = 1$, and $u(k - 1) = u(k - 2) = u(k - 3) = u(k - 4) = 0$ and $y(k - 1) = y(k - 2) = y(k - 3) = y(k - 4) = 0$. In other words, the system initially at rest is given a pulse for the input $u(k)$ at time k, i.e., $u(k) = 1$, and all past and future inputs are zero. The Toeplitz matrix $\mathcal{T}$ can be easily formed once its first column is computed.

10.4 System Markov Parameters

There are many types of inputs that are particularly important in theory as well as in practice. Among them, the pulse input is frequently used to generate a sequence of pulse responses for modal parameter identification, including system frequencies, damping, and mode shapes.

To observe the response to a pulse for a system with the input–output relationship described by Eq. (10.24), let $u(k) = 1$, $u(k - 1) = u(k - 2) =, \ldots, = u(k - p) = 0$ and $u(k + 1) = u(k + 2) =, \ldots, = u(k + s - 1) = 0$, i.e.,

$$
\begin{bmatrix} u(k-p) \\ \vdots \\ u(k-1) \\ u(k) \\ u(k+1) \\ \vdots \\ u(k+s-1) \end{bmatrix} = \begin{bmatrix} 0 \\ \vdots \\ 0 \\ 1 \\ 0 \\ \vdots \\ 0 \end{bmatrix}, \tag{10.27}
$$

where the dummy time indices k and s are arbitrary integers, but p is a prespecified integer. The integer p must be larger than or equal to a specific number related to the order of the system and the number of outputs, which will be addressed later. The conditions at the time index $k = 0$ are commonly referred to as the initial conditions. Here, for simplicity, the system is assumed to have a single input only. Assume that the system at time k is initially at rest in the sense that

$$\begin{bmatrix} y(k-p) \\ \vdots \\ y(k-2) \\ y(k-1) \end{bmatrix} = \begin{bmatrix} 0 \\ \vdots \\ 0 \\ 0 \end{bmatrix}. \tag{10.28}$$

Substituting the pulse input [Eq. (10.27)] and the initial condition [Eq. (10.28)] into Eq. (10.24) produces

$$\begin{bmatrix} y(k) \\ y(k+1) \\ \vdots \\ y(k+s-1) \end{bmatrix} = \begin{bmatrix} \beta_0 \\ \beta_0^{(1)} \\ \vdots \\ \beta_0^{(s-1)} \end{bmatrix}. \tag{10.29}$$

Equation (10.29) shows that the parameters β_0, $\beta_0^{(1)}, \ldots,$ and $\beta_0^{(s-1)}$ are nothing but the pulse-response of a system. The matrix $\mathcal{T}$ shown in Eq. (10.24) that is formed from the parameters, β_0, $\beta_0^{(1)}, \ldots,$ and $\beta_0^{(s-1)}$ (the pulse-response sequence) may thus be called the Toeplitz pulse-response matrix.

Equation (10.29) is true for a unit pulse applied to any input. For the system with multiple inputs, we apply to the system a unit pulse one at a time for each input at the time step k; the results can then be assembled into a sequence of pulse-response matrices β_0, $\beta_0^{(1)}, \ldots,$ and $\beta_0^{(s-1)}$. Each pulse-response matrix has a dimension of $m \times r$, where m is the number of outputs and r is the number of inputs.

What is the significance of these pulse-response matrices β_0, $\beta_0^{(1)}, \ldots, \beta_0^{(s-1)}$? To answer this question, let us repeat here the discrete-time state–space model:

$$x(k+1) = Ax(k) + Bu(k), \tag{10.30a}$$

$$y(k) = Cx(k) + Du(k). \tag{10.30b}$$

Solving for the output $y(k)$ with zero initial condition $x(k) = 0$ and the unit pulse $u(k) = 1$ shown in Eq. (10.27) yields

$$x(k) = 0 \quad \Longrightarrow \quad y(k) = D,$$
$$x(k+1) = B \quad \Longrightarrow \quad y(k+1) = CB,$$
$$x(k+2) = AB \quad \Longrightarrow \quad y(k+2) = CAB$$

$$\vdots \qquad\qquad \vdots \qquad\qquad \vdots$$

$$x(k+s-1) = A^{s-2}B \Longrightarrow y(k+s-1) = CA^{s-2}B. \tag{10.31}$$

Equations (10.31) are true for any single input, i.e., any column of the input matrix B. For the system with multiple inputs, we apply to the system a unit pulse one at a time for each column of B at the time step k; the results can then be assembled into a sequence of pulse-response matrices D, CB, CAB, ..., and $CA^{(s-2)}B$. The dimension of each pulse-response matrix is $m \times r$, where m is the number of outputs and r is the number of inputs.

If Eqs. (10.24) and (10.30) represent the same system, their pulse responses from Eqs. (10.29) and (10.31) should be identical, i.e.,

$$\beta_0 = D,$$
$$\beta_0^{(1)} = CB,$$
$$\beta_0^{(2)} = CAB,$$

$$\vdots$$

$$\beta_0^{(s-1)} = CA^{s-2}B. \tag{10.32}$$

The constant matrices

$$D, \ CB, \ CAB, \ldots, \ CA^{s-2}B$$

are referred to as system Markov parameters or, in short, Markov parameters. Markov parameters can be obtained from time-domain experimental data by use of Eq. (10.8) in conjunction with Eqs. (10.21) or (10.22). On the other hand, we may use the frequency-response function, as shown in Chapter 5 of the system identification book (Ref. [2]) to compute Markov parameters.

The Markov parameters are commonly used as the basis for identifying mathematical models for linear dynamical systems. Indeed, it is obvious from Eqs. (10.32) that the matrices A, B, C, and D are embedded in the Markov parameter sequence. Because the Markov parameter sequence is simply the pulse response of the system, it must be unique for a given system. We may show this by noting that any coordinate transformation of the state vector, say, $z(k) = Tx(k)$, yields the same Markov parameters, i.e.,

$$\beta_0^{(k)} = CT^{-1}(TAT^{-1})^{k-1}TB = CA^{k-1}B, \quad k = 1, 2, \ldots.$$

There are an infinite number of coordinate transformation matrices that produce the same Markov parameters.

EXAMPLE 10.7

Assume that the following state–space model can represent the system under test:

$$x(k+1) = Ax(k) + Bu(k),$$

$$y(k) = Cx(k) + Du(k),$$

where

$$A = \begin{bmatrix} 0 & 0 & 0 & -1 \\ 1 & 0 & 0 & -\frac{3}{2} \\ 0 & 1 & 0 & -\frac{5}{2} \\ 0 & 0 & 1 & -\frac{3}{2} \end{bmatrix},$$

$$B = \begin{bmatrix} -\frac{1}{2} \\ -\frac{17}{6} \\ -\frac{3}{2} \\ -\frac{8}{3} \end{bmatrix},$$

$$C = [0 \quad 0 \quad 0 \quad 1],$$

$$D = 1.$$

Given a pulse at time $k = 1$, i.e., $u(1) = 1$ and $u(k) = 0$ for $k > 1$, with the assumption that the initial condition $x(1) = [0 \ 0 \ 0 \ 0]^T$, the output response is

$$x(1) = \begin{bmatrix} 0 \\ 0 \\ 0 \\ 0 \end{bmatrix} \implies y(1) = D = 1,$$

$$x(2) = B = \begin{bmatrix} -\frac{1}{2} \\ -\frac{17}{6} \\ -\frac{3}{2} \\ -\frac{8}{3} \end{bmatrix} \implies y(2) = CB = -\frac{8}{3},$$

$$x(3) = AB = \begin{bmatrix} \frac{8}{3} \\ \frac{7}{2} \\ \frac{23}{6} \\ \frac{5}{2} \end{bmatrix} \implies y(3) = CAB = \frac{5}{2},$$

$$x(4) = A^2 B = \begin{bmatrix} -\frac{5}{2} \\ -\frac{13}{12} \\ -\frac{11}{4} \\ \frac{1}{12} \end{bmatrix} \implies y(4) = CA^2 B = \frac{1}{12}.$$

The pulse response is identical to that shown in Example 10.6, i.e., the first column of the Toeplitz matrix $\mathcal{T}$. It means that the state–space model used in this example and the finite-difference model used in Example 10.6 give the same input–output map.

10.5 Eigensystem Realization Algorithm

Given pulse-response histories (system Markov parameters), there are several methods available to compute a state–space model. In this section, a method commonly used in the control and modal testing of flexible structures is described, namely the eigensystem realization algorithm (ERA). The optimal nature and characteristics of the ERA-identified system model will be discussed.

Let us denote the Markov parameters as follows:

$$
\begin{aligned}
Y_0 &\overset{\text{def}}{=} D = \beta_0, \\
Y_1 &\overset{\text{def}}{=} CB = \beta_0^{(1)}, \\
Y_2 &\overset{\text{def}}{=} CAB = \beta_0^{(2)}, \\
Y_3 &\overset{\text{def}}{=} CA^2B = \beta_0^{(3)} \\
&\vdots \\
Y_k &\overset{\text{def}}{=} CA^{k-1}B = \beta_0^{(k)}.
\end{aligned}
\tag{10.33}
$$

The new definition is given to stress that the system Markov parameters can be obtained by several means, such as application of a pulse input to the system under test, conversion of frequency-response functions, etc. We may also first compute the finite-difference model, Eqs. (10.1), and then use the recursive formula, Eqs. (10.21) or (10.22), to obtain $Y_k = \beta_0^{(k)}$ for $k = 1, 2, \ldots, \infty$. From the Markov parameters, the embedded system matrices A, B, C, and D can be extracted.

The system realization begins by forming the generalized $pm \times \gamma r$ Hankel matrix (see Ref. [2]) composed of the Markov parameters from Eqs. (10.33):

$$
H(0) = \begin{bmatrix}
Y_1 & Y_2 & \cdots & Y_\gamma \\
Y_2 & Y_3 & \cdots & Y_{\gamma+1} \\
\vdots & \vdots & \ddots & \vdots \\
Y_p & Y_{p+1} & \cdots & Y_{p+\gamma-1}
\end{bmatrix},
\tag{10.34}
$$

where p and γ are integers such that $\gamma r \geq pm$. Recall again that m is the number of outputs and r is the number of inputs. Note that $Y_0 = D$ is not included in $H(0)$. The Hankel matrix is also known as the block data matrix.

The matrix $H(0)$ is of rank n (the order of the system) if $pm \geq n$. To confirm this point, substituting the system Markov parameters from Eqs. (10.33) into Eq. (10.34) and decomposing $H(0)$ into two matrices yield

$$
H(0) = \begin{bmatrix}
CB & CAB & \cdots & CA^{\gamma-1}B \\
CAB & CA^2B & \cdots & CA^\gamma B \\
\vdots & \vdots & \ddots & \vdots \\
CA^{p-1}B & CA^pB & \cdots & CA^{p+\gamma-2}B
\end{bmatrix} = \mathcal{P}_p \mathcal{Q}_\gamma,
\tag{10.35}
$$

where $\mathcal{P}_p$ and $\mathcal{Q}_\gamma$ are

$$\mathcal{P}_p = \begin{bmatrix} C \\ CA \\ CA^2 \\ \vdots \\ CA^{p-1} \end{bmatrix}, \tag{10.36a}$$

$$\mathcal{Q}_\gamma = [B \quad AB \quad A^2B \quad \cdots \quad A^{\gamma-1}B]. \tag{10.36b}$$

The block matrix $\mathcal{P}_p$ is the observability matrix, whereas the block matrix $\mathcal{Q}_\gamma$ is the controllability matrix. If the order of the system is n, then the minimum dimension of the state matrix is $n \times n$. If the system is controllable and observable, the block matrices $\mathcal{P}_p$ and $\mathcal{Q}_\gamma$ are of rank n. Therefore, the Hankel matrix $H(0)$ is of rank n by Eq. (10.35). *It becomes trivial to note that the maximum order of the identified system is pm.*

Based on the properties of the Hankel matrix composed of the Markov parameters (pulse-response samples), the ERA is used to compute the three matrices A, B, and C. The ERA process starts with the factorization of the Hankel matrix, defined in Eq. (10.34), by use of SVD:

$$H(0) = R \, \Sigma \, S^T, \tag{10.37}$$

where the columns of matrices R and S are orthonormal and Σ is a rectangular matrix,

$$\Sigma = \begin{bmatrix} \Sigma_n & 0 \\ 0 & 0 \end{bmatrix},$$

in which 0's are zero matrices with appropriate dimensions, and

$$\Sigma_n = \text{diag}[\sigma_1, \quad \sigma_2, \cdots, \sigma_i, \quad \sigma_{i+1}, \cdots, \sigma_n]$$

and monotonically nonincreasing σ_i $(i = 1, 2, \ldots, n)$,

$$\sigma_1 \geq \sigma_2 \geq \cdots \geq \sigma_i \geq \sigma_{i+1} \geq \cdots \sigma_n \geq 0.$$

Next, let R_n and S_n be the matrices formed by the first n columns of R and S, respectively. Hence the matrix $H(0)$ becomes

$$H(0) = R_n \Sigma_n S_n^T, \quad \text{where} \quad R_n^T R_n = I_n = S_n^T S_n. \tag{10.38}$$

Because of measurement noise, nonlinearity, and computer round-off, the Hankel $H(0)$ will usually be of full rank, which does not, in general, equal the true order of the system under test. It should not be the aim to obtain a system realization that exactly reproduces the noisy sequence of data. A realization that produces a smoothed version of the sequence and that closely represents the underlying linear dynamics of the system is more desirable.

If Eqs. (10.35) and (10.38) are examined as a whole, we can write the equality

$$H(0) = \left[R_n \Sigma_n^{1/2} \right] \left[\Sigma_n^{1/2} S_n^T \right] \cong \mathcal{P}_p \mathcal{Q}_\gamma, \tag{10.39}$$

where the approximation sign $\cong$ is used because noise and truncation of nonzero small singular values. One interpretation from Eq. (10.39) is that $\mathcal{P}_p$ is related to R_n and $\mathcal{Q}_\gamma$ is related to S_n^T. Indeed, one possible choice is $\mathcal{P}_p = R_n \Sigma_n^{1/2}$ and $\mathcal{Q}_\gamma = \Sigma_n^{1/2} S_n^T$. This choice appears to make both $\mathcal{P}_p$ and $\mathcal{Q}_\gamma$ balanced in the sense that

$$\mathcal{P}_p^T \mathcal{P}_p = \Sigma_n^{1/2} R_n^T R_n \Sigma_n^{1/2} = \Sigma_n,$$

$$\mathcal{Q}_\gamma \mathcal{Q}_\gamma^T = \Sigma_n^{1/2} S_n^T S_n \Sigma_n^{1/2} = \Sigma_n.$$

From Eqs. (10.36), we obtain

$B = $ the first r columns of $\mathcal{Q}_\gamma$,

$C = $ the first m rows of $\mathcal{P}_p$.

To compute A, we form the generalized $pm \times \gamma r$ Hankel matrix, $H(1)$, such that

$$H(1) = \begin{bmatrix} Y_2 & Y_3 & \cdots & Y_{\gamma+1} \\ Y_3 & Y_4 & \cdots & Y_{\gamma+2} \\ \vdots & \vdots & \ddots & \vdots \\ Y_{p+1} & Y_{p+2} & \cdots & Y_{p+\gamma} \end{bmatrix}. \tag{10.40}$$

Similar to Eq. (10.35), another form of $H(1)$ in terms of system matrices A, B, and C is

$$H(1) = \begin{bmatrix} CAB & CA^2B & \cdots & CA^\gamma B \\ CA^2B & CA^3B & \cdots & CA^{\gamma+1}B \\ \vdots & \vdots & \ddots & \vdots \\ CA^pB & CA^{p+1}B & \cdots & CA^{p+\gamma-1}B \end{bmatrix}$$

$$= \mathcal{P}_p A \mathcal{Q}_\gamma$$

$$= R_n \Sigma_n^{1/2} A \Sigma_n^{1/2} S_n^T. \tag{10.41}$$

One obvious solution for the state matrix A becomes

$$A = \Sigma_n^{-1/2} R_n^T H(1) S_n \Sigma_n^{-1/2}. \tag{10.42}$$

Although the above equation is valid, the reader is directed to Ref. [2] for a rigorous mathematical proof.

As a summary, we have described a procedure to compute the system matrices A, B, C, and D from system Markov parameters. The procedure requires computation of two Hankel matrices, $H(0)$ and $H(1)$. The identified system matrices A, B, C, and D are

$$A = \Sigma_n^{-1/2} R_n^T H(1) S_n \Sigma_n^{-1/2}, \tag{10.43a}$$

$B = $ the first r columns of $\Sigma_n^{1/2} S_n^T$, $\tag{10.43b}$

$C = $ the first m rows of $R_n \Sigma_n^{1/2}$, $\tag{10.43c}$

$D = Y_0.$ $\tag{10.43d}$

It is known that the identified matrices A, B, and C are not unique in the sense that they are coordinate dependent. Nevertheless, the state–space realization represented by A, B, C, and D is a minimum realization, i.e., the model is controllable and observable and its order is minimum.

Recall the general finite-difference model defined in Eq. (10.1). All the coefficients α_i and β_i are uniquely determined if the order of the system, n, is related to the order of the ARX model p according to $n = pm$, where m is the number of outputs. In this case, the sequence Y_k is uniquely determined from the input and the output data u and y. On the other hand, the sequence $Y_k = CA^{k-1}B$ for $k = 1, 2, \ldots, \infty$ represents the system pulse-responses for a given physical system under testing. Therefore, Y_k must be unique, regardless of whether or not the parameters $\alpha_1, \ldots, \alpha_p$ and $\beta_1, \ldots, \beta_p$ are unique.

EXAMPLE 10.8

Consider a rigid body of mass m with a force f acting along the direction of motion. The discrete-time model with the sampling time interval Δt for this system can be written as

$$x(k + 1) = Ax(k) + Bu(k),$$

where

$$A = \begin{bmatrix} 1 & \Delta t \\ 0 & 1 \end{bmatrix}, \qquad B = \begin{bmatrix} \frac{1}{2}\Delta t^2 \\ \Delta t \end{bmatrix}, \qquad u(k) = \frac{f(k)}{m}.$$

Assume that the displacement is sampled on every time interval Δt. The discrete measurement equation becomes

$$y(k) = Cx(k) \quad \text{with} \quad C = [1 \quad 0].$$

The pulse-response samples for this model are

$$Y_0 = D = 0,$$

$$Y_1 = CB = [1 \quad 0]\begin{bmatrix} \frac{1}{2}\Delta t^2 \\ \Delta t \end{bmatrix} = \frac{1}{2}\Delta t^2,$$

$$Y_2 = CAB = [1 \quad 0]\begin{bmatrix} 1 & \Delta t \\ 0 & 1 \end{bmatrix}\begin{bmatrix} \frac{1}{2}\Delta t^2 \\ \Delta t \end{bmatrix} = \frac{3}{2}\Delta t^2,$$

$$Y_3 = CA^2B = [1 \quad 0]\begin{bmatrix} 1 & 2\Delta t \\ 0 & 1 \end{bmatrix}\begin{bmatrix} \frac{1}{2}\Delta t^2 \\ \Delta t \end{bmatrix} = \frac{5}{2}\Delta t^2,$$

$$Y_4 = CA^3B = [1 \quad 0]\begin{bmatrix} 1 & 3\Delta t \\ 0 & 1 \end{bmatrix}\begin{bmatrix} \frac{1}{2}\Delta t^2 \\ \Delta t \end{bmatrix} = \frac{7}{2}\Delta t^2,$$

$$Y_5 = CA^4B = [1 \quad 0]\begin{bmatrix} 1 & 4\Delta t \\ 0 & 1 \end{bmatrix}\begin{bmatrix} \frac{1}{2}\Delta t^2 \\ \Delta t \end{bmatrix} = \frac{9}{2}\Delta t^2.$$

In practice, these pulse-response samples are obtained from measurements. To compute matrices A, B, and C from these pulse-response samples, form the

Hankel matrix $H(0)$,

$$H(0) = \begin{bmatrix} Y_1 & Y_2 \\ Y_2 & Y_3 \end{bmatrix} = \frac{\Delta t^2}{2} \begin{bmatrix} 1 & 3 \\ 3 & 5 \end{bmatrix},$$

and the shifted Hankel matrix $H(1)$,

$$H(1) = \begin{bmatrix} Y_2 & Y_3 \\ Y_3 & Y_4 \end{bmatrix} = \frac{\Delta t^2}{2} \begin{bmatrix} 3 & 5 \\ 5 & 7 \end{bmatrix}.$$

For a noise-free system of the order of 2, a 2×2 Hankel matrix is sufficient for system identification. Taking the SVD of the Hankel matrix $H(0)$ yields

$$H(0) = R \, \Sigma \, S^T$$

$$= \begin{bmatrix} 0.3346 & 0.9424 \\ 0.9424 & -0.3346 \end{bmatrix} \begin{bmatrix} 4.5616\Delta t^2 & 0 \\ 0 & 0.4384\Delta t^2 \end{bmatrix} \begin{bmatrix} 0.5531 & -0.8331 \\ 0.8331 & 0.5531 \end{bmatrix}^T.$$

Let $H(0)$ be written as

$$H(0) = [R\Sigma^{1/2}] \, [\Sigma^{1/2}S^T]$$
$$= \left\{ \begin{bmatrix} 0.7146 & 0.6240 \\ 2.0127 & -0.2215 \end{bmatrix} \Delta t \right\} \left\{ \begin{bmatrix} 1.1814 & 1.7793 \\ -0.5516 & 0.3663 \end{bmatrix} \Delta t \right\}$$
$$\underline{\underline{\Delta}} \, \{\mathcal{P}\} \, \{\mathcal{Q}\}.$$

The first row of $\mathcal{P}$ becomes the identified output matrix $\hat{C}$,

$$\hat{C} = [0.7146 \quad 0.6240] \, \Delta t,$$

and the first column of $\mathcal{Q}$ becomes the identified input matrix $\hat{B}$,

$$\hat{B} = \begin{bmatrix} 1.1814 \\ -0.5516 \end{bmatrix} \Delta t.$$

The state matrix can then be identified by

$$\hat{A} = \Sigma^{-1/2} R^T H(1) S \Sigma^{-1/2} = \begin{bmatrix} 1.3881 & -0.2528 \\ 0.5958 & 0.6119 \end{bmatrix}.$$

The identified state matrix has two repeated eigenvalues $\lambda = 1$, which mean zero, $\ln(1) = 0$, in the continuous-time model; i.e., rigid-body motion. The identified matrices $\hat{A}$, $\hat{B}$, and $\hat{C}$ are different from the original discrete matrices A, B and C. Nevertheless, both $\hat{A}$, $\hat{B}$, $\hat{C}$ and A, B, C represent the same dynamical systems in the sense that they provide identical pulse-response samples.

10.6　Observable Canonical-Form Realization

Given the input–output relationship such as the finite-difference model identified from experimental data, there are several methods of converting it to a state–space model. The method described in the last section uses the Markov parameters (pulse-responses) computed from the coefficient matrices of the finite-difference model to identify system matrices. It is based on the minimum realization theory that produces a controllable and observable model of minimum size. The other method that is commonly used is the observable canonical-form realization. It directly uses the coefficient matrices of the finite-element model without any further process of computing pulse-responses to form an observable model. The drawback is that the model thus obtained may not be controllable.

The finite-difference model that describes the input–output map of a system is

$$y(k) + \alpha_1 y(k-1) + \cdots + \alpha_p y(k-p)$$
$$= \beta_0 u(k) + \beta_1 u(k-1) + \cdots + \beta_p u(k-p). \tag{10.44}$$

The first step is to define the state variables. One method is presented in this section. Let us choose the state variables as

$$x_p(k) = y(k) - \beta_0 u(k), \tag{10.45a}$$

$$x_{p-1}(k) = y(k+1) - \beta_0 u(k+1) + \alpha_1 y(k) - \beta_1 u(k), \tag{10.45b}$$

$$x_{p-2}(k) = y(k+2) - \beta_0 u(k+2)$$
$$+ \alpha_1 y(k+1) - \beta_1 u(k+1) + \alpha_2 y(k) - \beta_2 u(k), \tag{10.45c}$$

$$\vdots \quad \vdots \qquad \vdots$$

$$x_1(k) = y(k+p-1) - \beta_0 u(k+p-1) + \alpha_1 y(k+p-2) - \beta_1 u(k+p-2)$$
$$+ \alpha_2 y(k+p-3) - \beta_2 u(k+p-3)$$

$$\vdots$$

$$+ \alpha_{p-1} y(k) - \beta_{p-1} u(k), \tag{10.45d}$$

where each vector $x_i(k)$, $i = 1, 2, \ldots, p$, has the length of m, which is the number of outputs. The set of equations (10.45) yields

$$y(k) = x_p(k) + \beta_0 u(k), \tag{10.46a}$$

$$x_{p-1}(k) = x_p(k+1) + \alpha_1 y(k) - \beta_1 u(k), \tag{10.46b}$$

$$x_{p-2}(k) = x_{p-1}(k+1) + \alpha_2 y(k) - \beta_2 u(k) \tag{10.46c}$$

$$\vdots \quad \vdots \qquad \vdots$$

$$x_1(k) = x_2(k+1) + \alpha_{p-1} y(k) - \beta_{p-1} u(k), \tag{10.46d}$$

$$x_1(k+1) = \alpha_p y(k) - \beta_p u(k), \tag{10.46e}$$

where the last equation is obtained by use of the last equation in Eqs. (10.45) with the help of Eq. (10.44). The above equations can be arranged in matrix form as

$$x(k+1) = Ax(k) + Bu(k), \tag{10.47a}$$

$$y(k) = Cx(k) + Du(k), \tag{10.47b}$$

where the state vector $x(k)$, the state matrix A, the input matrix B, the output matrix C, and the direct-transmission matrix are

$$x(k) = \begin{bmatrix} x_1(k) \\ x_2(k) \\ x_3(k) \\ \vdots \\ x_{p-1}(k) \\ x_p(k) \end{bmatrix}, \qquad (10.48a)$$

$$A = \begin{bmatrix} 0 & 0 & 0 & \cdots & 0 & -\alpha_p \\ I & 0 & 0 & \cdots & 0 & -\alpha_{p-1} \\ 0 & I & 0 & \cdots & 0 & -\alpha_{p-2} \\ \vdots & \vdots & \vdots & \ddots & \vdots & \vdots \\ 0 & 0 & 0 & \cdots & 0 & -\alpha_2 \\ 0 & 0 & 0 & \cdots & I & -\alpha_1 \end{bmatrix}, \qquad (10.48b)$$

$$B = \begin{bmatrix} \beta_p - \alpha_p\beta_0 \\ \beta_{p-1} - \alpha_{p-1}\beta_0 \\ \beta_{p-2} - \alpha_{p-2}\beta_0 \\ \vdots \\ \beta_2 - \alpha_2\beta_0 \\ \beta_1 - \alpha_1\beta_0 \end{bmatrix}, \qquad (10.48c)$$

$$C = [0 \quad 0 \quad 0 \quad \cdots \quad 0 \quad I], \qquad (10.48d)$$

$$D = \beta_0. \qquad (10.48e)$$

Recall that p is the number of time steps, m is the number of outputs, and r is the number of inputs. The state matrix A becomes an $mp \times mp$ matrix, the input matrix B an $mp \times r$ matrix, and the output matrix C an $m \times mp$ matrix. A state–space model in the form of Eqs. (10.48) is said to be in the canonical form.

The observability matrix of the canonical-form realization is

$$\mathcal{P}_p = \begin{bmatrix} C \\ CA \\ CA^2 \\ \vdots \\ CA^{p-1} \end{bmatrix} = \begin{bmatrix} 0 & 0 & 0 & \cdots & 0 & 0 & I \\ 0 & 0 & 0 & \cdots & 0 & I & -\alpha_1 \\ 0 & 0 & 0 & \cdots & I & -\alpha_1 & -\alpha_2 + \alpha_1\alpha_1 \\ \vdots & \vdots & \vdots & \ddots & \vdots & \vdots & \vdots \\ 0 & I & X & \cdots & X & X & X \\ I & X & X & \cdots & X & X & X \end{bmatrix},$$

where X denotes possible nonzero elements. The matrix $\mathcal{P}_p$ is nonsingular, implying that the observability matrix $\mathcal{P}_p$ has a rank of mp and the system is observable. Are they controllable as well? First, observe from Eqs. (10.48) that

$$
\begin{aligned}
D &= \beta_0, \\
CB &= \beta_1 - \alpha_1 \beta_0 &&= \beta_0^{(1)}, \\
CAB &= \beta_2 - \alpha_2 \beta_0 - \alpha_1 \beta_0^{(1)} &&= \beta_0^{(2)}, \\
CA^2B &= \beta_2 - \alpha_3 \beta_0 - \alpha_2 \beta_0^{(1)} - \alpha_1 \beta_0^{(2)} &&= \beta_0^{(3)}, \\
\vdots \quad \vdots \qquad & \qquad\qquad \vdots && \quad \vdots \quad \vdots
\end{aligned}
\tag{10.49}
$$

where the last equality is obtained with Eqs. (10.22). Equations (10.49) show that the canonical-form model, Eqs. (10.48), produces the same sequence of system Markov parameters as the finite-difference model, Eq. (10.44). The triplet $[A, B, C]$ thus provides a valid system realization. Is it a minimum realization? *A model is a minimum realization if and only if it is controllable and observable.* The order of this model is mp, i.e., the dimension of A shown in Eqs. (10.48). If the number p is chosen such that mp is larger than the order of the system, then the triplet $[A, B, C]$ is not a minimum realization. Because it is proved that the canonical form, Eq. (10.48), is observable, it cannot be controllable for a nonminimum realization. In general, the order of a system under test is not known a priori. The number mp tends to be chosen significantly larger than the effective order of the system to accommodate the measurement noise and system uncertainties. "Effective" here means the part of the model that can be excited by the inputs and measured by the outputs. A state–space model in the form of Eqs. (10.48) is thus said to be in the observable canonical form.

Once we have the system matrices A, B, C, and D we may be interested in looking for an observer gain matrix for state estimation so that it can be used for state-feedback control, as discussed in Chap. 9. One particular observer matrix is introduced relating to the deadbeat controller that will be discussed in Chap. 11. Let us define the matrix G as

$$
G = \begin{bmatrix} \alpha_p \\ \alpha_{p-1} \\ \alpha_{p-2} \\ \vdots \\ \alpha_2 \\ \alpha_1 \end{bmatrix},
\tag{10.50}
$$

which makes the matrix $A + GC$ deadbeat at the p time step, i.e.,

$$(A + GC)^p = 0,$$

where 0 is an $mp \times mp$ zero matrix. The matrix G is called the deadbeat-observer gain matrix for the state–space model described by Eqs. (10.47). It is worth noting that if the identified observable model is not controllable, some of the system poles may not be relocated by any feedback controller.

EXAMPLE 10.9

Consider the following finite-difference (ARX) model,

$$y(k) = -\frac{3}{2}y(k-1) - \frac{5}{2}y(k-2) - \frac{3}{2}y(k-3) - y(k-4)$$

$$+ u(k) - \frac{7}{6}u(k-1) + u(k-2) - \frac{4}{3}u(k-3) + \frac{1}{2}u(k-4),$$

which can be rearranged to produce the following state–space equations,

$$x(k+1) = Ax(k) + Bu(k),$$

$$y(k) = Cx(k) + Du(k),$$

where, from Eqs. (10.48),

$$A = \begin{bmatrix} 0 & 0 & 0 & -\alpha_4 \\ 1 & 0 & 0 & -\alpha_3 \\ 0 & 1 & 0 & -\alpha_2 \\ 0 & 0 & 1 & -\alpha_1 \end{bmatrix} = \begin{bmatrix} 0 & 0 & 0 & -1 \\ 1 & 0 & 0 & -\frac{3}{2} \\ 0 & 1 & 0 & -\frac{5}{2} \\ 0 & 0 & 1 & -\frac{3}{2} \end{bmatrix},$$

$$B = \begin{bmatrix} \beta_4 - \alpha_4\beta_0 \\ \beta_3 - \alpha_3\beta_0 \\ \beta_2 - \alpha_2\beta_0 \\ \beta_1 - \alpha_1\beta_0 \end{bmatrix} = \begin{bmatrix} -\frac{1}{2} \\ -\frac{17}{6} \\ -\frac{3}{2} \\ -\frac{8}{3} \end{bmatrix},$$

$$C = [0 \quad 0 \quad 0 \quad 1],$$

$$D = \beta_0 = 1.$$

To verify the result, let us compute the pulse response:

$$[D \quad CB \quad CAB \quad CA^2B \quad \cdots] = [1 \quad -\frac{8}{3} \quad \frac{5}{2} \quad \frac{1}{12} \quad \cdots].$$

This is identical to the pulse response generated from the finite-difference model (see Example 10.6).

10.7 Relationship with an Observer Model

In this section, we will derive the relationship between the state–space observer model and the finite-difference model. A state–space model of a linear system describes the system input and output through an intermediate quantity called the state vector. However, the state vector itself is, in general, not accessible for direct measurement. A state estimator, also known as an observer, is used to provide an estimate of the system state from input and output measurements.

10.7.1   State–Space Observer Model

The observer equation shown in Eq. (9.17) has the form

$$\hat{x}(k+1) = (A + GC)\hat{x}(k) + (B + GD)u(k) - Gy(k), \tag{10.51}$$

$$\hat{y}(k) = C\hat{x}(k) + Du(k), \tag{10.52}$$

where $\hat{x}(k)$ and $\hat{y}(k)$ denote the estimated state and output, respectively. The matrix G is the observer gain matrix to be determined. Let state–space observer equation (10.51) be rewritten in a slightly different form. Define

$$\bar{A} = A + GC, \tag{10.53a}$$

$$\bar{B} = [B + GD \quad -G], \tag{10.53b}$$

$$v(k) = \begin{bmatrix} u(k) \\ y(k) \end{bmatrix}. \tag{10.53c}$$

The observer state equation becomes

$$\hat{x}(k+1) = \bar{A}\hat{x}(k) + \bar{B}v(k). \tag{10.54}$$

Equations (10.54) and (10.52) constitute a discrete-time state–space observer model of a dynamical system. This matrix difference equation is driven by both the input $u(k)$ and the output $y(k)$, and its solution yields the estimated state $\hat{x}(k)$. Equation (10.54) is identical in form to the state-space equation

$$x(k+1) = Ax(k) + Bu(k).$$

However, the eigenvalues of $\bar{A}$ are moved as a consequence of the additional term GC and the number of columns in $\bar{B}$ is increased relative to those in B by the number of outputs m. Because the matrix G can be arbitrarily chosen, $\bar{A}$ may be made as asymptotically stable as desired. However, there is an optimal choice, which will be discussed later.

EXAMPLE 10.10

Consider the discrete-time model

$$A = \begin{bmatrix} \cos\gamma & \sin\gamma \\ -\sin\gamma & \cos\gamma \end{bmatrix},$$

$$B = \begin{bmatrix} 1 - \cos\gamma \\ \sin\gamma \end{bmatrix},$$

$$C = [-1 \quad 0],$$

$$D = 1,$$

where γ is a constant value. The goal is to find a matrix G such that $A + GC$ is as asymptotically stable as desired. Let the matrix G be chosen as

$$G = \begin{bmatrix} 2\cos\gamma \\ -\sin\gamma + \dfrac{\cos^2\gamma}{\sin\gamma} \end{bmatrix}.$$

The matrices $\bar{A}$ and $\bar{B}$ in Eqs. (10.53) become

$$\bar{A} = A + GC$$

$$= \begin{bmatrix} \cos\gamma & \sin\gamma \\ -\sin\gamma & \cos\gamma \end{bmatrix} + \begin{bmatrix} 2\cos\gamma \\ -\sin\gamma + \dfrac{\cos^2\gamma}{\sin\gamma} \end{bmatrix} \begin{bmatrix} -1 & 0 \end{bmatrix}$$

$$= \begin{bmatrix} -\cos\gamma & \sin\gamma \\ -\dfrac{\cos^2\gamma}{\sin\gamma} & \cos\gamma \end{bmatrix},$$

$$\bar{B} = [B + GD \quad -G]$$

$$= \begin{bmatrix} \left(\begin{matrix} 1 - \cos\gamma \\ \sin\gamma \end{matrix}\right) + \left(\begin{matrix} 2\cos\gamma \\ -\sin\gamma + \dfrac{\cos^2\gamma}{\sin\gamma} \end{matrix}\right) & -\left(\begin{matrix} 2\cos\gamma \\ -\sin\gamma + \dfrac{\cos^2\gamma}{\sin\gamma} \end{matrix}\right) \end{bmatrix}$$

$$= \begin{bmatrix} 1 + \cos\gamma & -2\cos\gamma \\ \dfrac{\cos^2\gamma}{\sin\gamma} & \sin\gamma - \dfrac{\cos^2\gamma}{\sin\gamma} \end{bmatrix}.$$

What is the significance of the above $\bar{A}$ and $\bar{B}$ in comparison with the A and B? First, all the eigenvalues of $\bar{A}$ are zero. Second, $\bar{A}^2 = \bar{A}^3 = \cdots = \bar{A}^n = \cdots = 0$, which we can prove by noting that

$$\bar{A}^2 = \bar{A}\bar{A} = \begin{bmatrix} -\cos\gamma & \sin\gamma \\ -\dfrac{\cos^2\gamma}{\sin\gamma} & \cos\gamma \end{bmatrix} \begin{bmatrix} -\cos\gamma & \sin\gamma \\ -\dfrac{\cos^2\gamma}{\sin\gamma} & \cos\gamma \end{bmatrix} = \begin{bmatrix} 0 & 0 \\ 0 & 0 \end{bmatrix}.$$

Therefore the matrix $\bar{A}$ is asymptotically stable in the sense that it takes only a power of two for $\bar{A}$ to decay completely. On the other hand, the matrix A is not asymptotically stable, because

$$A^n = \begin{bmatrix} \cos n\gamma & \sin n\gamma \\ -\sin n\gamma & \cos n\gamma \end{bmatrix},$$

which does not go to zero regardless of how large an n is chosen.

10.7.2 Relationship with State–Space System Matrices

To establish the relationship between the finite-difference model and the observer model, we need to solve for the output $\hat{y}(k)$ from Eqs. (10.51) and (10.52) with zero initial condition, $\hat{x}(1) = 0$, in terms of the current and previous inputs $u(1), u(2), \ldots, u(k)$

and previous outputs $y(1)$, $y(2), \ldots,$ $y(k-1)$. At the initial point with $k=1$, Eqs. (10.51) and (10.52) produce

$$\hat{x}(1) = 0,$$

$$\hat{y}(1) = Du(1).$$

At the second time step with $k=2$, Eqs. (10.51) and (10.52) give

$$\hat{x}(2) = (B + GD)u(1) - Gy(1),$$

$$\hat{y}(2) = C(B + GD)u(1) + Du(2) - CGy(1).$$

Continuing the process for the third time step with $k=3$ yields

$$\hat{x}(3) = (A + GC)(B + GD)u(1) + (B + GD)u(2)$$
$$- (A + GC)Gy(1) - Gy(2),$$

$$\hat{y}(3) = C(A + GC)(B + GD)u(1) + C(B + GD)u(2) + Du(3)$$
$$- C(A + GC)Gy(1) - CGy(2).$$

Finally, by induction, we obtain

$$\hat{x}(k) = \sum_{i=1}^{k-1} (A + GC)^{i-1}(B + GD)u(k - i)$$

$$- \sum_{i=1}^{k-1} (A + GC)^{i-1} Gy(k - i), \tag{10.55a}$$

$$\hat{y}(k) = \sum_{i=1}^{k-1} C(A + GC)^{i-1}(B + GD)u(k - i) + Du(k)$$

$$- \sum_{i=1}^{k-1} C(A + GC)^{i-1} Gy(k - i). \tag{10.55b}$$

Let the matrix G be chosen such that $\bar{A} = A + GC$ is sufficiently stable to make $\bar{A}^p$ negligible. In this case, the estimated $\hat{y}(k)$ closely approaches the measured $y(k)$ for $k > p$ because the estimation error, Eq. (9.20), approaches zero. Therefore, for large k, the estimated $\hat{y}(k)$ becomes

$$\hat{y}(k) = \sum_{i=1}^{p} C(A + GC)^{i-1}(B + GD)u(k - i) + Du(k)$$

$$- \sum_{i=1}^{p} C(A + GC)^{i-1} Gy(k - i). \tag{10.56}$$

The relationship between the state–space observer model and the finite-difference model can be easily established by a comparison of Eq. (10.1) with Eq. (10.56). Indeed, the coefficients α_i and β_i for $i = 1, 2, \ldots, p$ shown in Eq. (10.1) are equivalent to

$$\alpha_i = C(A + GC)^{i-1}G, \quad i = 1, 2, \ldots, p, \tag{10.57a}$$

$$\beta_i = C(A + GC)^{i-1}(B + GD), \quad i = 1, 2, \ldots, p, \tag{10.57b}$$

$$\beta_0 = D. \tag{10.57c}$$

The coefficient matrices a_i and β_i are also called observer Markov parameters (OMPs), because they are the system Markov parameters of the observer model described by Eqs. (10.52) and (10.54) (see Problem 10.5). Equations (10.57) are always true as long as the system is observable.

The finite-difference model describes only the relationship between the input and the output under the assumption that the initial condition is zero or that the system is in the condition of a steady state. In practice, if the test for direct input and output measurements is sufficiently long to allow the transient response to decay, the error that is due to a nonzero initial condition becomes minimum.

EXAMPLE 10.11

Let us use the results obtained in Example 10.10 to compute the corresponding finite-difference model. From Eq. (10.57a), the coefficients α_i are computed as follows:

$$\alpha_1 = CG = \begin{bmatrix} -1 & 0 \end{bmatrix} \begin{bmatrix} 2\cos\gamma \\ -\sin\gamma + \dfrac{\cos^2\gamma}{\sin\gamma} \end{bmatrix} = -2\cos\gamma,$$

$$\alpha_2 = C(A + GC)G$$

$$= \begin{bmatrix} -1 & 0 \end{bmatrix} \begin{bmatrix} -\cos\gamma & \sin\gamma \\ -\dfrac{\cos^2\gamma}{\sin\gamma} & \cos\gamma \end{bmatrix} \begin{bmatrix} 2\cos\gamma \\ -\sin\gamma + \dfrac{\cos^2\gamma}{\sin\gamma} \end{bmatrix} = 1,$$

$$\alpha_3 = \alpha_4 = \cdots = a_\infty = 0.$$

and similarly, from Eqs. (10.57b) and (10.57c), the coefficients β_i are

$$\beta_0 = 1,$$

$$\beta_1 = C(B + GD) = \begin{bmatrix} -1 & 0 \end{bmatrix} \begin{bmatrix} 1 + \cos\gamma \\ \dfrac{\cos^2\gamma}{\sin\gamma} \end{bmatrix} = -1 - \cos\gamma,$$

$$\beta_2 = C(A + GC)(B + GD)$$

$$= \begin{bmatrix} -1 & 0 \end{bmatrix} \begin{bmatrix} -\cos\gamma & \sin\gamma \\ -\dfrac{\cos^2\gamma}{\sin\gamma} & \cos\gamma \end{bmatrix} \begin{bmatrix} 1 + \cos\gamma \\ \dfrac{\cos^2\gamma}{\sin\gamma} \end{bmatrix} = \cos\gamma,$$

$$\beta_3 = \beta_4 = \cdots = b_\infty = 0.$$

The linear difference model, Eq. (10.56), can be written as

$$y(k) - (2\cos\gamma)\,y(k-1) + y(k-2)$$
$$= u(k) - (1 + \cos\gamma)\,u(k-1) + (\cos\gamma)\,u(k-2).$$

It is clear that the current output $y(k)$ depends on only the past two outputs, $y(k-1)$ and $y(k-2)$ and the current and past two inputs, $u(k)$, $u(k-1)$ and $u(k-2)$. It should be pointed out that this expression is valid only if a proper observer gain is used for computing the OMPs.

10.7.3 Computation of Observer Gain

The key quantity of relating the finite-difference model to the observer model is the observer gain G. Given the system matrices A, B, C, and D and the gain matrix G, the coefficient matrices α_i and β_i can be easily computed by Eqs. (10.57). On the other hand, computing the gain matrix G if the coefficient matrices α_i and β_i are given is discussed in this section.

To identify the observer gain G, first define the sequence of parameters

$$Y_k^o = C A^{k-1} G, \quad k = 1, 2, 3, \ldots, \tag{10.58}$$

and then compute them in terms of the OMP parameter matrices α_i and β_i. From Eq. (10.57a), the first parameter in the sequence is simply

$$Y_1^o = CG = \alpha_1. \tag{10.59}$$

The next parameter in the sequence from Eq. (10.57a) is obtained by considering α_2:

$$\alpha_2 = C\bar{A}G = (CAG + CGCG)$$
$$= Y_2^o + \alpha_1 Y_1^o,$$

which yields

$$Y_2^o = \alpha_2 - \alpha_1 Y_1^o. \tag{10.60}$$

Similarly,

$$\alpha_3 = C\bar{A}^2 G$$
$$= (CA^2G + CGCAG + C\bar{A}GCG)$$
$$= Y_3^o + \alpha_1 Y_2^o + \alpha_2 Y_1^o,$$

which yields

$$Y_3^o = \alpha_3 - \alpha_1 Y_2^o - \alpha_2 Y_1^o. \tag{10.61}$$

By induction, the general relationship is

$$Y_1^o = CG = \alpha_1,$$

$$Y_k^o = \alpha_k - \sum_{i=1}^{k-1} \alpha_i Y_{k-i}^o \quad \text{for} \quad k = 2, \ldots, p,$$

$$Y_k^o = -\sum_{i=1}^{p} \alpha_i Y_{k-i}^o \quad \text{for} \quad k = p+1, \ldots, \infty. \tag{10.62}$$

We have thus obtained the sequence $Y_k^o = C A^{k-1} G$, where C and A can be realized by an identification method shown in Section (10.5) from the Markov parameter sequence

$Y_k = CA^{k-1}B$. The sequence $Y_k^o = CA^{k-1}G$ is commonly called the observer gain Markov parameters in parallel with the system Markov parameter sequence $Y_k = CA^{k-1}B$. The observer gain G can be computed from

$$G = (\mathcal{P}^T\mathcal{P})^{-1}\mathcal{P}^T Y^o, \tag{10.63}$$

where

$$\mathcal{P} = \begin{bmatrix} C \\ CA \\ CA^2 \\ \vdots \\ CA^N \end{bmatrix}, \tag{10.64}$$

$$Y^o = \begin{bmatrix} Y_1^o \\ Y_2^o \\ Y_3^o \\ \vdots \\ Y_{N+1}^o \end{bmatrix} = \begin{bmatrix} CG \\ CAG \\ CA^2G \\ \vdots \\ CA^N G \end{bmatrix} = \mathcal{P}G. \tag{10.65}$$

Note that the integer N has to be chosen large enough to make sure that the observability matrix $\mathcal{P}$ has a rank of n, which is the order of the identified system.

EXAMPLE 10.12

Assume that the identified system matrices A and C are

$$A = \begin{bmatrix} \cos\gamma & \sin\gamma \\ -\sin\gamma & \cos\gamma \end{bmatrix},$$

$$C = [-1 \quad 0],$$

where γ is a constant value. The identified parameters α_1 and α_2 are

$$\alpha_1 = -2\cos\gamma,$$

$$\alpha_2 = 1.$$

The observer gain Markov parameters from Eqs. (10.59) and (10.60) become

$$Y_1^o = \alpha_1 = -2\cos\gamma.$$

$$Y_2^o = \alpha_2 - \alpha_1 Y_1^o = 1 - 4\cos^2\gamma.$$

Form the matrices

$$\mathcal{P} = \begin{bmatrix} C \\ CA \end{bmatrix} = \begin{bmatrix} -1 & 0 \\ -\cos\gamma & -\sin\gamma \end{bmatrix},$$

$$Y^o = \begin{bmatrix} -2\cos\gamma \\ 1 - 4\cos^2\gamma \end{bmatrix}.$$

The observer gain G can be computed from Eq. (10.63) as follows:

$$G = (\mathcal{P}^T \mathcal{P})^{-1} \mathcal{P}^T Y^o = \mathcal{P}^{-1} Y^o$$

$$= \begin{bmatrix} 2\cos\gamma \\ \cos 2\gamma \, \csc\gamma \end{bmatrix} = \begin{bmatrix} 2\cos\gamma \\ -\sin\gamma + \dfrac{\cos^2\gamma}{\sin\gamma} \end{bmatrix}.$$

The gain is identical to the one shown in Example 10.10.

10.8 Concluding Remarks

The area of system identification encompasses a multitude of approaches, perspectives, and techniques whose interrelationships and relative merits can be difficult to sort out. As a result, it is difficult for a nonspecialist to extract the fundamental concepts. It may take considerable effort to gain enough intuition about a particular technique to be able to use it effectively in practice. In this chapter, we have briefly described some system identification basics, particularly the mathematical relationships of inputs and outputs, and the conversions of input–output models to state–space models appropriate to modern control design methods. We emphasize identification techniques in the time domain, including those from system realization theory and observer-based identification methods. A list of references is given at the end of this chapter for the readers who are interested in learning more about system identification.

10.9 Problems

10.1 The following finite-difference model,

$$y_t(k) + 2y_t(k-1) + y_t(k-2) = u(k-1) + u(k-2),$$

is given, where the subscript t signifies that the associated quantities are true output values. Let the input vector u be

$$u(1) = 1, \qquad u(k) = 0 \quad \text{for} \quad k > 1.$$

Assume that the system is initially at rest.

(a) Compute and plot the time history $y_t(k)$ with $k = 1, 2, \ldots, 9$. What is the vibration frequency (cycles per time step) of the output?

(b) Add the following noise vector

$$y_e = 10^{-1} \times [1 \quad -0.3 \quad -0.5 \quad -0.1 \quad -1.4 \quad -0.7 \quad 1.8 \quad -2.9 \quad 0.1]$$

to the output vector

$$y_t = [y_t(1) \quad y_t(2) \quad \cdots \quad y_t(9)]$$

to form the measured output vector $y = y_t + y_e$. Plot the measured output vector y.

(c) Use the time histories u and y to compute the estimated output vector $\hat{y}$ as defined in Example 10.1, i.e.,

$$\hat{y}(k) = -2y(k-1) - y(k-2) + u(k-1) + u(k-2).$$

(d) Solve the roots z_1 and z_2 of the following polynomial:

$$1 + 2z^{-1} + z^{-2} = 0.$$

Let

$$z = e^{j2\pi\omega\gamma},$$

where γ is the sample time interval, $j = \sqrt{-1}$, and ω is the vibration frequency (cycles/s). Determine ω_1 and ω_2 from the above equation with $\gamma = 1$.

10.2 Repeat Problem 10.1 with the following finite-difference model:

$$y_t(k) - 2y_t(k-1) + y_t(k-2) = u(k-1) + u(k-2).$$

Discuss the differences of the results obtained from Problems 10.1 and 10.2.

10.3 The following is a set of input data $u(k)$ and output data $y(k)$ for $k = 1, 2, \ldots, 7$:

$$
\begin{bmatrix} u(1) \\ u(2) \\ u(3) \\ u(4) \\ u(5) \\ u(6) \\ u(7) \end{bmatrix} = \begin{bmatrix} 1 \\ 0 \\ 0 \\ 0 \\ 0 \\ 0 \\ 0 \end{bmatrix}, \qquad \begin{bmatrix} y(1) \\ y(2) \\ y(3) \\ y(4) \\ y(5) \\ y(6) \\ y(7) \end{bmatrix} = \begin{bmatrix} 0 \\ 1 \\ -1 \\ 1 \\ -1 \\ 1 \\ -1 \end{bmatrix}.
$$

Assume that the system is initially at rest and the ARX order is $p = 2$. Use Eqs. (10.5) and (10.6) to solve for the parameter matrix θ.

10.4 Given any set of input signals $u(k)$, $k = 1, 2, \ldots, 10$, compute a set of output data sequence according to the following equation:

$$y(k) = u(k) + \sum_{i=1}^{k} [\cos(i) - \cos(i-1)] u(k-i).$$

Assume that the system is initially at rest. Use Eqs. (10.5) and (10.6) to solve for the parameter matrix θ for $p = 2$ and $p = 4$.

10.5 What are the system Markov parameters of the observer model described by Eqs. (10.52) and (10.54)? Prove that the coefficient matrices $[\beta_i \ -\alpha_i]$ are the system Markov parameters of the observer model.

10.6 Assume that we are given a sequence of pulse-responses (Markov parameters)

$$Y_1, \ Y_2, \ Y_7, \ Y_8, \ Y_9, \ Y_{10}, \ Y_{12}, \ Y_{13}, \ Y_{14}, \ Y_{15}, \ Y_{16}, \ Y_{17}, \ Y_{20}, \ Y_{21}, \ Y_{22}, \ Y_{23}, \ Y_{24}.$$

Note that some data are either missing or too noisy to be used. Let A be the state matrix, B the input matrix, and C the output matrix. Each Markov parameter $Y_k = CA^{k-1}B$ is a 3×2 matrix, i.e., the number of outputs is 3 and the number of inputs is 2.

(a) Write two 9×6 Hankel matrices $H(0)$ and $H(1)$ in terms of the Markov parameters, which can be used for system identification. Explain why.

(b) What is the maximum order of the identified model. Is the identified model unique?

10.10 Appendix: Efficient Computation of Data Correlation Matrix

Because of the nature of data shifting to form V defined in Eq. (10.5e), there is an efficient way of computing correlation matrices VV^T. Let us first define the matrix $\tilde{V}$ as

$$
\tilde{V} = \begin{bmatrix} y \\ V \end{bmatrix} = \begin{bmatrix} v(p+1) & v(p+2) & \cdots & v(\ell) \\ v(p) & v(p+1) & \cdots & v(\ell-1) \\ \vdots & \vdots & \ddots & \vdots \\ v(2) & v(3) & \cdots & v(\ell-p+1) \\ v(1) & v(2) & \cdots & v(\ell-p) \end{bmatrix}, \tag{10.66}
$$

where y and V are defined in Eqs. (10.5b) and (10.5e), respectively. Note that each quantity v in $\tilde{V}$ is a $(m+r) \times 1$ column vector. The product of $\tilde{V}\tilde{V}^T$ can be written as

$$
\tilde{V}\tilde{V}^T = \begin{bmatrix} yy^T & yV^T \\ Vy^T & VV^T \end{bmatrix}
$$

$$
= \begin{bmatrix} \sum\limits_{\tau=p+1}^{\ell} v(\tau)v^T(\tau) & \sum\limits_{\tau=p+1}^{\ell} v(\tau)v^T(\tau-1) & \sum\limits_{\tau=p+1}^{\ell} v(\tau)v^T(\tau-2) & \cdots \\[2ex] \sum\limits_{\tau=p}^{\ell-1} v(\tau)v^T(\tau+1) & \sum\limits_{\tau=p}^{\ell-1} v(\tau)v^T(\tau) & \sum\limits_{\tau=p}^{\ell-1} v(\tau)v^T(\tau-1) & \cdots \\[2ex] \sum\limits_{\tau=p-1}^{\ell-2} v(\tau)v^T(\tau+2) & \sum\limits_{\tau=p-1}^{\ell-2} v(\tau)v^T(\tau+1) & \sum\limits_{\tau=p-1}^{\ell-2} v(\tau)v^T(\tau) & \cdots \\[2ex] \vdots & \vdots & \vdots & \ddots \\[2ex] \sum\limits_{\tau=1}^{\ell-p} v(\tau)v^T(\tau+p) & \sum\limits_{\tau=1}^{\ell-p} v(\tau)v^T(\tau+p-1) & \sum\limits_{\tau=1}^{\ell-p} v(\tau)v^T(\tau+p-2) & \cdots \end{bmatrix}.
$$

$$\tag{10.67}$$

The relationships among yV^T, VV^T, and $\tilde{V}\tilde{V}^T$ are

$$
yV^T = \text{first } m \text{ rows and last } p(r+m)+r \text{ columns of } \tilde{V}\tilde{V}^T, \tag{10.68a}
$$

$$
VV^T = \text{last } p(r+m)+r \text{ rows and columns of } \tilde{V}\tilde{V}^T. \tag{10.68b}
$$

These two equations indicate that matrices yV^T and VV^T can be computed from $\tilde{V}\tilde{V}^T$.

From the pattern appearing in Eq. (10.67), the reader should not have any difficulty in filling out all other elements that are not shown. For example, all the diagonal elements are identical except for their upper and lower limits and also for all the subdiagonal elements. As a result, the second diagonal element may be computed from the first diagonal element by

$$\sum_{\tau=p}^{\ell-1} v(\tau)v^T(\tau) = \sum_{\tau=p+1}^{\ell} v(\tau)v^T(\tau) + v(p)v^T(p) - v(\ell)v^T(\ell). \tag{10.69}$$

By induction, other diagonal submatrices can be computed recursively by

$$\sum_{\tau=p-i}^{\ell-1-i} v(\tau)v^T(\tau) = \sum_{\tau=p+1-i}^{\ell-i} v(\tau)v^T(\tau)$$
$$+ v(p-i)v^T(p-i) - v(\ell-i)v^T(\ell-i) \tag{10.70}$$

for $i = 0, 1, \ldots, p-1$. This recursive formula indicates that each quantity is computed from its previous quantity. Similarly, the first upper off-diagonal submatrix on the second $m + r$ rows may be calculated from the first upper off-diagonal submatrix on the first $m + r$ rows by

$$\sum_{\tau=p}^{\ell-1} v(\tau)v^T(\tau-1) = \sum_{\tau=p+1}^{\ell} v(\tau)v^T(\tau-1)$$
$$+ v(p)v^T(p-1) - v(\ell)v^T(\ell-1). \tag{10.71}$$

Therefore the first upper off-diagonal $(m + r) \times (m + r)$ submatrices can be computed recursively by

$$\sum_{\tau=p-i}^{\ell-1-i} v(\tau)v^T(\tau-1) = \sum_{\tau=p+1-i}^{\ell-i} v(\tau)v^T(\tau-1)$$
$$+ v(p-i)v^T(p-i-1) - v(\ell-i)v^T(\ell-i-1) \tag{10.72}$$

for $i = 0, 1, \ldots, p-2$. Furthermore, the second upper off-diagonal $(m + r) \times (m + r)$ submatrices are calculated recursively by

$$\sum_{\tau=p-i}^{\ell-1-i} v(\tau)v^T(\tau-2) = \sum_{\tau=p+1-i}^{\ell-i} v(\tau)v^T(\tau-2)$$
$$+ v(p-i)v^T(p-i-2) - v(\ell-i)v^T(\ell-i-2) \tag{10.73}$$

for $i = 0, 1, \ldots, p-3$. Similarly, the third, the fourth, up to the $(p-1)$th upper off-diagonal quantities can be calculated recursively as soon as their first ones are known. The first ones mean the first $m + r$ rows of the product $\tilde{V}\tilde{V}^T$.

Computing the product $\tilde{V}\tilde{V}^T$ becomes recursive as soon as the first $m + r$ rows are computed. For a long data record, i.e, $\ell \gg 1$, this recursive procedure is very efficient in computing time. Because the product $\tilde{V}\tilde{V}^T$ is symmetric, only the upper half or the lower half need be computed.

BIBLIOGRAPHY

[1] Graupe, D., *Identification of Systems*, 2nd ed., Krieger, Malabar, FL, 1975.
[2] Juang, J.-N., *Applied System Identification*, Prentice-Hall, Englewood Cliffs, NJ, 1994.
[3] Ljung, L., *System Identification: Theory for the User,* Prentice-Hall, Englewood Cliffs, NJ, 1987.

11

Predictive Control

11.1 Introduction

This chapter describes several computational algorithms to compute the predictive control law that has some feature of adaptive control. All algorithms make use of the multi-step-ahead output prediction as derived in Chap. 10 based on the finite-difference model. The generalized predictive control (GPC) algorithm (Ref. [1–4]) is based on system output predictions over a finite horizon known as the prediction horizon. In determining the future control inputs, it is assumed that control is applied only over a finite horizon known as the control horizon. The GPC is computed with the Toeplitz matrix formed from the step-response time history of the system in conjunction with a cost function with weighted input and output. The control input is obtained by minimization of the cost function. There are three design parameters involved: the control weight, the prediction horizon, and the control horizon. A proper combination of these parameters is required in order to guarantee stability of the predictive control law.

In contrast to the GPC approach, another approach is the deadbeat predictive control (DPC) (Ref. [5–9]). The DPC feedback law is supposed to bring the output response to rest after a few specific time steps. Similar to GPC, DPC has a control design parameter and an identification parameter related to the order of the system. The control design parameter, which is similar to the GPC control horizon, gives the number of time steps for the system to become deadbeat (rest). The DPC guarantees closed-loop stability for a controllable system.

Several GPC and DPC algorithms are discussed in this chapter. Numerical examples are given to illustrate the concepts of these algorithms.

11.2 Generalized Predictive Control

The system identification techniques described in Chap. 10 determine the observer Markov parameters (OMPs) (i.e., ARX coefficient matrices) based on input and output data. The OMP may be estimated with batch least squares, recursive least squares, or any other appropriate system identification technique (Ref. [10]). If the OMP of the system is known, the future outputs may be predicted with a recursive relationship as shown in Eqs. (10.22) and (10.23) to yield

$$y_s(k) = \mathcal{T}u_s(k) + \mathcal{B}u_p(k-p) - \mathcal{A}y_p(k-p). \tag{11.1}$$

Here the vectors $y_s(k)$ and $y_p(k - p)$ are defined as

$$y_s(k) = \begin{bmatrix} y(k) \\ y(k+1) \\ \vdots \\ y(k+s-1) \end{bmatrix},$$

$$y_p(k - p) = \begin{bmatrix} y(k-p) \\ y(k-p+1) \\ \vdots \\ y(k-1) \end{bmatrix},$$

and $u_s(k)$ and $u_p(k - p)$ are similar to $y_s(k)$ and $y_p(k - p)$, with y replaced with u. Matrices T and A are

$$T = \begin{bmatrix} \beta_0 & & \\ \beta_0^{(1)} & \beta_0 & \\ \vdots & \vdots & \ddots \\ \beta_0^{(s-1)} & \beta_0^{(s-2)} & \cdots & \beta_0 \end{bmatrix},$$

(11.2)

$$A = \begin{bmatrix} \alpha_p & \alpha_{(p-1)} & \cdots & \alpha_1 \\ \alpha_p^{(1)} & \alpha_{p-1}^{(1)} & \cdots & \alpha_1^{(1)} \\ \vdots & \vdots & \ddots & \vdots \\ \alpha_p^{(s-1)} & \alpha_{p-1}^{(s-1)} & \cdots & \alpha_1^{(s-1)} \end{bmatrix}.$$

(11.3)

Matrix B is similar to A, with α replaced with β. In Eq. (11.1), p is the integer related to the system order and s is the integer related to the prediction horizon $h_p = s - 1$. The quantity $y_s(k)$ is the vector containing the current and predicted future outputs, whereas $u_s(k)$ is the vector containing the current and future control inputs yet to be determined. Also $y_p(k - p)$ is the vector containing the past system outputs, and $u_p(k - p)$ is the vector containing the past control inputs.

The GPC algorithm is based on system output predictions over a finite horizon known as the prediction horizon. To predict the future outputs, some assumptions need to be made about the future control inputs. In determining the future control inputs, it is assumed that control is applied only over a finite horizon known as the control horizon. Beyond the control horizon, the control input is assumed to be zero. In the GPC algorithm, the control horizon is always equal to or less than the prediction horizon. In addition to the horizons, a control penalty is introduced to limit the control effort and stabilize the closed-loop system. The cost function to be minimized in the GPC algorithm is

$$J(k) = \frac{1}{2}\{[y_s(k) - \tilde{y}_s(k)]^T[y_s(k) - \tilde{y}_s(k)] + u_s^T(k)\lambda u_s(k)\}.$$

(11.4)

In Eq. (11.4), $y_s(k)$ is the predicted output vector, $\tilde{y}_s(k)$ is the desired output vector, $u_s(k)$ is the control input vector, and λ is the control penalty scalar. For simplicity, both prediction horizon and control horizon are set to $s-1$ and the control penalty λ is assumed to be a positive scalar rather than a matrix. Minimizing Eq. (11.4) with respect to $u_s(k)$ yields

$$\frac{\partial J(k)}{\partial u_s(k)} = \left[\frac{\partial y_s(k)}{\partial u_s(k)}\right]^T [y_s(k) - \tilde{y}_s(k)] + \lambda u_s(k) \equiv 0. \tag{11.5}$$

Note from Eq. (11.1) that

$$\frac{\partial y_s(k)}{\partial u_s(k)} = \begin{bmatrix} \dfrac{\partial y(k)}{\partial u(k)} & \dfrac{\partial y(k)}{\partial u(k+1)} & \cdots & \dfrac{\partial y(k)}{\partial u(k+s-1)} \\[2mm] \dfrac{\partial y(k+1)}{\partial u(k)} & \dfrac{\partial y(k+1)}{\partial u(k+1)} & \cdots & \dfrac{\partial y(k+1)}{\partial u(k+s-1)} \\ \vdots & \vdots & \ddots & \vdots \\ \dfrac{\partial y(k+s-1)}{\partial u(k)} & \dfrac{\partial y(k+s-1)}{\partial u(k+1)} & \cdots & \dfrac{\partial y(k+s-1)}{\partial u(k+s-1)} \end{bmatrix} = \mathcal{T}. \tag{11.6}$$

Solving for $u_s(k)$ from Eq. (11.5) with the aid of Eq. (11.1) and Eq. (11.6) will give the control sequence to be applied to the system:

$$u_s(k) = -[(\mathcal{T}^T\mathcal{T} + \lambda I)^{-1}\mathcal{T}^T][\mathcal{B}u_p(k-p) - \mathcal{A}y_p(k-p) - \tilde{y}_s(k)], \tag{11.7}$$

where I is an identity matrix. The first r values of the control sequence are applied to the r control inputs, the remainder is discarded, and a new control sequence is calculated for the next time step. To carry out the above process, the desired outputs $\tilde{y}(k), \tilde{y}(k+1), \ldots, \tilde{y}(k+s-1)$ must be given.

In the regulation problem, the desired system output is zero. Taking the first r rows with zero desired outputs results in

$$\begin{aligned} u(k) &= \text{the first } r \text{ rows of } [-(\mathcal{T}^T\mathcal{T} + \lambda I)^{-1}\mathcal{T}^T] \\ &\quad \times [\mathcal{B}u_p(k-p) - \mathcal{A}y_p(k-p)] \\ &= \alpha_1^c y(k-1) + \alpha_2^c y(k-2) + \cdots + \alpha_p^c y(k-p) \\ &\quad + \beta_1^c u(k-1) + \beta_2^c u(k-2) + \cdots + \beta_p^c u(k-p). \end{aligned} \tag{11.8}$$

The constant coefficient matrices $\alpha_1^c, \alpha_2^c, \ldots, \alpha_p^c$ are computed by

$$\begin{bmatrix} \alpha_1^c & \alpha_2^c & \cdots & \alpha_p^c \end{bmatrix} = \text{the first } r \text{ rows of } [(\mathcal{T}^T\mathcal{T} + \lambda I)^{-1}\mathcal{T}^T]\mathcal{A}$$

and $\beta_1^c, \beta_2^c, \cdots, \beta_p^c$ by

$$\begin{bmatrix} \beta_1^c & \beta_2^c & \cdots & \beta_p^c \end{bmatrix} = \text{the first } r \text{ rows of } [-(\mathcal{T}^T\mathcal{T} + \lambda I)^{-1}\mathcal{T}^T]\mathcal{B}.$$

The control force $u(k)$ at the current time step k is a weighted sum of the past input and output data back to the time step $k-p$. When $\lambda = 0$, the closed-loop system will be unstable for nonminimum phase systems because the matrix $\mathcal{T}$ is rank deficient. The quantity λ must be carefully tuned to make the system stable.

The formulation given in Eq. (11.8) assumes that the control horizon h_c is equal to the prediction horizon h_p. Nevertheless, the control horizon may be chosen to be less than the prediction horizon, resulting in a more stable and sluggish regulator. This is achieved if the matrix $\mathcal{T}$ in Eq. (11.8) is reduced to become

$$
\mathcal{T}_c =
\begin{bmatrix}
\beta_0 & & & \\
\beta_0^{(1)} & \beta_0 & & \\
\vdots & \vdots & \ddots & \\
\beta_0^{h_c} & \beta_0^{(h_c-1)} & \cdots & \beta_0 \\
\beta_0^{(h_c+1)} & \beta_0^{(h_c)} & \cdots & \beta_0^{(1)} \\
\vdots & \vdots & \vdots & \vdots \\
\beta_0^{(s-1)} & \beta_0^{(s-2)} & \cdots & \beta_0^{(s-h_c-1)}
\end{bmatrix}.
\tag{11.9}
$$

The control sequence determined with Eq. (11.9) in Eq. (11.8) is for a shorter control horizon, i.e., $h_c < s - 1$. Beyond the control horizon the control input is assumed to be zero.

EXAMPLE 11.1

Consider the discrete-time model

$$
A = \begin{bmatrix} \cos\gamma & \sin\gamma \\ -\sin\gamma & \cos\gamma \end{bmatrix},
$$

$$
B = \begin{bmatrix} 1 - \cos\gamma \\ \sin\gamma \end{bmatrix},
$$

$$
C = [-1 \quad 0],
$$

$$
D = d,
$$

where γ is a constant value. Note that the discrete-time model represents a mass–spring system with unit mass and unit stiffness at the sampling interval γ. From Example 10.10, the deadbeat-observer gain matrix is

$$
G = \begin{bmatrix} 2\cos\gamma \\ -\sin\gamma + \dfrac{\cos^2\gamma}{\sin\gamma} \end{bmatrix}.
$$

The matrices $\bar{A}$ and $\bar{B}$ in Eq. (10.53) become

$$
\begin{aligned}
\bar{A} &= A + GC \\
&= \begin{bmatrix} \cos\gamma & \sin\gamma \\ -\sin\gamma & \cos\gamma \end{bmatrix} + \begin{bmatrix} 2\cos\gamma \\ -\sin\gamma + \dfrac{\cos^2\gamma}{\sin\gamma} \end{bmatrix} [-1 \quad 0] \\
&= \begin{bmatrix} -\cos\gamma & \sin\gamma \\ -\dfrac{\cos^2\gamma}{\sin\gamma} & \cos\gamma \end{bmatrix},
\end{aligned}
$$

$$\bar{B} = [B + GD, \ -G]$$

$$= \left(\begin{bmatrix} 1 - \cos\gamma \\ \sin\gamma \end{bmatrix} + d \begin{bmatrix} 2\cos\gamma \\ -\sin\gamma + \dfrac{\cos^2\gamma}{\sin\gamma} \end{bmatrix}, \ - \begin{bmatrix} 2\cos\gamma \\ -\sin\gamma + \dfrac{\cos^2\gamma}{\sin\gamma} \end{bmatrix} \right)$$

$$= \begin{bmatrix} 1 + 2d\cos\gamma - \cos\gamma & -2\cos\gamma \\ \sin\gamma - d\sin\gamma + d\dfrac{\cos^2\gamma}{\sin\gamma} & \sin\gamma - \dfrac{\cos^2\gamma}{\sin\gamma} \end{bmatrix}.$$

Note that $\bar{A}^2 = \bar{A}^3 = \cdots = \bar{A}^n = \cdots = 0$, as proved in Example 10.10.

The coefficients α_i for the corresponding finite-difference model are computed by Eq. (10.57a), as follows,

$$\alpha_1 = CG = \begin{bmatrix} -1 & 0 \end{bmatrix} \begin{bmatrix} 2\cos\gamma \\ -\sin\gamma + \dfrac{\cos^2\gamma}{\sin\gamma} \end{bmatrix} = -2\cos\gamma,$$

$$\alpha_2 = C(A + GC)G$$

$$= \begin{bmatrix} -1 & 0 \end{bmatrix} \begin{bmatrix} -\cos\gamma & \sin\gamma \\ -\dfrac{\cos^2\gamma}{\sin\gamma} & \cos\gamma \end{bmatrix} \begin{bmatrix} 2\cos\gamma \\ -\sin\gamma + \dfrac{\cos^2\gamma}{\sin\gamma} \end{bmatrix} = 1,$$

$$\alpha_3 = \alpha_4 = \cdots = \alpha_\infty = 0,$$

and similarly, from Eqs. (10.57b) and (10.57c), the coefficients β_i are

$$\beta_0 = d,$$

$$\beta_1 = C(B + GD) = \begin{bmatrix} -1 & 0 \end{bmatrix} \begin{bmatrix} 1 + 2d\cos\gamma - \cos\gamma \\ \sin\gamma - d\sin\gamma + d\dfrac{\cos^2\gamma}{\sin\gamma} \end{bmatrix}$$

$$= \cos\gamma - 1 - 2d\cos\gamma,$$

$$\beta_2 = C(A + GC)(B + GD)$$

$$= \begin{bmatrix} -1 & 0 \end{bmatrix} \begin{bmatrix} -\cos\gamma & \sin\gamma \\ -\dfrac{\cos^2\gamma}{\sin\gamma} & \cos\gamma \end{bmatrix} \begin{bmatrix} 1 + 2d\cos\gamma - \cos\gamma \\ \sin\gamma - d\sin\gamma + d\dfrac{\cos^2\gamma}{\sin\gamma} \end{bmatrix}$$

$$= \cos\gamma - 1 + d,$$

$$\beta_3 = \beta_4 = \cdots = \beta_\infty = 0.$$

The linear finite-difference model, Eq. (10.56), can then be written as

$$y(k) - (2\cos\gamma)\,y(k-1) + y(k-2)$$
$$= u(k) + (\cos\gamma - 1 - 2d\cos\gamma)\,u(k-1) + (\cos\gamma - 1 + d)\,u(k-2).$$

In practice, the coefficients $\alpha_1, \alpha_2, \beta_0, \beta_1$, and β_2 for the finite-difference model should be obtained from input and output time history as described in Chap. 10.

Application of Eqs. (10.19) and (10.20) for $j = 1$ yields

$$\alpha_1^{(1)} = \alpha_2 - \alpha_1\alpha_1 = 1 - 4\cos^2\gamma,$$

$$\alpha_2^{(1)} = -\alpha_1\alpha_2 = 2\cos\gamma,$$

$$\beta_1^{(1)} = \beta_2 - \alpha_1\beta_1 = (\cos\gamma - 1)(1 + 2\cos\gamma) + d - 4d\cos^2\gamma,$$
$$\beta_2^{(1)} = -\alpha_1\beta_2 = 2\cos\gamma(\cos\gamma - 1 + d).$$

Using Eq. (10.21) for $j = 1$ yields

$$\beta_0^{(1)} = \cos\gamma - 1.$$

The constant matrices $\mathcal{T}$, $\mathcal{A}$, and $\mathcal{B}$ shown in Eq. (11.1) become

$$\mathcal{T} = \begin{bmatrix} \beta_0 & 0 \\ \beta_0^{(1)} & \beta_0 \end{bmatrix} = \begin{bmatrix} d & 0 \\ \cos\gamma - 1 & d \end{bmatrix},$$

$$\mathcal{A} = \begin{bmatrix} \alpha_2 & \alpha_1 \\ \alpha_2^{(1)} & \alpha_1^{(1)} \end{bmatrix} = \begin{bmatrix} 1 & -2\cos\gamma \\ 2\cos\gamma & 1 - 4\cos^2\gamma \end{bmatrix},$$

$$\mathcal{B} = \begin{bmatrix} \beta_2 & \beta_1 \\ \beta_2^{(1)} & \beta_1^{(1)} \end{bmatrix}$$

$$= \begin{bmatrix} \cos\gamma - 1 + d & \cos\gamma - 1 + 2d\cos\gamma \\ 2\cos\gamma(\cos\gamma - 1 + d) & (\cos\gamma - 1)(1 + 2\cos\gamma) + d - 4d\cos^2\gamma \end{bmatrix}.$$

It is interesting to note that the coefficient matrices $\mathcal{T}$, $\mathcal{A}$, and $\mathcal{B}$ are functions of $\cos\gamma$ and d only, without involving $\sin\gamma$. For simplicity without losing generality, let us assume that

$$\gamma = (2i + 1)\pi/2 \quad \Longrightarrow \quad \cos\gamma = 0,$$

where i is an arbitrary positive integer. The output-prediction equation with $h_p = 1$ (prediction horizon) becomes

$$\begin{bmatrix} y(k) \\ y(k+1) \end{bmatrix} = \begin{bmatrix} d & 0 \\ -1 & d \end{bmatrix} \begin{bmatrix} u(k) \\ u(k+1) \end{bmatrix} - \begin{bmatrix} 1 & 0 \\ 0 & 1 \end{bmatrix} \begin{bmatrix} y(k-2) \\ y(k-1) \end{bmatrix}$$
$$+ \begin{bmatrix} d-1 & -1 \\ 0 & d-1 \end{bmatrix} \begin{bmatrix} u(k-2) \\ u(k-1) \end{bmatrix}.$$

Application of Eq. (11.7) with control weighting λ and $h_c = 1$ (control horizon) thus yields

$$\begin{bmatrix} u(k) \\ u(k+1) \end{bmatrix} = \begin{bmatrix} \dfrac{d(d^2 + \lambda)}{(d^2 + \lambda)^2 + \lambda} & \dfrac{-\lambda}{(d^2 + \lambda)^2 + \lambda} \\[2ex] \dfrac{d^2}{(d^2 + \lambda)^2 + \lambda} & \dfrac{d(d^2 + \lambda)}{(d^2 + \lambda)^2 + \lambda} \end{bmatrix} \begin{bmatrix} y(k-2) \\ y(k-1) \end{bmatrix}$$

$$+ \begin{bmatrix} \dfrac{(d-1)d(d^2 + \lambda)}{(d^2 + \lambda)^2 + \lambda} & \dfrac{d^3 + 2\lambda d - \lambda}{(d^2 + \lambda)^2 + \lambda} \\[2ex] \dfrac{(d-1)d^2}{(d^2 + \lambda)^2 + \lambda} & \dfrac{d\{\lambda - d[(d-1)d + a - 1]\}}{(d^2 + \lambda)^2 + \lambda} \end{bmatrix} \begin{bmatrix} u(k-2) \\ u(k-1) \end{bmatrix}.$$

It is quite obvious that the above control equation has no solution if both λ and d are zero. However, as long as λ is not zero, the equation has a solution. Let us assume that

$$d = 0, \qquad \lambda \neq 0.$$

The control force at $u(k)$ becomes

$$u(k) = \frac{-1}{\lambda + 1}[y(k - 1) + u(k - 1)].$$

Note again that this equation is not valid for $\lambda = 0$. Without the presence of the output $y(k - 1)$, the control force $u(k)$ will decay eventually because

$$u(k) = \left[\frac{-1}{\lambda + 1}\right]^{k-1} u(1) \;\to\; 0 \quad \text{for} \quad \lambda > 0 \quad \text{and} \quad k \to \infty.$$

Such a controller is commonly referred to as a stable controller (see Problem 11.6 for general cases).

COMPUTATIONAL STEPS

The generalized predictive control algorithm is summarized as follows:

(1) Use any system identification (batch or recursive) technique to determine the open-loop OMPs (ARX) $\alpha_1, \ldots, \alpha_p$, and $\beta_0, \beta_1, \ldots, \beta_p$, before the control action is turned on. The integer p must be properly chosen such that $pm \geq n$, where n is the order of the system and m is the number of outputs.
(2) Compute the system Markov parameters (pulse-response sequence) with the recursive formula, Eq. (10.21), and form the Toeplitz matrix $\mathcal{T}$ shown in Eq. (11.2). The prediction horizon h_p must be chosen larger than p.
(3) Form matrices $\mathcal{B}$ and $\mathcal{A}$ shown in Eq. (11.3) with their elements computed with the recursive formula, Eqs. (10.19) and (10.20).
(4) Choose a nonzero positive value for the control weighting λ and insert it into Eq. (11.8) to compute the feedback-control parameters $\alpha_1^c, \alpha_2^c, \ldots, \alpha_p^c$ and $\beta_1^c, \beta_2^c, \ldots, \beta_p^c$.

In theory, the smaller the value of λ, the larger the control magnitude, and hence the better the performance of GPC. However, there is usually a practical limit on control magnitude and so the performance and control magnitude should be compromised, which is controlled by λ. An iterative procedure may be necessary to tune the control weighting to meet the performance requirements.

11.3 Deadbeat Predictive Control

The GPC controller is very attractive because it can be easily implemented from input and output data. One drawback is that the control weighting must be chosen before implementation. One way is to develop a logical procedure to determine the weighting. Another way is to eliminate the control weighting if possible without compromising the performance requirements. The deadbeat controller introduced in this section is a GPC-like controller without control weighting.

To motivate the design of a different feedback controller, consider the following question: What should the future input signal $u(k)$, $u(k+1)$, $\ldots$, $u(k+q-1)$ be to make the future output sequence $y(k+q)$, $y(k+q+1)$, $\ldots$, ∞ equal to zero (deadbeat)? Here we have assumed that the control action starts at time step k. Before time step k, the system is open loop.

Let the control action be turned on at time step k and ended at $k+q$. In other words, the control action occurs only from $u(k)$ to $u(k+q-1)$, beyond which the control action is zero, i.e., $u(k+q) = u(k+q+1) = \cdots = 0$. Under this condition, Eq. (11.1) produces the following equation:

$$y_p(k+q) = T'u_q(k) + B'u_p(k-p) - A'y_p(k-p), \tag{11.10}$$

where $y_p(k+q)$ and $u_q(k)$ are

$$y_p(k+q) = \begin{bmatrix} y(k+q) \\ y(k+q+1) \\ \vdots \\ y(k+q+p-1) \end{bmatrix}, \tag{11.11a}$$

$$u_q(k) = \begin{bmatrix} u(k) \\ u(k+1) \\ \vdots \\ u(k+q-1) \end{bmatrix}, \tag{11.11b}$$

and T', B', and A' are

$$T' = \begin{bmatrix} \beta_0^{(q)} & \beta_0^{(q-1)} & \cdots & \beta_0^{(1)} \\ \beta_0^{(q+1)} & \beta_0^{(q)} & \cdots & \beta_0^{(2)} \\ \vdots & \vdots & \ddots & \vdots \\ \beta_0^{(q+p-1)} & \beta_0^{(q+p-2)} & \cdots & \beta_0^{(p)} \end{bmatrix}, \tag{11.12a}$$

$$B' = \begin{bmatrix} \beta_p^{(q)} & \beta_{p-1}^{(q)} & \cdots & \beta_1^{(q)} \\ \beta_p^{(q+1)} & \beta_{p-1}^{(q+1)} & \cdots & \beta_1^{(q+1)} \\ \vdots & \vdots & \ddots & \vdots \\ \beta_p^{(q+p-1)} & \beta_{p-1}^{(q+p-1)} & \cdots & \beta_1^{(q+p-1)} \end{bmatrix}, \tag{11.12b}$$

$$A' = \begin{bmatrix} \alpha_p^{(q)} & \alpha_{p-1}^{(q)} & \cdots & \alpha_1^{(q)} \\ \alpha_p^{(q+1)} & \alpha_{p-1}^{(q+1)} & \cdots & \alpha_1^{(q+1)} \\ \vdots & \vdots & \ddots & \vdots \\ \alpha_p^{(q+p-1)} & \alpha_{p-1}^{(q+p-1)} & \cdots & \alpha_1^{(q+p-1)} \end{bmatrix}. \tag{11.12c}$$

Equation (11.10) is a reduced version of Eq. (11.1) with its first q equations and the equations beyond $q + p - 1$ removed. The matrix T' of dimension $pm \times qr$ is formed from the pulse response (system Markov parameters). Note that m is the number of outputs, p is the order of the ARX model (finite-difference model), r is the number of inputs, and q is the number of control steps. If we flip the columns in the left/right direction and preserve the rows of T', it becomes a Hankel matrix of the pulse response, i.e.,

$$H = \begin{bmatrix} \beta_0^{(1)} & \beta_0^{(2)} & \cdots & \beta_0^{(q)} \\ \beta_0^{(2)} & \beta_0^{(3)} & \cdots & \beta_0^{(q+1)} \\ \vdots & \vdots & \ddots & \vdots \\ \beta_0^{(p)} & \beta_0^{(p+1)} & \cdots & \beta_0^{(q+p-1)} \end{bmatrix}. \tag{11.13}$$

The Hankel matrix is known to have maximum rank of n, which is the order of the system if $pm \geq n$, where m is the number of outputs, p is a selectable integer, and n is the system order. The integer p must be selected such that $pm \geq n$ but the rank of the Hankel matrix would not be increased by choosing a large p. That is why the number of rows for T' is chosen to be pm, although any number greater than pm may be used to form Eq. (11.10). The integer q must also be chosen such that $qr \geq n$ to make sure that the Hankel matrix has rank n.

The output vector $y_p(k + q)$ in Eq. (11.10) includes the output sequence from the time step $k + q$ to $k + q + p - 1$. It depends on the input vector $u_q(k)$, consisting of the input sequence from the time step k to $k + q - 1$, which is one step before the step $k + q$ for the first output in $y_p(k + q)$. It also relies on $u_p(k - p)$ and $y_p(k - p)$, consisting of the input and output sequences from the time step $k - p$ to $k - 1$. The significance of Eq. (11.10) is that the input and output relation has been rewritten so that the output at time $k + q$ and beyond can be computed from the input sequence from $k - p$ to $k + q - 1$ and the output sequence from $k - p$ to $k - 1$. In other words, the output sequence from k to $k + q - 1$ is not required to be known for the prediction of the output at the time $k + q$ and beyond. This prediction characteristic can be capitalized on for the feedback design shown below.

From Eq. (11.10), it is clear that the following equality,

$$u_q(k) = -[T']^\dagger [\mathcal{B}' u_p(k - p) - \mathcal{A}' y_p(k - p)], \tag{11.14}$$

will bring $y_p(k + q)$ to rest, i.e.,

$$y_p(k + q) = \begin{bmatrix} y(k + q) \\ y(k + q + 1) \\ \vdots \\ y(k + q + p - 1) \end{bmatrix} = 0_{pm \times 1},$$

where $0_{pm \times 1}$ is a pm$\times$1 zero vector. The first r rows of Eq. (11.14) thus give

$$\begin{aligned} u(k) &= -\text{first } r \text{ rows of } \{[T']^\dagger\} [\mathcal{B}' u_p(k - p) - \mathcal{A}' y_p(k - p)] \\ &= \alpha_1^c y(k - 1) + \alpha_2^c y(k - 2) + \cdots + \alpha_p^c y(k - p) \\ &\quad + \beta_1^c u(k - 1) + \beta_2^c u(k - 2) + \cdots + \beta_p^c u(k - p), \end{aligned} \tag{11.15}$$

where the superscript c signifies the control parameters. The feedback-control parameters $\alpha_1^c, \ldots, \alpha_p^c$ and $\beta_1^c, \ldots, \beta_p^c$ are to be used to compute the current control signal $u(k)$ with the past p input and output measurements. The control action is supposed to bring the output to zero for all time steps larger than $k + q$. Along with the desired zero input $u(k + q)$ and beyond, the system should be at rest, i.e., deadbeat, beyond time step $k + q$. That is in theory. In practice, when the system has input and output uncertainties, the control action can only bring the output down to the level of the uncertainties.

EXAMPLE 11.2

For the purpose of comparing GPC and DPC, let us repeat Example 11.1 by using a DPC controller. First we need to compute the elements in the matrices T', A', and B' shown in Eq. (11.12). Given the constant values α_1, α_2, $\alpha_1^{(1)}$, $\alpha_2^{(1)}$, β_1, β_2, $\beta_1^{(1)}$, $\beta_2^{(1)}$, and β_0, from Example 11.1, application of Eqs. (10.19) and (10.20) for $j = 2$ and $j = 3$ yields

$$\alpha_1^{(2)} = \alpha_2^{(1)} - \alpha_1^{(1)}\alpha_1 = -2[\cos\gamma + \cos(3\gamma)],$$

$$\alpha_2^{(2)} = -\alpha_1^{(1)}\alpha_2 = -1 + 4\cos^2\gamma,$$

$$\alpha_1^{(3)} = \alpha_2^{(2)} - \alpha_1^{(2)}\alpha_1 = -1 - 4\cos\gamma\cos(3\gamma),$$

$$\alpha_2^{(3)} = -\alpha_1^{(2)}\alpha_2 = 2[\cos\gamma + \cos(3\gamma)];$$

$$\beta_1^{(2)} = \beta_2^{(1)} - \alpha_1^{(1)}\beta_1 = -\cos(2\gamma) + \cos(3\gamma) - 2d[\cos\gamma + \cos(3\gamma)],$$

$$\beta_2^{(2)} = -\alpha_1^{(1)}\beta_2 = (1 - d - \cos\gamma)(1 - 4\cos^2\gamma),$$

$$\beta_1^{(3)} = \beta_2^{(2)} - \alpha_1^{(2)}\beta_1$$
$$\qquad = -d - \cos(3\gamma) + \cos(4\gamma) - 2d[\cos(2\gamma) + \cos(4\gamma)],$$

$$\beta_2^{(3)} = -\alpha_1^{(2)}\beta_2 = 2(1 - d - \cos\gamma)[\cos\gamma + \cos(3\gamma)].$$

Using Eq. (10.21) for $j = 2$ and $j = 3$ yields

$$\beta_0^{(2)} = \beta_1^{(1)} - \alpha_1^{(1)}\beta_0 = -\cos\gamma + \cos(2\gamma),$$

$$\beta_0^{(3)} = \beta_1^{(2)} - \alpha_1^{(2)}\beta_0 = -\cos(2\gamma) + \cos(3\gamma).$$

For simplicity, let us set

$$d = 0.$$

The constant matrices T', A', and B' in Eq. (11.12) become

$$T' = \begin{bmatrix} \beta_0^{(2)} & \beta_0^{(1)} \\ \beta_0^{(3)} & \beta_0^{(2)} \end{bmatrix} = \begin{bmatrix} -\cos\gamma + \cos(2\gamma) & \cos\gamma - 1 \\ -\cos(2\gamma) + \cos(3\gamma) & -\cos\gamma + \cos(2\gamma) \end{bmatrix},$$

$$A' = \begin{bmatrix} \alpha_2^{(2)} & \alpha_1^{(2)} \\ \alpha_2^{(3)} & \alpha_1^{(3)} \end{bmatrix} = \begin{bmatrix} -1 + 4\cos^2\gamma & -2[\cos\gamma + \cos(3\gamma)] \\ 2[\cos\gamma + \cos(3\gamma)] & -1 - 4\cos\gamma\cos(3\gamma) \end{bmatrix},$$

$$\mathcal{B}' = \begin{bmatrix} \beta_2^{(2)} & \beta_1^{(2)} \\ \beta_2^{(3)} & \beta_1^{(3)} \end{bmatrix}$$

$$= \begin{bmatrix} (1 - \cos\gamma)(1 - 4\cos^2\gamma) & -\cos(2\gamma) + \cos(3\gamma) \\ 2(1 - \cos\gamma)[\cos\gamma + \cos(3\gamma)] & -\cos(3\gamma) + \cos(4\gamma) \end{bmatrix}.$$

The determinant of matrix $\mathcal{T}'$ is

$$\|\mathcal{T}'\| = 16\cos^2\frac{\gamma}{2}\sin^4\frac{\gamma}{2} = 4\sin^2\gamma\,\sin^2\frac{\gamma}{2}.$$

As long as the constant γ is chosen such that

$$\gamma \neq i\pi,$$

where i is zero or a positive integer, the matrix $\mathcal{T}'$ is nonsingular and thus invertible. The system matrices A, B, C, and D shown in Example 11.1 represent a single-degree-of-freedom mass–spring system with unit mass and unit spring constant producing the system frequency at

$$\omega = \sqrt{\frac{\text{spring constant}}{\text{mass}}} = \sqrt{\frac{1}{1}} = 1 \text{ rad/s} = \frac{1}{2\pi} \text{ Hz}.$$

To identify the system frequency, the sampling frequency must be higher than twice the system frequency, i.e., $1/\pi$ Hz. This means that the sampling interval must be shorter than π.

Assume that

$$\gamma = \pi/2 \implies \cos\gamma = 0, \quad \sin\gamma = 1.$$

The output-prediction equation, Eq. (11.10), for deadbeat-control designs becomes

$$\begin{bmatrix} y(k+2) \\ y(k+3) \end{bmatrix} = \begin{bmatrix} -1 & -1 \\ 1 & -1 \end{bmatrix}\begin{bmatrix} u(k) \\ u(k+1) \end{bmatrix} - \begin{bmatrix} -1 & 0 \\ 0 & -1 \end{bmatrix}\begin{bmatrix} y(k-2) \\ y(k-1) \end{bmatrix}$$

$$+ \begin{bmatrix} 1 & 1 \\ 0 & 1 \end{bmatrix}\begin{bmatrix} u(k-2) \\ u(k-1) \end{bmatrix}.$$

To make $y(k+2)$ and $y(k+3)$ equal to zero, $u(k)$ and $u(k+1)$ must satisfy the following equation:

$$\begin{bmatrix} -1 & -1 \\ 1 & -1 \end{bmatrix}\begin{bmatrix} u(k) \\ u(k+1) \end{bmatrix} = -\begin{bmatrix} y(k-2) \\ y(k-1) \end{bmatrix} - \begin{bmatrix} 1 & 1 \\ 0 & 1 \end{bmatrix}\begin{bmatrix} u(k-2) \\ u(k-1) \end{bmatrix}$$

or

$$\begin{bmatrix} u(k) \\ u(k+1) \end{bmatrix} = \frac{1}{2}\begin{bmatrix} 1 & -1 \\ 1 & 1 \end{bmatrix}\begin{bmatrix} y(k-2) \\ y(k-1) \end{bmatrix} + \frac{1}{2}\begin{bmatrix} 1 & 0 \\ 1 & 2 \end{bmatrix}\begin{bmatrix} u(k-2) \\ u(k-1) \end{bmatrix}.$$

The control force $u(k)$ becomes

$$u(k) = \frac{1}{2}[y(k-2) - y(k-1) + u(k-2)].$$

This result is indeed different from the one computed in Example 11.1.

COMPUTATIONAL STEPS

The indirect method for predictive control design is summarized as follows:

(1) Use any system identification (batch or recursive) technique to determine the open-loop OMPs $\alpha_1, \ldots, \alpha_p$, and $\beta_0, \beta_1, \ldots, \beta_p$, before the control action is turned on. The integer p must be properly chosen such that $pm \geq n$, where n is the order of the system and m is the number of outputs.

(2) Compute the system Markov parameters (pulse-response sequence) with the recursive formula, Eq. (10.21), and form the Toeplitz matrix T' shown in Eq. (11.12).

(3) Form matrices $\mathcal{A}'$ and $\mathcal{B}'$ shown in Eq. (11.12) with their elements computed with the recursive formula, Eqs. (10.19) and (10.20).

(4) Make sure that T' is invertible and use Eq. (11.15) to compute the feedback-control parameters $\alpha_1^c, \alpha_2^c, \ldots, \alpha_p^c$ and $\beta_1^c, \beta_2^c, \ldots, \beta_p^c$. If the condition number of T' is very poor, resulting in an inaccurate inverse of T', the value of the integer p should be reduced.

11.4 Direct Algorithm for GPC and DPC

The GPC and DPC algorithms derived above are based on Eq. (11.1) with the assumption that the parameters $\alpha_1, \alpha_2, \ldots, \alpha_p$ and $\beta_0, \beta_1, \ldots, \beta_p$ are given a priori. The direct algorithm derived in this section uses the input and output data directly without explicitly involving the parameters $\alpha_1, \alpha_2, \ldots, \alpha_p$ and $\beta_0, \beta_1, \ldots, \beta_p$. One may be interested in computing the feedback-control parameters $\alpha_1^c, \alpha_2^c, \ldots, \alpha_p^c$ and $\beta_1^c, \beta_2^c, \ldots, \beta_p^c$ shown in Eq. (11.15) directly from input and output data, that is to bypass the first three steps presented in the previous sections for GPC and DPC.

11.4.1 Direct GPC Algorithm

The direct GPC algorithm starts with Eqs. (10.25) and (10.26) and forms the following input and output matrices:

$$Y_s(k) = [\, y_s(k) \quad y_s(k+1) \quad \cdots \quad y_s(k+N-1) \,]$$

$$= \begin{bmatrix} y(k) & y(k+1) & \cdots & y(k+N-1) \\ y(k+1) & y(k+2) & \cdots & y(k+N) \\ \vdots & \vdots & \ddots & \vdots \\ y(k+s-1) & y(k+s) & \cdots & y(k+s+N-2) \end{bmatrix}, \quad (11.16a)$$

$$U_s(k) = [u_s(k) \quad u_s(k+1) \quad \cdots \quad u_s(k+N-1)]$$

$$= \begin{bmatrix} u(k) & u(k+1) & \cdots & u(k+N-1) \\ u(k+1) & u(k+2) & \cdots & u(k+N) \\ \vdots & \vdots & \ddots & \vdots \\ u(k+s-1) & u(k+s) & \cdots & u(k+s+N-2) \end{bmatrix}, \qquad (11.16b)$$

where N is an integer. Similar to $Y_s(k)$ and $U_s(k)$, the matrices $Y_p(k-p)$ and $U_p(k-p)$ are defined when s is replaced with p and k with $k-p$. The data matrices $U_s(k)$ and $Y_s(k)$ include the input and the output data up to the time step $k+s+N-2$, whereas $U_p(k-p)$ and $Y_p(k-p)$ have data up to the time step $k+N-2$.

Application of Eq. (11.1) yields

$$Y_s(k) = \mathcal{T} U_s(k) + \mathcal{B} U_p(k-p) - \mathcal{A} Y_p(k-p) \qquad (11.17)$$

or

$$Y_s(k) = [\mathcal{T} \quad \mathcal{B} \quad \mathcal{A}] \begin{bmatrix} U_s(k) \\ U_p(k-p) \\ -Y_p(k-p) \end{bmatrix}. \qquad (11.18)$$

Let the integers s and N be chosen large enough such that the matrix $U_s(k)$ of dimension $sr \times N$ with $sr \le N$ has rank sr, the matrix $U_p(k-p)$ of dimension $pr \times N$ with $pr \le N$ has rank pr, and the matrix $Y_p(k-p)$ of dimension $pm \times N$ with $pm \le N$ has rank pm. Again, r and m denote the number of inputs and outputs, respectively. Equation (11.17) produces the following least-squares solution:

$$[\mathcal{T} \quad \mathcal{B} \quad \mathcal{A}] = Y_s(k) \begin{bmatrix} U_s(k) \\ U_p(k-p) \\ -Y_p(k-p) \end{bmatrix}^{\dagger}, \qquad (11.19)$$

where $\dagger$ means the pseudoinverse.

The matrices $\mathcal{T}$, $\mathcal{B}$, and $\mathcal{A}$ solved in Eq. (11.19) can be used in Eq. (11.8) to compute the control parameters $\alpha_1^c, \alpha_2^c, \ldots, \alpha_p^c$ and $\beta_1^c, \beta_2^c, \ldots, \beta_p^c$ for a GPC controller. In theory, they must produce the same GPC controller with the assumption that the input and the output data are noise free.

11.4.2 Direct DPC Algorithm

From the triple $[\mathcal{T}, \mathcal{B}, \mathcal{A}]$, it is easy to extract the triple $[\mathcal{T}', \mathcal{B}', \mathcal{A}']$ defined in Eq. (11.12) for computing the DPC control parameters $\alpha_1^c, \alpha_2^c, \ldots, \alpha_p^c$ and β_1^c, $\beta_2^c, \ldots, \beta_p^c$ in Eq. (11.15). Equation (11.17) has some redundant equations that may be eliminated to directly compute the triple $[\mathcal{T}', \mathcal{B}', \mathcal{A}']$ without computing $[\mathcal{T}, \mathcal{B}, \mathcal{A}]$. Indeed, let us set

$$s = q + p$$

and delete the first qm rows of Eq. (11.19). Equation (11.19) reduces to

$$[\mathcal{T}'' \quad \mathcal{B}' \quad \mathcal{A}'] = Y_p(k+q) \begin{bmatrix} U_{q+p}(k) \\ U_p(k-p) \\ -Y_p(k-p) \end{bmatrix}^\dagger, \tag{11.20}$$

where $\mathcal{T}''$, $\mathcal{B}'$, and $\mathcal{A}'$ should be identical to those obtained by deletion of the first qm rows of $\mathcal{T}$, $\mathcal{B}$, and $\mathcal{A}$, respectively. The matrices $\mathcal{B}'$ and $\mathcal{A}'$ are identical to those defined in Eq. (11.12). The matrix $\mathcal{T}''$ has more columns than $\mathcal{T}'$, defined in Eq. (11.12), i.e.,

$$\mathcal{T}' = \mathcal{T}''(:, 1 : qr), \tag{11.21}$$

where $\mathcal{T}''(:, 1 : qr)$ contains all rows and columns from 1 to qr of $\mathcal{T}''$. Now the data matrices become

$$Y_p(k+q) = [y_p(k+q) \quad \cdots \quad y_p(k+q+N-1)]$$

$$= \begin{bmatrix} y(k+q) & \cdots & y(k+q+N-1) \\ y(k+q+1) & \cdots & y(k+q+N) \\ \vdots & \ddots & \vdots \\ y(k+q+p-1) & \cdots & y(k+q+p+N-2) \end{bmatrix}, \tag{11.22a}$$

$$U_{q+p}(k) = [u_{q+p}(k) \quad \cdots \quad u_{q+p}(k+N-1)]$$

$$= \begin{bmatrix} u(k) & \cdots & u(k+N-1) \\ u(k+1) & \cdots & u(k+N) \\ \vdots & \ddots & \vdots \\ u(k+q+p-1) & \cdots & u(k+q+p+N-2) \end{bmatrix}. \tag{11.22b}$$

At this moment, all input and output data are measured from the open-loop system, before any control action begins.

From the triple $[\mathcal{T}'', \mathcal{B}', \mathcal{A}']$, the control law from Eq. (11.15) can be applied to compute the control gain parameters,

$$\begin{aligned} u(k) &= -\text{first } r \text{ rows of } \{[\mathcal{T}''(:, 1 : qr)]^\dagger\} [\mathcal{B}' u_p(k-p) - \mathcal{A}' y_p(k-p)] \\ &= \alpha_1^c y(k-1) + \alpha_2^c y(k-2) + \cdots + \alpha_p^c y(k-p) \\ &\quad + \beta_1^c u(k-1) + \beta_2^c u(k-2) + \cdots + \beta_p^c u(k-p) \end{aligned} \tag{11.23}$$

EXAMPLE 11.3
Given the input time history $u(k)$ for $k = 1, 2, \ldots, 15$,

$$u = [1 \quad 0 \quad 0 \quad 0 \quad -1 \quad 0 \quad 2 \quad -2 \quad 1 \quad -1 \quad 0 \quad 2 \quad 0 \quad 1 \quad 1],$$

and the output time history $y(k)$ for $k = 1, 2, \ldots, 15$,

$$y = [0 \quad -1 \quad -1 \quad 1 \quad 1 \quad 0 \quad 0 \quad -2 \quad 0 \quad 3 \quad 0 \quad -2 \quad -2 \quad 0 \quad 1]$$

The output time history was generated from the discrete-time model described by the system matrices A, B, C, and D shown in Example 11.1 with $\gamma = \pi/2$ and $d = 0$.

Let $s = 4$ and $k = 3$ be inserted into Eq. (11.16) to form the following input and output matrices:

$$
Y_4(3) = \begin{bmatrix}
y(3) & y(4) & \cdots & y(12) \\
y(4) & y(5) & \cdots & y(13) \\
\vdots & \vdots & \ddots & \vdots \\
y(6) & y(7) & \cdots & y(15)
\end{bmatrix}
$$

$$
= \begin{bmatrix}
-1 & 1 & 1 & 0 & 0 & -2 & 0 & 3 & 0 & -2 \\
1 & 1 & 0 & 0 & -2 & 0 & 3 & 0 & -2 & -2 \\
1 & 0 & 0 & -2 & 0 & 3 & 0 & -2 & -2 & 0 \\
0 & 0 & -2 & 0 & 3 & 0 & -2 & -2 & 0 & 1
\end{bmatrix},
$$

$$
U_4(3) = \begin{bmatrix}
u(3) & u(4) & \cdots & u(12) \\
u(4) & u(5) & \cdots & u(13) \\
\vdots & \vdots & \ddots & \vdots \\
u(6) & u(7) & \cdots & u(15)
\end{bmatrix}
$$

$$
= \begin{bmatrix}
0 & 0 & -1 & 0 & 2 & -2 & 1 & -1 & 0 & 2 \\
0 & -1 & 0 & 2 & -2 & 1 & -1 & 0 & 2 & 0 \\
-1 & 0 & 2 & -2 & 1 & -1 & 0 & 2 & 0 & 1 \\
0 & 2 & -2 & 1 & -1 & 0 & 2 & 0 & 1 & 1
\end{bmatrix}.
$$

Similarly, we can also form the following two matrices:

$$
Y_2(1) = \begin{bmatrix}
y(1) & y(2) & \cdots & y(10) \\
y(2) & y(3) & \cdots & y(11)
\end{bmatrix}
$$

$$
= \begin{bmatrix}
0 & -1 & -1 & 1 & 1 & 0 & 0 & -2 & 0 & 3 \\
-1 & -1 & 1 & 1 & 0 & 0 & -2 & 0 & 3 & 0
\end{bmatrix},
$$

$$
U_2(1) = \begin{bmatrix}
u(1) & u(2) & \cdots & u(10) \\
u(2) & u(3) & \cdots & u(11)
\end{bmatrix}
$$

$$
= \begin{bmatrix}
1 & 0 & 0 & 0 & -1 & 0 & 2 & -2 & 1 & -1 \\
0 & 0 & 0 & -1 & 0 & 2 & -2 & 1 & -1 & 0
\end{bmatrix}.
$$

Application of Eq. (11.18) yields

$$
Y_4(3) = \begin{bmatrix} T & B & A \end{bmatrix} \begin{bmatrix} U_4(3) \\ U_2(1) \\ -Y_2(1) \end{bmatrix}.
$$

This equation produces

$$[\mathcal{T} \ \vdots \ \mathcal{B} \ \vdots \ \mathcal{A}] = Y_4(3) \begin{bmatrix} U_4(3) \\ U_2(1) \\ -Y_2(1) \end{bmatrix}^\dagger$$

$$= \begin{bmatrix} 0 & 0 & 0 & 0 & \vdots & -1 & -1 & \vdots & 1 & 0 \\ -1 & 0 & 0 & 0 & \vdots & 0 & -1 & \vdots & 0 & 1 \\ -1 & -1 & 0 & 0 & \vdots & 1 & 1 & \vdots & -1 & 0 \\ 1 & -1 & -1 & 0 & \vdots & 0 & 1 & \vdots & 0 & -1 \end{bmatrix}.$$

Note that for the matrix

$$[\mathcal{T} \ \vdots \ \mathcal{B} \ \vdots \ \mathcal{A}]$$

of dimension 4×8 to be unique, the matrix

$$\begin{bmatrix} U_4(3) \\ U_2(1) \\ -Y_2(1) \end{bmatrix}$$

of dimensions 8×10 must have the rank of 8. It means that the number of columns must be equal to or larger than 8. Indeed, we may get the same solution with only 13 data points rather than 15 points used in this example. However, more data points will enhance the accuracy in the least-squares sense, in particular for the data with considerable uncertainties.

GPC Controller
For a linear time-invariant system, $\mathcal{T}$, $\mathcal{B}$, and $\mathcal{A}$ are constant matrices in the sense that they are valid for any time step. For a GPC controller, we may take the first two rows of the computed matrices $\mathcal{T}$, $\mathcal{B}$, and $\mathcal{A}$ to form the following output prediction equation:

$$\begin{bmatrix} y(k) \\ y(k+1) \end{bmatrix} = \begin{bmatrix} 0 & 0 \\ -1 & 0 \end{bmatrix} \begin{bmatrix} u(k) \\ u(k+1) \end{bmatrix} + \begin{bmatrix} -1 & -1 \\ 0 & -1 \end{bmatrix} \begin{bmatrix} u(k-2) \\ u(k-1) \end{bmatrix}$$
$$- \begin{bmatrix} 1 & 0 \\ 0 & 1 \end{bmatrix} \begin{bmatrix} y(k-2) \\ y(k-1) \end{bmatrix}.$$

This is identical to the output prediction equation shown in Example 11.1 with $d = 0$. With the control weighting λ, first note the following equality:

$$\left\{ \begin{bmatrix} 0 & 0 \\ -1 & 0 \end{bmatrix}^T \begin{bmatrix} 0 & 0 \\ -1 & 0 \end{bmatrix} + \begin{bmatrix} \lambda & 0 \\ 0 & \lambda \end{bmatrix} \right\}^{-1} \begin{bmatrix} 0 & 0 \\ -1 & 0 \end{bmatrix}^T = \begin{bmatrix} 0 & \frac{-1}{\lambda+1} \\ 0 & 0 \end{bmatrix}.$$

The GPC controller will then become

$$
\begin{bmatrix} u(k) \\ u(k+1) \end{bmatrix} = \begin{bmatrix} 0 & \dfrac{-1}{\lambda+1} \\ 0 & 0 \end{bmatrix} \left\{ \begin{bmatrix} 1 & 1 \\ 0 & 1 \end{bmatrix} \begin{bmatrix} u(k-2) \\ u(k-1) \end{bmatrix} + \begin{bmatrix} y(k-2) \\ y(k-1) \end{bmatrix} \right\},
$$

which implies that

$$
u(k) = \frac{-1}{\lambda+1}[u(k-1) + y(k-1)].
$$

DPC Controller

Taking the last two rows of the computed matrix T, B, and A and assuming that $u(k+3) = 0$ and $u(k+4) = 0$, we may form the following output-prediction equation for the DPC control algorithm:

$$
\begin{bmatrix} y(k+3) \\ y(k+4) \end{bmatrix} = \begin{bmatrix} -1 & -1 \\ 1 & -1 \end{bmatrix} \begin{bmatrix} u(k) \\ u(k+1) \end{bmatrix} + \begin{bmatrix} 1 & 1 \\ 0 & 1 \end{bmatrix} \begin{bmatrix} u(k-2) \\ u(k-1) \end{bmatrix}
$$
$$
+ \begin{bmatrix} 1 & 0 \\ 0 & 1 \end{bmatrix} \begin{bmatrix} y(k-2) \\ y(k-1) \end{bmatrix}.
$$

To make $y(k+3) = y(k+4) = 0$, the DPC controller must satisfy the following equation:

$$
\begin{bmatrix} u(k) \\ u(k+1) \end{bmatrix} = \begin{bmatrix} 1 & 1 \\ -1 & 1 \end{bmatrix}^{-1} \left\{ \begin{bmatrix} 1 & 1 \\ 0 & 1 \end{bmatrix} \begin{bmatrix} u(k-2) \\ u(k-1) \end{bmatrix} + \begin{bmatrix} y(k-2) \\ y(k-1) \end{bmatrix} \right\}
$$

or

$$
\begin{bmatrix} u(k) \\ u(k+1) \end{bmatrix} = \frac{1}{2}\begin{bmatrix} 1 & 0 \\ 1 & 2 \end{bmatrix} \begin{bmatrix} u(k-2) \\ u(k-1) \end{bmatrix} + \frac{1}{2}\begin{bmatrix} 1 & -1 \\ 1 & 1 \end{bmatrix} \begin{bmatrix} y(k-2) \\ y(k-1) \end{bmatrix}.
$$

The first equation yields the DPC controller

$$
u(k) = \frac{1}{2}[u(k-2) + y(k-2) - y(k-1)].
$$

COMPUTATIONAL STEPS

The following computational steps are involved in the direct algorithms for predictive control designs:

(1) Form the data matrices $Y_s(k)$ and $U_s(k)$, and $Y_p(k-p)$ and $U_p(k-p)$ defined in Eq. (11.16). The integer p must be chosen such that $pm \geq n$, where m is the number of outputs and n is the anticipated system order. The integer q is also chosen such that the Hankel matrix defined in Eq. (11.13) has rank n.
(2) Compute the least-squares solution, Eq. (11.19), to determine T, B, and A for the direct GPC algorithm, or Eq. (11.20), to determine T'', B', and A' for the direct DPC algorithm.

(3) Use Eq. (11.8) for GPC, or Eq. (11.23) for DPC, to compute the feedback control-parameters $\alpha_1^c, \alpha_2^c \ldots, \alpha_p^c$ and $\beta_1^c, \beta_2^c, \ldots, \beta_p^c$.

The direct algorithm may seem simpler than the indirect algorithm, which involves the computation of the OMPs. However, the direct method requires larger matrix manipulation in computing T, $\mathcal{B}$, and $\mathcal{A}$ from Eq. (11.19) or T'', $\mathcal{B}'$, and $\mathcal{A}'$ from Eq. (11.20).

11.5 State–Space Representation

Some researchers may be interested in knowing the corresponding state–space representation for the predictive control techniques described earlier. There are cases in which a state–space model is very useful in conducting controller designs. It provides them with flexibilities for real-time implementation.

Let us recall the following finite-difference model that may be identified from experimental data:

$$
\begin{aligned}
y(k) + \alpha_1 y(k-1) + \cdots + \alpha_p y(k-p) \\
= \beta_0 u(k) + \beta_1 u(k-1) + \cdots + \beta_p u(k-p).
\end{aligned}
\tag{11.24}
$$

There are several methods of converting it to a state–space model. We have shown one method in Section 10.6 of Chap. 10. Here another method is introduced for derivation of a predictive control method in state–space representation.

From Eq. (11.1) with $s = p$, let the state vector $x(k)$ of dimension $pm \times 1$ be defined as

$$
x(k) = y_p(k) - T u_p(k) = \mathcal{B} u_p(k-p) - \mathcal{A} y_p(k-p),
\tag{11.25}
$$

where $\mathcal{B}$ is a $pm \times pr$ matrix and $\mathcal{A}$ is a $pm \times pm$ matrix:

$$
\mathcal{B} =
\begin{bmatrix}
\beta_p & \beta_{p-1} & \cdots & \beta_1 \\
\beta_p^{(1)} & \beta_{p-1}^{(1)} & \cdots & \beta_1^{(1)} \\
\vdots & \vdots & \ddots & \vdots \\
\beta_p^{(p-1)} & \beta_{p-1}^{(p-1)} & \cdots & \beta_1^{(p-1)}
\end{bmatrix},
\tag{11.26a}
$$

$$
\mathcal{A} =
\begin{bmatrix}
\alpha_p & \alpha_{p-1} & \cdots & \alpha_1 \\
\alpha_p^{(1)} & \alpha_{p-1}^{(1)} & \cdots & \alpha_1^{(1)} \\
\vdots & \vdots & \ddots & \vdots \\
\alpha_p^{(p-1)} & \alpha_{p-1}^{(p-1)} & \cdots & \alpha_1^{(p-1)}
\end{bmatrix}.
\tag{11.26b}
$$

Equation (11.25) signifies the relationship between the state vector and the input and output data. It implies that the state at time step k can be estimated from the past p input and output data. This provides the basis for predictive control designs for a system represented by a state–space model.

11.5.1 Observable Canonical-Form Realization

The question now is how to build a state–space model based on the definition of the state vector. To answer this question, first partition the state vector as

$$x(k) = \begin{bmatrix} x_1(k) \\ x_2(k) \\ \vdots \\ x_{p-1}(k) \\ x_p(k) \end{bmatrix}. \tag{11.27}$$

Based on the definition of $\mathcal{T}$ in Eq. (11.2) and $x(k)$ in Eq. (11.25), each component $x_i(k)$ for $i = 1, 2, \ldots, p$ is a $m \times 1$ vector that has the following expression:

$$x_1(k) = y(k) - \beta_0 u(k), \tag{11.28a}$$

$$x_2(k) = y(k + 1) - \beta_0 u(k + 1) - \beta_0^{(1)} u(k), \tag{11.28b}$$

$$x_3(k) = y(k + 2) - \beta_0 u(k + 2) - \beta_0^{(1)} u(k + 1) - \beta_0^{(2)} u(k) \tag{11.28c}$$

$$\vdots$$

$$x_p(k) = y(k + p - 1)$$
$$\qquad - \beta_0 u(k + p - 1) - \beta_0^{(1)} u(k + p - 2) - \cdots - \beta_0^{(p-1)} u(k). \tag{11.28d}$$

The set of equations (11.28) yields

$$x_1(k + 1) = x_2(k) + \beta_0^{(1)} u(k), \tag{11.29a}$$

$$x_2(k + 1) = x_3(k) + \beta_0^{(2)} u(k), \tag{11.29b}$$

$$x_3(k + 1) = x_4(k) + \beta_0^{(3)} u(k) \tag{11.29c}$$

$$\vdots$$

$$x_p(k + 1) = y(k + p)$$
$$\qquad - \beta_0 u(k + p) - \beta_0^{(1)} u(k + p - 1) - \cdots - \beta_0^{(p-1)} u(k + 1)$$
$$= -\alpha_1 x_p(k) - \alpha_2 x_{p-1}(k) + \cdots - \alpha_p x_1(k) + \beta_0^{(p)} u(k), \tag{11.29d}$$

where the last equation is obtained with Eqs. (10.22) and (11.24). The reader should not have any difficulty to deriving Eq. (11.29d) (see Problem 11.1). Equation (11.29) can be arranged in matrix form as

$$x(k + 1) = Ax(k) + Bu(k), \tag{11.30a}$$

$$y(k) = Cx(k) + Du(k), \tag{11.30b}$$

where the state matrix A and the input matrix B are

$$A = \begin{bmatrix} 0 & I & 0 & \cdots & 0 & 0 \\ 0 & 0 & I & \cdots & 0 & 0 \\ \vdots & \vdots & \vdots & \ddots & \vdots & \vdots \\ 0 & 0 & 0 & \cdots & 0 & I \\ -\alpha_p & -\alpha_{p-1} & -\alpha_{p-2} & \cdots & -\alpha_2 & -\alpha_1 \end{bmatrix}, \tag{11.31a}$$

$$B = \begin{bmatrix} \beta_0^{(1)} \\ \beta_0^{(2)} \\ \vdots \\ \beta_0^{(p-1)} \\ \beta_0^{(p)} \end{bmatrix}, \tag{11.31b}$$

and the output matrix C and the direct-transmission matrix D are

$$C = [I \quad 0 \quad \cdots \quad 0 \quad 0], \tag{11.32a}$$
$$D = \beta_0. \tag{11.32b}$$

Recall that p is the number of available OMPs, m is the number of outputs, and r is the number of inputs. The state vector x becomes an $mp \times 1$ vector, the state matrix A an $mp \times mp$ matrix, the input matrix B an $mp \times r$ matrix, and the output matrix C an $m \times mp$ matrix. A state–space model in the form of Eq. (11.30) is said to be in the canonical-form.

The observability matrix of the canonical-form realization is

$$Q = \begin{bmatrix} C \\ CA \\ CA^2 \\ \vdots \\ CA^{p-1} \end{bmatrix} = \begin{bmatrix} I & 0 & 0 & \cdots & 0 \\ 0 & I & 0 & \cdots & 0 \\ 0 & 0 & I & \cdots & 0 \\ \vdots & \vdots & \vdots & \ddots & \vdots \\ 0 & 0 & 0 & \cdots & I \end{bmatrix}. \tag{11.33}$$

The matrix Q is an identity matrix that is obviously nonsingular. It implies that the observability matrix Q has a rank of mp and thus all states in the state vector x are observable. Are they controllable as well? First, form the controllability matrix

$$H = [B \quad AB \quad \cdots \quad A^{s-1}B]$$
$$= \begin{bmatrix} \beta_0^{(1)} & \beta_0^{(2)} & \cdots & \beta_0^{(s)} \\ \beta_0^{(2)} & \beta_0^{(3)} & \cdots & \beta_0^{(s+1)} \\ \vdots & \vdots & \ddots & \vdots \\ \beta_0^{(p)} & \beta_0^{(p+1)} & \cdots & \beta_0^{(s+p-1)} \end{bmatrix}, \tag{11.34}$$

where Eq. (10.21) has been used to form this matrix. The controllability matrix H

is a $pm \times rs$ Hankel matrix formed from system Markov parameters (pulse-response sequence). The maximum rank of H is n, which is the order of the system. Assume that the integer s is chosen large enough, i.e., $rs \geq pm$. If $pm = n$, the rank of H is identical to that of Q. As a result, the state–space representation, Eqs. (11.30), is a minimum realization from given OMPs $\alpha_1, \alpha_2, \ldots, \alpha_p, \beta_0, \beta_1, \ldots, \beta_p$. A state–space representation is a minimum realization if and only if it is controllable and observable, i.e., the state matrix is minimum order.

The maximum order of the model, Eq. (11.30), is pm, which is the dimension of the realized state matrix A. If the number p is chosen such that pm is larger than the order of the system, then the triplet $[A, B, C]$ is not a minimum realization. This is because the corresponding canonical form, Eqs. (11.30), is observable (the rank of Q is pm), but not controllable (the rank of H is less than pm). In this case, some of the states in the state vector x are not controllable. The state–space model given by Eqs. (11.30) is said to be in an observable canonical form. A different form of observable canonical-form realization is discussed in Section 10.6 of Chap. 10.

One may be interested in knowing the observer that makes the state matrix become deadbeat in a finite number of time steps. First, recall the matrices $\alpha_1, \alpha_1^{(1)}, \alpha_1^{(2)}, \ldots,$ $\alpha_1^{(p-1)}$ defined in Eq. (10.23). The following observer gain matrix

$$
G = \begin{bmatrix} \alpha_1 \\ \alpha_1^{(1)} \\ \vdots \\ \alpha_1^{(p-2)} \\ \alpha_1^{(p-1)} \end{bmatrix}
\tag{11.35}
$$

will result in

$$
(A + GC)^p = \begin{bmatrix} \alpha_1 & I & 0 & \cdots & 0 & 0 \\ \alpha_1^{(1)} & 0 & I & \cdots & 0 & 0 \\ \vdots & \vdots & \vdots & \ddots & \vdots & \vdots \\ \alpha_1^{(p-2)} & 0 & 0 & \cdots & 0 & I \\ \alpha_1^{(p-1)} - \alpha_p & -\alpha_{p-1} & -\alpha_{p-2} & \cdots & -\alpha_2 & -\alpha_1 \end{bmatrix}^p
$$

$$
= 0_{pm \times pm},
\tag{11.36}
$$

where $0_{pm \times pm}$ is a pm by pm zero matrix. In other words, the observer gain G will bring the observer state matrix $A + GC$ to zero in p steps. The reader is suggested to prove Eq. (11.36) as a problem exercise (see Problem 11.2). For a system with significant uncertainties, the deadbeat observer will converge to the steady-state Kalman filter under certain conditions (Ref. [10]) related to the data length and the choice of p.

Knowing the state–space model represented by the four matrices A, B, C, and D defined in Eqs. (11.31) and (11.32) and the observer gain matrix G defined in

Eq. (11.35), the observer equations for the state estimation are

$$\hat{x}(k+1) = A\hat{x}(k) + Bu(k) + G[\hat{y}(k) - y(k)], \tag{11.37a}$$

$$\hat{y}(k) = C\hat{x}(k) + Du(k), \tag{11.37b}$$

where $\hat{x}(k)$ and $\hat{y}(k)$ are estimated state and output at the time step k. The matrix G is used to estimate the state vector x. The initial state vector may be assumed to be any arbitrary vector such as a zero vector. Theoretically, a deadbeat observer should converge to the true state vector in p steps. Note that we may use either Eq. (11.25) or Eq. (11.37) to estimate the state vector.

11.5.2 GPC Controller

Given a state–space representation, there are many ways to design a feedback law to control the system. Common methods include optimal control designs, pole-placement techniques, virtual passive techniques, etc. The control law for full state feedback is commonly expressed by [see Eq. (9.75) in Section 9.6 of Chap. 9]

$$u(k) = \mathcal{F}\hat{x}(k). \tag{11.38}$$

For the GPC controller, the control gain matrix $\mathcal{F}$ can be computed from Eq. (11.8):

$$\mathcal{F} = -\text{the first } r \text{ rows of } [(T^T T + \lambda I)^{-1} T^T], \tag{11.39}$$

where

$$T = \begin{bmatrix} D & & & \\ CB & D & & \\ \vdots & \vdots & \ddots & \\ CA^{(s-1)}B & CA^{(s-2)}B & \cdots & D \end{bmatrix}, \tag{11.40}$$

which is computed from the state–space system matrices A, B, C, and D. It is easy to prove that the Toeplitz matrix computed from Eq. (11.40) is identical to the one shown in Eq. (11.2) (see Problem 11.3). For other general forms of the state–space model, Eqs. (11.8) and (11.39) may still be used to design a full-state-feedback controller as long as an observer is available to estimate the state accurately. Note that the observable canonical-form realization shown in Eqs. (11.31) and (11.32) may not be controllable. We may use a model reduction technique to reduce the size of system matrices to make it observable and controllable. However, the reduced model may not preserve the canonical form defined in Eqs. (11.31) and (11.32).

11.5.3 DPC Controller

For the DPC controller, the control gain matrix may be computed with the state–space approach as follows. With some algebraic manipulations, Eqs. (11.30) produce

$$x(k + 1) = Ax(k) + Bu(k),$$

$$x(k + 2) = A^2x(k) + [AB \quad B]\begin{bmatrix} u(k) \\ u(k + 1) \end{bmatrix}$$

$$\vdots \quad \vdots \quad \vdots$$

$$x(k + q) = A^q x(k) + T' u_q(k), \tag{11.41}$$

where

$$u_q(k) = \begin{bmatrix} u(k) \\ u(k + 1) \\ \vdots \\ u(k + q - 1) \end{bmatrix}, \tag{11.42}$$

and

$$T' = [\, A^{q-1}B \quad A^{q-2}B \quad \cdots \quad B \,]. \tag{11.43}$$

The matrix T' is an $n \times qr$ controllability matrix, where n is the order of the system and r is the number of inputs. The integer q must be chosen such that $qr \geq n$ to ensure that the matrix T' has rank of n. Note that the T' shown in both Eqs. (11.12a) and (11.43) are identical. Equation (11.43) is calculated from the state matrix A and input matrix B, whereas Eq. (11.12a) may be computed directly from input and output data.

Equation (11.41) shows that the state $x(k + q)$ at time $k + q$ becomes zero when the input series $u(k)$, $u(k + 1)$, ..., $u(k + q - 1)$ is given by

$$u_q(k) = -[T']^\dagger A^q x(k) \quad \Longrightarrow \quad x(k + q) = 0, \tag{11.44}$$

which clearly implies that the input $u(k)$ at time k is

$$u(k) = \mathcal{F}x(k)$$
$$= -\{\text{first } r \text{ rows of } [T']^\dagger\} A^q x(k). \tag{11.45}$$

Equation (11.45) gives a state-feedback controller that drives the state $x(k)$ at time step k to zero after q time steps. The control laws obtained from Eq. (11.15) and Eq. (11.45) are identical (see Problem 11.4).

Some researchers may prefer to use the state–space representation described by the system matrices A, B, C, D, the observer gain matrix G, and the control gain matrix $\mathcal{F}$ for real-time implementation. The control gain $\mathcal{F}$ can be computed with any other existing methods, such as the pole-placement techniques, optimal control methods, etc.

EXAMPLE 11.4

The input and output time histories

$$u = [1 \quad 0 \quad 0 \quad 0 \quad -1 \quad 0 \quad 2 \quad -2 \quad 1 \quad -1 \quad 0 \quad 2 \quad 0 \quad 1 \quad 1],$$
$$y = [0 \quad -1 \quad -1 \quad 1 \quad 1 \quad 0 \quad 0 \quad -2 \quad 0 \quad 3 \quad 0 \quad -2 \quad -2 \quad 0 \quad 1].$$

are the same as those given in Example 11.3 that produce the following three constant matrices:

$$
\mathcal{T} = \begin{bmatrix} \beta_0 & 0 & 0 & 0 \\ \beta_0^{(1)} & \beta_0 & 0 & 0 \\ \beta_0^{(2)} & \beta_0^{(1)} & \beta_0 & 0 \\ \beta_0^{(3)} & \beta_0^{(2)} & \beta_0^{(1)} & \beta_0 \end{bmatrix} = \begin{bmatrix} 0 & 0 & 0 & 0 \\ -1 & 0 & 0 & 0 \\ -1 & -1 & 0 & 0 \\ 1 & -1 & -1 & 0 \end{bmatrix},
$$

$$
\mathcal{A} = \begin{bmatrix} \alpha_2 & \alpha_1 \\ \alpha_2^{(1)} & \alpha_1^{(1)} \\ \alpha_2^{(2)} & \alpha_1^{(2)} \\ \alpha_2^{(3)} & \alpha_1^{(3)} \end{bmatrix} = \begin{bmatrix} 1 & 0 \\ 0 & 1 \\ -1 & 0 \\ 0 & -1 \end{bmatrix},
$$

$$
\mathcal{B} = \begin{bmatrix} \beta_2 & \beta_1 \\ \beta_2^{(1)} & \beta_1^{(1)} \\ \beta_2^{(2)} & \beta_1^{(2)} \\ \beta_2^{(3)} & \beta_1^{(3)} \end{bmatrix} = \begin{bmatrix} -1 & -1 \\ 0 & -1 \\ 1 & 1 \\ 0 & 1 \end{bmatrix}.
$$

From Eqs. (11.31), the state matrix A and input matrix B become

$$
A = \begin{bmatrix} 0 & 1 \\ -\alpha_2 & -\alpha_1 \end{bmatrix} = \begin{bmatrix} 0 & 1 \\ -1 & 0 \end{bmatrix},
$$

$$
B = \begin{bmatrix} \beta_0^{(1)} \\ \beta_0^{(2)} \end{bmatrix} = \begin{bmatrix} -1 \\ -1 \end{bmatrix}
$$

From Eqs. (11.32), the output matrix C and the direct-transmission matrix D become

$$
C = [1 \quad 0],
$$

$$
D = \beta_0 = 0.
$$

The observability matrix of the canonical-form realization is

$$
Q = \begin{bmatrix} C \\ CA \end{bmatrix} = \begin{bmatrix} 1 & 0 \\ 0 & 1 \end{bmatrix}.
$$

The matrix Q is an identity matrix that is obviously nonsingular and thus all states in the state vector x are observable. The controllability matrix is

$$
H = [B \quad AB] = \begin{bmatrix} -1 & -1 \\ -1 & 1 \end{bmatrix}.
$$

It is also a nonsingular matrix because its determinant is 2. The identified model is observable and controllable. The observer gain matrix is

$$
G = \begin{bmatrix} \alpha_1 \\ \alpha_1^{(1)} \end{bmatrix} = \begin{bmatrix} 0 \\ 1 \end{bmatrix}.
$$

The observer equation is

$$\begin{bmatrix} \hat{x}_1(k+1) \\ \hat{x}_2(k+1) \end{bmatrix} = \begin{bmatrix} 0 & 1 \\ -1 & 0 \end{bmatrix} \begin{bmatrix} \hat{x}_1(k) \\ \hat{x}_2(k) \end{bmatrix} - \begin{bmatrix} 1 \\ 1 \end{bmatrix} u(k) + \begin{bmatrix} 0 \\ 1 \end{bmatrix} [\hat{y}(k) - y(k)]$$

$$\hat{y}(k) = \begin{bmatrix} 1 & 0 \end{bmatrix} \begin{bmatrix} \hat{x}_1(k) \\ \hat{x}_2(k) \end{bmatrix}.$$

Let us assume an arbitrary vector for the initial value of the estimated state vector $\hat{x}$, say,

$$\begin{bmatrix} \hat{x}_1(1) \\ \hat{x}_2(1) \end{bmatrix} = \begin{bmatrix} 5 \\ 10 \end{bmatrix}.$$

For $k = 1$, first note from the input and the output data that $u(1) = 1$ and $y(1) = 0$:

$$\hat{y}(1) = \begin{bmatrix} 1 & 0 \end{bmatrix} \begin{bmatrix} 5 \\ 10 \end{bmatrix} = 5,$$

$$\begin{bmatrix} \hat{x}_1(2) \\ \hat{x}_2(2) \end{bmatrix} = \begin{bmatrix} 0 & 1 \\ -1 & 0 \end{bmatrix} \begin{bmatrix} 5 \\ 10 \end{bmatrix} - \begin{bmatrix} 1 \\ 1 \end{bmatrix} 1 + \begin{bmatrix} 0 \\ 1 \end{bmatrix} [5 - 0] = \begin{bmatrix} 9 \\ -1 \end{bmatrix}.$$

For $k = 2$ with $u(2) = 0$ and $y(2) = -1$,

$$\hat{y}(2) = \begin{bmatrix} 1 & 0 \end{bmatrix} \begin{bmatrix} 9 \\ -1 \end{bmatrix} = 9,$$

$$\begin{bmatrix} \hat{x}_1(3) \\ \hat{x}_2(3) \end{bmatrix} = \begin{bmatrix} 0 & 1 \\ -1 & 0 \end{bmatrix} \begin{bmatrix} 9 \\ -1 \end{bmatrix} - \begin{bmatrix} 1 \\ 1 \end{bmatrix} 0 + \begin{bmatrix} 0 \\ 1 \end{bmatrix} [9 + 1] = \begin{bmatrix} -1 \\ 1 \end{bmatrix}.$$

For $k = 3$ with $u(3) = 0$ and $y(3) = -1$,

$$\hat{y}(3) = \begin{bmatrix} 1 & 0 \end{bmatrix} \begin{bmatrix} -1 \\ 1 \end{bmatrix} = -1,$$

$$\begin{bmatrix} \hat{x}_1(4) \\ \hat{x}_2(4) \end{bmatrix} = \begin{bmatrix} 0 & 1 \\ -1 & 0 \end{bmatrix} \begin{bmatrix} -1 \\ 1 \end{bmatrix} - \begin{bmatrix} 1 \\ 1 \end{bmatrix} 0 + \begin{bmatrix} 0 \\ 1 \end{bmatrix} [-1 + 1] = \begin{bmatrix} 1 \\ 1 \end{bmatrix}.$$

It clearly shows that $\hat{y}(3) = y(3)$. Indeed, the estimated output after the time step $k = 2$ is identical to the real output, i.e., $\hat{y}(k) = y(k)$ for $k > 2$. It implies that the gain matrix G is a two-step deadbeat observer, i.e.,

$$(A + GC)^2 = \left\{ \begin{bmatrix} 0 & 1 \\ -1 & 0 \end{bmatrix} + \begin{bmatrix} 0 \\ 1 \end{bmatrix} \begin{bmatrix} 1 & 0 \end{bmatrix} \right\}^2 = \begin{bmatrix} 0 & 1 \\ 0 & 0 \end{bmatrix}^2 = 0_{2 \times 2}.$$

GPC Controller
For the GPC controller, first form the matrix $\mathcal{T}$ from Eq. (11.40):

$$\mathcal{T} = \begin{bmatrix} D & 0 \\ CB & D \end{bmatrix} = \begin{bmatrix} 0 & 0 \\ -1 & 0 \end{bmatrix}.$$

Using Eq. (11.39) to compute the control gain matrix yields

$$\mathcal{F} = -\text{the first } r \text{ rows of } [(\mathcal{T}^T\mathcal{T} + \lambda I)^{-1}\mathcal{T}^T]$$

$$= -\text{the first } r \text{ rows of } \begin{bmatrix} 0 & \dfrac{-1}{\lambda+1} \\ 0 & 0 \end{bmatrix}$$

$$= \begin{bmatrix} 0 & \dfrac{1}{\lambda+1} \end{bmatrix},$$

where λ is the control weighting. Note that this solution is valid only for a nonzero λ. The control law becomes

$$u(k) = \mathcal{F}\hat{x}(k) = \begin{bmatrix} 0 & \dfrac{1}{\lambda+1} \end{bmatrix}\begin{bmatrix} \hat{x}_1(k) \\ \hat{x}_2(k) \end{bmatrix} = \frac{1}{\lambda+1}\hat{x}_2(k).$$

This is the formulation for actual implementation with the state–space approach. One question may arise regarding the stability of the GPC controller. The closed-loop state–space equation is

$$x(k+1) = Ax(k) + Bu(k) = [A + B\mathcal{F}]x(k),$$

where

$$A + B\mathcal{F} = \begin{bmatrix} 0 & 1 \\ -1 & 0 \end{bmatrix} + \begin{bmatrix} -1 \\ -1 \end{bmatrix}\begin{bmatrix} 0 & \dfrac{1}{\lambda+1} \end{bmatrix}$$

$$= \begin{bmatrix} 0 & \dfrac{\lambda}{\lambda+1} \\ -1 & \dfrac{-1}{\lambda+1} \end{bmatrix}.$$

The eigenvalues of the closed-loop state matrix $A + B\mathcal{F}$ are

$$z_1 = \frac{-1 - \sqrt{1 - 4\lambda(1+\lambda)}}{2(1+\lambda)}, \qquad z_2 = \frac{-1 + \sqrt{1 - 4\lambda(1+\lambda)}}{2(1+\lambda)}.$$

Note that λ must be a positive value. Because the identified state–space model is in the discrete-time domain, the closed-loop system is stable if and only if the eigenvalues of its closed-loop state matrix are within the unit circle, i.e., its magnitude must be less than one. The characteristics of the two eigenvalues can be understood by examination of the following three values of λ.

CASE 1:

$$\lambda \to 0 \implies z_1 \to -1, \qquad z_2 \to 0.$$

CASE 2:

$$\lambda \to \infty \implies z_1 \to -\sqrt{-1}, \qquad z_2 \to \sqrt{-1}.$$

CASE 3:

$$\lambda = \frac{-1 + \sqrt{2}}{2} \implies z_1 = z_2 = \frac{-1}{2}.$$

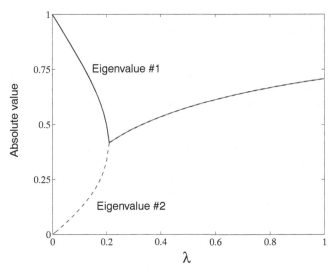

Figure 11.1. Absolute values of the close-loop system eigenvalues.

When $\lambda < [(-1 + \sqrt{2})/2]$, both eigenvalues are real and negative. On the other hand, when $\lambda > [(-1 + \sqrt{2}/2]$, both eigenvalues are a pair of complex values. The closed-loop system for both Case 1 and Case 2 are marginally stable because at lease one of the eigenvalues is close to the unit circle. Figure 11.1 shows the absolute values of both eigenvalues versus the control weighting. Case 3 seems to be the optimal case for this example. Regardless of what the weighting λ is, the closed-loop system is stable for $\lambda > 0$.

One may be curious how it compares with the other approach presented in Example 11.3. From Eq. (11.25) for $p = 2$, the state vector can be expressed in terms of past input and output data as follows:

$$
\begin{bmatrix} \hat{x}_1(k) \\ \hat{x}_2(k) \end{bmatrix} = \begin{bmatrix} \beta_2 & \beta_1 \\ \beta_2^{(1)} & \beta_1^{(1)} \end{bmatrix} \begin{bmatrix} u(k-2) \\ u(k-1) \end{bmatrix} - \begin{bmatrix} \alpha_2 & \alpha_1 \\ \alpha_2^{(1)} & \alpha_1^{(1)} \end{bmatrix} \begin{bmatrix} y(k-2) \\ y(k-1) \end{bmatrix}
$$

$$
= \begin{bmatrix} -1 & -1 \\ 0 & -1 \end{bmatrix} \begin{bmatrix} u(k-2) \\ u(k-1) \end{bmatrix} - \begin{bmatrix} 1 & 0 \\ 0 & 1 \end{bmatrix} \begin{bmatrix} y(k-2) \\ y(k-1) \end{bmatrix}.
$$

For $k > 2$, $\hat{x}(k)$ should be identical to the true state vector $x(k)$. From the above equation, we obtain

$$
\hat{x}_1(k) = -u(k-2) - u(k-1) - y(k-2),
$$
$$
\hat{x}_2(k) = -u(k-1) - y(k-1).
$$

Therefore the GPC control law may also be written as

$$
u(k) = \mathcal{F}x(k) = \frac{1}{\lambda + 1} \cdot \hat{x}_2(k) = \frac{-1}{\lambda + 1} [u(k-1) + y(k-1)],
$$

which is identical to the result shown in Example 11.3.

DPC Controller

For the DPC controller, first form the matrix T' from Eq. (11.43) for $q = 2$:

$$T' = [AB \quad B] = \begin{bmatrix} -1 & -1 \\ 1 & -1 \end{bmatrix}.$$

Using Eq. (11.45) to compute the control gain matrix yields

$$\mathcal{F} = -\text{the first } r \text{ rows of } [T']^{-1} A^2$$

$$= -\text{the first } r \text{ rows of } \frac{1}{2} \begin{bmatrix} -1 & 1 \\ -1 & -1 \end{bmatrix} \begin{bmatrix} -1 & 0 \\ 0 & -1 \end{bmatrix}$$

$$= \frac{1}{2} [-1 \quad 1].$$

The DPC control law becomes

$$u(k) = \mathcal{F}\hat{x}(k) = \frac{1}{2} [-1 \quad 1] \begin{bmatrix} \hat{x}_1(k) \\ \hat{x}_2(k) \end{bmatrix} = \frac{1}{2}[-\hat{x}_1(k) + \hat{x}_2(k)].$$

This is the DPC formulation for actual implementation with the state–space approach. The closed-loop state matrix for the DPC controller is

$$A + B\mathcal{F} = \begin{bmatrix} 0 & 1 \\ -1 & 0 \end{bmatrix} + \begin{bmatrix} -1 \\ -1 \end{bmatrix} \begin{bmatrix} -\frac{1}{2} & \frac{1}{2} \end{bmatrix}$$

$$= \frac{1}{2} \begin{bmatrix} 1 & 1 \\ -1 & -1 \end{bmatrix}.$$

The eigenvalues of the closed-loop state matrix $A + B\mathcal{F}$ are

$$z_1 = z_2 = 0.$$

Both are at the origin of the unit circle, implying that this is the most stable case for a closed-loop discrete-time system. Indeed, it will take only two steps for the system to become deadbeat, i.e., the closed-loop state vector becomes a zero vector after $k > 2$, because

$$x(3) = [A + B\mathcal{F}]^2 x(1) = 0,$$

where

$$[A + B\mathcal{F}]^2 = \frac{1}{4} \begin{bmatrix} 1 & 1 \\ -1 & -1 \end{bmatrix}^2 = \begin{bmatrix} 0 & 0 \\ 0 & 0 \end{bmatrix}.$$

In practice, this case may not happen as expected because the identified model would be different from the actual model because of measurement noise and system uncertainties.

Similar to the GPC case, the DPC control law may also be implemented with past input and output data, as shown in Example 11.3. First note that

$$-\hat{x}_1(k) + \hat{x}_2(k) = u(k - 2) + y(k - 2) - y(k - 1).$$

Therefore, the DPC control law may also be written as

$$u(k) = \mathcal{F}\hat{x}(k) = \frac{1}{2}[u(k-2) + y(k-2) - y(k-1)],$$

which is again identical to the result shown in Example 11.3.

Although the two-step deadbeat controller seems very attractive, the resulting control force may likely exceed the limit allowed from the control actuator. We may reduce the control force by increasing the deadbeat time step q. The matrix $\mathcal{T}'$ from Eq. (11.43) with $q = 3$ is

$$\mathcal{T}' = [\, A^2B \quad AB \quad B\,] = \begin{bmatrix} 1 & -1 & -1 \\ 1 & 1 & -1 \end{bmatrix}.$$

The control gain matrix from Eq. (11.45) becomes

$$\begin{aligned} \mathcal{F} &= -\text{the first } r \text{ rows of } [\mathcal{T}']^{\dagger}\, A^3 \\ &= \frac{1}{4}[\, -1 \quad 1\,]. \end{aligned}$$

The closed-loop state matrix for the DPC controller is

$$\begin{aligned} A + B\mathcal{F} &= \begin{bmatrix} 0 & 1 \\ -1 & 0 \end{bmatrix} + \begin{bmatrix} -1 \\ -1 \end{bmatrix}\begin{bmatrix} -\frac{1}{4} & \frac{1}{4} \end{bmatrix} \\ &= \frac{1}{4}\begin{bmatrix} 1 & 3 \\ -3 & -1 \end{bmatrix}. \end{aligned}$$

The eigenvalues of the closed-loop state matrix $A + B\mathcal{F}$ are

$$z_1 = \sqrt{\frac{-1}{2}}, \qquad z_2 = -\sqrt{\frac{-1}{2}}.$$

Both are on the imaginary axis of the unit circle. The magnitude $\sqrt{1/2}$ of the two eigenvalues indicates that the closed-loop system is quite stable. It should be noted that $[A + B\mathcal{F}]^3 \neq 0_{2\times 2}$.

EXAMPLE 11.5

Consider the three-degree-of-freedom spring–mass–damper system shown in Fig. 11.2. The corresponding second-order ordinary differential equation is

$$M\ddot{w} + \Xi\dot{w} + Kw = u,$$

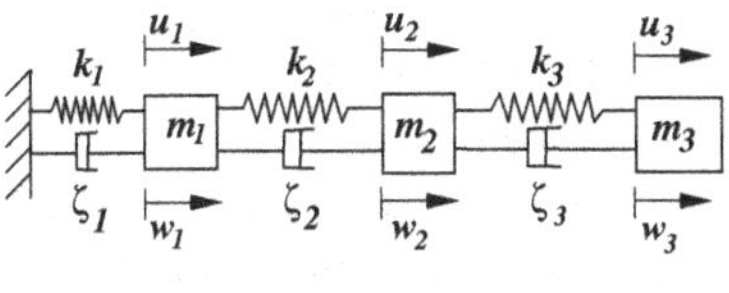

Figure 11.2. A simple spring–mass–damper system.

where

$$M = \begin{bmatrix} m_1 & 0 & 0 \\ 0 & m_2 & 0 \\ 0 & 0 & m_3 \end{bmatrix},$$

$$\Xi = \begin{bmatrix} \zeta_1 + \zeta_2 & -\zeta_2 & 0 \\ -\zeta_2 & \zeta_2 + \zeta_3 & -\zeta_3 \\ 0 & -\zeta_3 & \zeta_3 \end{bmatrix},$$

$$K = \begin{bmatrix} k_1 + k_2 & -k_2 & 0 \\ -k_2 & k_2 + k_3 & -k_3 \\ 0 & -k_3 & k_3 \end{bmatrix},$$

$$w = \begin{bmatrix} w_1 \\ w_2 \\ w_3 \end{bmatrix}, \quad u = \begin{bmatrix} u_1 \\ u_2 \\ u_3 \end{bmatrix}.$$

where m_i, k_i, and ζ_i, $i = 1, 2, 3$, are the mass, spring stiffness, and damping coefficients, respectively. For this system, the order of the equivalent state–space representation is 6 ($n = 6$). The input force applied to each mass is denoted by u_i, $i = 1, 2, 3$. The variables w_i, $i = 1, 2, 3$ are the positions of the three masses measured from their equilibrium positions. In the simulation, $m_1 = m_2 = m_3 = 1$ Kg, $k_1 = k_2 = k_3 = 1000$ N/m, and $\zeta_1 = \zeta_2 = \zeta_3 = 0.1$ N-s/m. The system is sampled at 20 Hz ($\Delta t = 0.05$ s). Let the measurements y be the accelerations of the three masses, $\ddot{w}_i$, $i = 1, 2, 3$.

Let us consider a single-control-input, single-disturbance-input, and single-output case in which the control input to the system is the force on the first mass (i.e., $u_c = u_1$), the disturbance input is at the second mass (i.e., $u_d = u_2$), and the output is the acceleration of the third mass (i.e., $y = \ddot{w}_3$) (noncolocated actuator–sensor). Therefore, the smallest order for the output prediction shown in Eq. (11.1) is $p = 6$, and the smallest value for the prediction horizon is $h_p = 5$ (i.e., $s = 6$) if no disturbance input is present.

Assume that the disturbance input is a single tone of 4 Hz but unknown, i.e., the disturbance is not measurable. A random signal is generated for the control input along with the single-tone disturbance signal to excite the system to generate the output time history. Note that the output data contain the system information as well as the disturbance information. The open-loop control input and the measured output data are used to compute the control parameters.

The GPC algorithm has four parameters to adjust. The system order p, the prediction horizon h_p, the control horizon h_c, and the control weight λ all must be chosen. As in most control systems, we must balance performance with system stability and actuator authority. Generally, we should set the prediction horizon $h_p + 1$ at least equal to or greater than the system order p.

For a system with unknown disturbance or system uncertainties, increasing the model size will produce a more accurate prediction of future outputs. To accommodate the unknown single-tone disturbance, let us increase the order p by 2 and, correspondingly, the prediction horizon also by 2, $s = 8$ in Eq. (11.1). The reason for increasing the order by 2 is to model the single tone (Ref. [4]). Furthermore, assume that the control horizon is the same as the prediction horizon and the GPC control weighting is $\lambda = 0.1$.

A total of 50 output data points are generated and used to compute the GPC control gain coefficients. Inserting OMPs obtained from the input and output data into Eqs. (11.1) and (11.8) with the design parameters

$$p = s = 8, \qquad \lambda = 0.1$$

yields the GPC controller:

$$
\begin{aligned}
u_c(k) = {} & 0.7960u_c(k-1) + 0.0674u_c(k-2) - 0.7230u_c(k-3) \\
& + 0.9486u_c(k-4) - 0.1017u_c(k-5) - 0.2234u_c(k-6) \\
& - 0.0426u_c(k-7) - 0.0027u_c(k-8) \\
& + 0.4735y(k-1) + 0.1750y(k-2) + 0.5586y(k-3) \\
& + 0.4157y(k-4) + 0.0756y(k-5) + 0.9947y(k-6) \\
& + 0.2217y(k-7) + 0.0162y(k-8).
\end{aligned}
$$

Figure 11.3 shows the time histories of control force (solid line), disturbance input (dashed–dotted lines), open-loop response (dashed–dotted line), and closed-loop response (solid line). The first 50 (2.5-s) open-loop data points were used

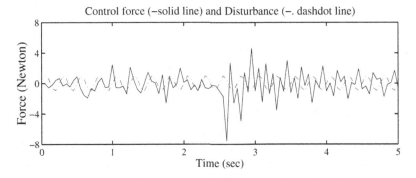

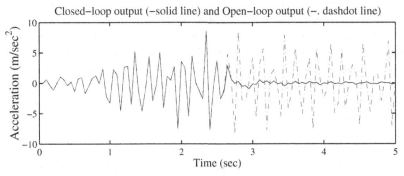

Figure 11.3. Single-output open-loop and closed-loop time histories.

for computing the GPC control gain coefficients. The GPC controller was then turned on after 2.5 s to produce the closed-loop response. Note that the single-tone disturbance is always on. The peaks of the open-loop response excited by the unknown disturbance are considerably reduced by the control feedback.

Next, consider the case in which there are additional measurements available for feedback control (unequal number of inputs and outputs). In addition to the acceleration of the third mass, acceleration measurement of the second mass is also available, i.e., $y_1 = \ddot{w}_3$ and $y_2 = \ddot{w}_2$. The minimum order of the finite-difference model is $p = 3 + 1$, which includes the additional one to accommodate for the unknown disturbance of a single tone. The identified model becomes $pm = 8$, where m is the number of outputs, which is identical to the previous single-input and single-output case. For comparison, the prediction horizon and control horizon are set to $h_p = h_c = 3$, which makes $s = 4$ in Eq. (11.1). Let the system be excited by the same random input and disturbance signal as the earlier case shown in Fig. (11.3). The first 50 data points were recorded and used to compute the GPC controller. From Eqs. (11.1) and (11.8) with the design parameters

$$p = s = 4, \qquad \lambda = 0.1,$$

the controller in this case becomes

$$u_c(k) = 1.1473 u_c(k-1) + 0.2307 u_c(k-2) - 1.2335 u_c(k-3)$$

$$+ 0.5898 u_c(k-4) - [\, 2.2054 \quad 0.6357 \,] \begin{bmatrix} y_1(k-1) \\ y_2(k-1) \end{bmatrix}$$

$$+ [\, 0.5560 \quad 2.4087 \,] \begin{bmatrix} y_1(k-2) \\ y_2(k-2) \end{bmatrix}$$

$$+ [\, -0.4357 \quad 0.3109 \,] \begin{bmatrix} y_1(k-3) \\ y_2(k-3) \end{bmatrix}$$

$$- [\, 0.0886 \quad 1.2134 \,] \begin{bmatrix} y_1(k-4) \\ y_2(k-4) \end{bmatrix}.$$

Note that with the additional measurements, fewer time steps (and fewer controller gains) are required. This is a reflection of the fact that complete state estimation can now be achieved faster with the additional sensors. Figure 11.4 shows the time histories of control force (solid line), disturbance input (dashed–dotted line), first output open-loop response (solid line) and closed-loop response (dashed–dotted line), and second output open-loop response (solid line) and closed-loop response (dashed–dotted line). The peaks of the open-loop response for both outputs are considerably reduced by the control feedback.

11.6 Concluding Remarks

Two control techniques were introduced, the generalized predictive control (GPC) and the deadbeat predictive control (DPC). Several computational algorithms were described to determine the GPC and DPC control laws directly from input and output

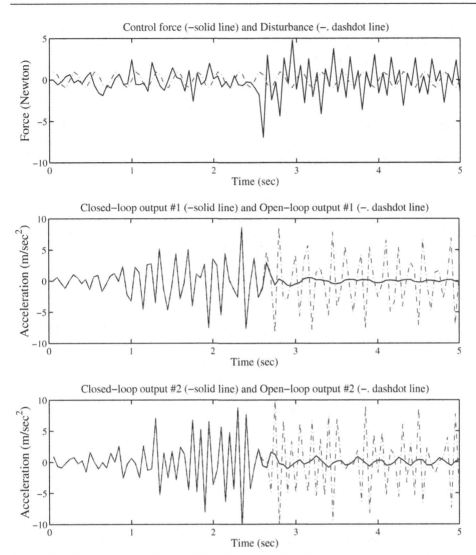

Figure 11.4. Two-output open-loop and closed-loop time histories.

data. These algorithms are simple and easy to compute and thus are good candidates for real-time implementation in a micro processor. Two steps are involved, system identification and controller design. The fundamental step in system identification is to calculate the coefficient matrices, such as the Toeplitz matrix formed by the pulse-response time history, for the output predictor. The most time-consuming task in the controller design is to compute a matrix inverse formed from the Toeplitz matrix.

The state–space representation of the GPC and DPC provides a clear insight into the fundamental structure of the predictive control law. The connection between the state–space control law and the predictive control law is clearly established. Because the control gains are designed from the input–output models, they may be adaptively tuned from on-line input and output measurements. As a result, these controllers should be able to handle systems with slowly time-varying dynamics, provided that input and output data are sufficiently rich to allow reasonable system identification.

We have focused on the computational algorithms directly by using input and output data to compute GPC and DPC controllers in this chapter. There are many important issues including robustness and stability associated with the measurement noise, unknown disturbances, and system uncertainties. These issues are quite essential in actual implementation for control of a real system. It is beyond the scope of this chapter to do an in-depth discussion of these issues. A list of references is provided at the end of this chapter on the subject of predictive control. Example 11.5 is given for the reader to observe how one may use the GPC controller to deal with the case in which a system is excited by a control signal and an unknown disturbance.

11.7 Problems

11.1 Use Eqs. (10.22) and (11.24) with the aid of Eq. (11.28) for the definition of $x_i(k)$ for $i = 1, 2, \ldots, p$ to prove Eq. (11.29d), i.e.,

$$y(k+p) - \beta_0 u(k+p) - \beta_0^{(1)} u(k+p-1) - \cdots - \beta_0^{(p-1)} u(k+1)$$

$$= -\alpha_1 x_p(k) - \alpha_2 x_{p-1}(k) + \cdots - \alpha_p x_1(k) + \beta_0^{(p)} u(k).$$

11.2 Based on the relationship of Eqs. (10.19) or Eqs. (10.23), prove that Eq. (11.36) is true:

$$(A + GC)^p = \begin{bmatrix} \alpha_1 & I & 0 & \cdots & 0 & 0 \\ \alpha_1^{(1)} & 0 & I & \cdots & 0 & 0 \\ \vdots & \vdots & \vdots & \ddots & \vdots & \vdots \\ \alpha_1^{(p-2)} & 0 & 0 & \cdots & 0 & I \\ \alpha_1^{(p-1)} - \alpha_p & -\alpha_{p-1} & -\alpha_{p-2} & \cdots & -\alpha_2 & -\alpha_1 \end{bmatrix}^p,$$

$$= 0_{pm \times pm}$$

where m is the number of outputs, p is the order of the finite-difference model, pm is the order of the state matrix A, and $0_{pm \times pm}$ is a $pm \times pm$ zero matrix. For simplicity, one may start with $p = 2, 3, \ldots$ to search for an approach to solve for the general case.

11.3 Using Eqs. (11.31) and (11.32) for the definition of state–space system matrices A, B, C, and D with the aid of Eq. (10.21) or Eq. (10.22), prove that the following equalities are true:

$$T' = [\, A^{q-1}B \quad A^{q-2}B \quad \cdots \quad B \,]$$

$$= \begin{bmatrix} \beta_0^{(q)} & \beta_0^{(q-1)} & \cdots & \beta_0^{(1)} \\ \beta_0^{(q+1)} & \beta_0^{(q)} & \cdots & \beta_0^{(2)} \\ \vdots & \vdots & \ddots & \vdots \\ \beta_0^{(q+p-1)} & \beta_0^{(q+p-2)} & \cdots & \beta_0^{(p)} \end{bmatrix}.$$

$$
\mathcal{T} = \begin{bmatrix} \beta_0 & & & \\ \beta_0^{(1)} & \beta_0 & & \\ \vdots & \vdots & \ddots & \\ \beta_0^{(s-1)} & \beta_0^{(s-2)} & \cdots & \beta_0 \end{bmatrix}
$$

$$
= \begin{bmatrix} D & & & \\ CB & D & & \\ \vdots & \vdots & \ddots & \\ CA^{(s-1)}B & CA^{(s-2)}B & \cdots & D \end{bmatrix}.
$$

11.4 Let matrices $\mathcal{A}$ of $(q+p)m \times pm$ and $\mathcal{B}$ of $(q+p)m \times pr$ be

$$
\mathcal{A} = \begin{bmatrix} \alpha_p & \alpha_{(p-1)} & \cdots & \alpha_1 \\ \alpha_p^{(1)} & \alpha_{p-1}^{(1)} & \cdots & \alpha_1^{(1)} \\ \vdots & \vdots & \ddots & \vdots \\ \alpha_p^{(q+p-1)} & \alpha_{p-1}^{(q+p-1)} & \cdots & \alpha_1^{(q+p-1)} \end{bmatrix},
$$

$$
\mathcal{B} = \begin{bmatrix} \beta_p & \beta_{(p-1)} & \cdots & \beta_1 \\ \beta_p^{(1)} & \beta_{p-1}^{(1)} & \cdots & \beta_1^{(1)} \\ \vdots & \vdots & \ddots & \vdots \\ \beta_p^{(q+p-1)} & \beta_{p-1}^{(q+p-1)} & \cdots & \beta_1^{(q+p-1)} \end{bmatrix},
$$

and the state matrix A of $pm \times pm$ be

$$
A = \begin{bmatrix} 0 & I & \cdots & 0 \\ 0 & 0 & \cdots & 0 \\ \vdots & \vdots & \ddots & \vdots \\ 0 & 0 & \cdots & I \\ -\alpha_p & -\alpha_{p-1} & \cdots & -\alpha_1 \end{bmatrix},
$$

where both p and q are integers and $qr \geq pm$.

(a) Based on the recursive relationships defined in Eqs.(10.19)–(10.23), show that

$$
\mathcal{A}(qm+1 : qm+pm, :) = A^q \mathcal{A}(1 : pm, :),
$$

$$
\mathcal{B}(qm+1 : qm+pm, :) = A^q \mathcal{B}(1 : pm, :),
$$

The last pm rows starting from the row $qm+1$ of $\mathcal{A}$ and $\mathcal{B}$ are the first pm rows of $\mathcal{A}$ and $\mathcal{B}$ premultiplied by A^q. This result provides an interesting connection between the state matrix A and the submatrices of $\mathcal{A}$ and $\mathcal{B}$.

(b) Prove that the control laws obtained from Eqs. (11.15) and (11.45) are identical. One may simply derive Eq. (11.15) from Eq. (11.45)

11.5 Use the state vector defined in Eq. (10.45), i.e.,

$$x_p(k) = y(k) - \beta_0 u(k),$$
$$x_{p-1}(k) = y(k+1) - \beta_0 u(k+1) + \alpha_1 y(k) - \beta_1 u(k)$$

$$\vdots \ \vdots \qquad \vdots$$

$$x_1(k) = y(k+p-1) - \beta_0 u(k+p-1)$$
$$+ \alpha_1 y(k+p-2) - \beta_1 u(k+p-2) + \cdots + \alpha_{p-1} y(k) - \beta_{p-1} u(k),$$

where each vector $x_i(k)$, $i = 1, 2, \ldots, p$, has the length of m (the number of outputs). What is the corresponding formulation in terms of past input and output vectors, $u_p(k-p)$ and $y_p(k-p)$? The formulation should be expressed in the form similar to that of Eq. (11.25). The definition of the state vector produces the canonical-form realization shown in Eqs. (10.48):

$$A = \begin{bmatrix} 0 & 0 & 0 & \cdots & 0 & -\alpha_p \\ I & 0 & 0 & \cdots & 0 & -\alpha_{p-1} \\ \vdots & \vdots & \vdots & \ddots & \vdots & \vdots \\ 0 & 0 & 0 & \cdots & I & -\alpha_1 \end{bmatrix}, \quad B = \begin{bmatrix} \beta_p - \alpha_p \beta_0 \\ \beta_{p-1} - \alpha_{p-1}\beta_0 \\ \vdots \\ \beta_1 - \alpha_1 \beta_0 \end{bmatrix},$$

$$C = [0 \ \ 0 \cdots \ I \ ], \qquad D = \beta_0.$$

Discuss the merits between this canonical-form realization and the one defined in Eqs. (11.31) and (11.32).

11.6 Given the following control law,

$$u(k) = \alpha_1^c y(k-1) + \alpha_2^c y(k-2) + \cdots + \alpha_p^c y(k-p)$$
$$+ \beta_1^c u(k-1) + \beta_2^c u(k-2) + \cdots + \beta_p^c u(k-p),$$

prove that the controller itself is asymptotically stable if and only if the magnitude of each eigenvalue of the following matrix is less than one:

$$\begin{bmatrix} 0 & 0 & 0 & \cdots & 0 & \alpha_p^c \\ I & 0 & 0 & \cdots & 0 & \alpha_{p-1}^c \\ \vdots & \vdots & \vdots & \ddots & \vdots & \vdots \\ 0 & 0 & 0 & \cdots & I & \alpha_1^c \end{bmatrix}.$$

BIBLIOGRAPHY

[1] Bialasiewicz, J. T., Horta, L. G., and Phan, M., "Identified Predictive Control," in *Proceedings of the American Control Conference,* IEEE, Piscataway, NJ, 1994.

[2] Clarke, D. W., Mohtadi, C., and Tuffs, P. S., "Generalized Predictive Control—Part I. The Basic Algorithm, " *Automatica,* Vol. 23, No. 2, pp. 137–148, 1987.

[3] Clarke, D. W., Mohtadi, C., and Tuffs, P. S., "Generalized Predictive Control—Part II. Extensions and Interpretations, " *Automatica,* Vol. 23, No. 2, pp. 149–160, 1987.

[4] Juang, J.-N. and Eure, K. W., "Predictive Feedback and Feedforward Control for Systems with Unknown Disturbances," NASA/TM-1998-208744, December 1998.

[5] Eldem, V. and Selbuz, H., "On the General Solution of the State Deadbeat Control Problem," *IEEE Trans. Autom. Control,* Vol. 39, pp. 1002–1006, 1994.

[6] Leden, B. "Multivariable deadbeat control," *Automatica,* Vol. 13, pp. 185–188, 1977.

[7] Marrari, M. R., Emami-Naeini, A., and Franklin, G. F., "Output Deadbeat Control of Discrete-Time Multivariable Systems," *IEEE Trans. Autom. Control*, Vol. 34, pp. 644–648, 1989.

[8] Phan, M. G., and Juang, J.-N., "Predictive Controllers for Feedback Stabilization," *J. Guid, Control Dyn.*, Vol. 21, pp. 747–753, 1988.

[9] Schlegel, M., "Parameterization of the Class of Deadbeat Controllers," *IEEE Trans. Autom. Control*, Vol. 27, pp. 727–729, 1982.

[10] Juang, J.-N., *Applied System Identification*, Prentice-Hall, Englewood Cliffs, NJ, 1994.

SUGGESTED READING

Astrom, K. J, and Wittenmark, B., *Adaptive Control*, Addison-Wesley, Reading MA, 1989.

De Keyser, R. M. C. and Van Cauwenberghe, A. R., "A Self-Tuning Multi-Step Predictor Application," *Automatica*, Vol. 17, No. 1, pp. 167–174, 1979.

Goodwin, G. C. and Sin, K. S., *Adaptive Filtering, Prediction, and Control*, Prentice-Hall, Englewood Cliffs, NJ, 1984.

Mosca, E., *Optimal, Predictive, and Adaptive Control*, Prentice-Hall, Englewood Cliffs, NJ, 1995.

Peterka, V., "Predictor-Based Self-Tuning Control," *Automatica*, Vol. 20, No. 1, pp. 39–50, 1984.

Richalet, J., Rault, A., Testud, J. L., and Papon, J., "Model Predictive Heuristic Control: Applications to Industrial Processes," *Automatica*, Vol. 14, No. 5, pp. 413–428, 1978.

Soeterboek, R., *Predictive Control: A Unified Approach*, Prentice-Hall, Englewood Cliffs, NJ, 1992.

Ydstie, B. E., "Extended Horizon Adaptive Control," in *Proceedings of the 9th IFAC World Congress*, Vol. VII, Elsevier Science Ltd., New York, pp. 133–138, 1984.

Index

333